Pathogenesis of the Avian Embryo

PATHOGENESIS OF THE AVIAN EMBRYO

AN ANALYSIS OF CAUSES OF MALFORMATIONS AND PRENATAL DEATH

ALEXIS L. ROMANOFF
Professor Emeritus of Chemical Embryology
Cornell University

with the collaboration of

ANASTASIA J. ROMANOFF

WILEY-INTERSCIENCE, a Division of John Wiley & Sons, Inc.
New York • London • Sydney • Toronto

Library of Congress Cataloging in Publication Data

Romanoff, Alexis Lawrence, 1892–
Pathogenesis of the avian embryo.

Bibliography: p.
1. Embryology—Birds. 2. Fetus—Abnormities and deformities. 3. Abnormalities (Animals) 4. Chick embryo. I. Romanoff, Anastasia J. (Sayenko) 1905– II. Title.

QL698.R57   598.2′2′16   72-3249
ISBN 0-471-73240-0

Printed in the United States of America

10 9 8 7 6 5 4 3 2 1

DEDICATED

to

those whose endeavors in scientific research have
contributed to the present knowledge of
Pathology of the Avian Embryo

Preface

This book is a detailed study of the effects of various deleterious agents on the incidence of specific syndromes and of prenatal mortality in the embryo. The text is based primarily on the analysis of experimental data selected from the world scientific literature.

The author's main objective is to consolidate in one volume the widely scattered scientific findings on teratological manifestations in the avian embryo—experimentally the most extensively studied species. These basic facts should provide better understanding of all the possible causes of congenital anomalies and prenatal death. Also, such a mass of information, especially the quantitative data on birds, can be used in the diagnosis of many diseased conditions, in comparative studies of mammalian material, and, in general, as a guide in the formulation of preventive measures against the occurrence of congenital malformations in humans.

Presentation of the well-documented results in an organized form, with a comprehensive, cross-referenced index, should make the contents of the volume quickly accessible to students, teachers, and investigators in the field of developmental biology, as well as to many other biologists, particularly medical and veterinary, interested in the subject of teratology.

This volume is the fourth and last in the series of monographic studies of the avian egg and embryo. In size it is comparable to the previous volume *Biochemistry of the Avian Embryo: A Quantitative Analysis of Prenatal Development* (1967), a highly condensed discussion. The scope of literature reviewed is about the same as in the two earlier extensive treatises *The Avian Egg* (1949) and *The Avian Embryo: Structural and Functional Development* (1960).

Special Features of the Text

The preparation of this volume is based on the analytical study of the scientific material available in more than 6000 published papers, most of them recent. The significant experimental data on pathological changes resulting from the application of various teratogens were carefully evaluated, frequently recalculated, and condensed for suitable text tables. In some cases they were incorporated into large numbered tables or plotted graphically for clearer presentation of important biological relationships.

In this study the chief emphasis has been placed on the quantitative evaluation of the incidence of various morbid syndromes and embryonic mortality which disclose characteristic effects of the tested teratogen. Whenever possible, the occurrence of each specific anomaly is presented numerically, expressed in percentages either of the total number of treated fertile eggs or of the total malformed (dead and alive) embryos.

The selected data on gross anatomical changes of the avian embryos are derived primarily from the experimental findings *in vivo* or *in situ*, with only occasional supporting data obtained *in vitro*. The conditions in the tissue culture are seldom fully comparable with the environment of the living organism (see Fig. 183, in Appendix). Exceptions have been made in the studies of growth changes demonstrated by mitosis and exchange of respiratory gases.

Of the 21 numbered tables, nine are summary tables (Tables 2, 3, 8, 9, 11, 12, 13, 14, and 15) which because of their excessive length or general information are placed at the end of each corresponding chapter.

The 192 figures, consisting of freehand sketches and graphic illustrations (prepared by the author), are intended to demonstrate qualitatively and quantitatively the nature and extent of damage to the embryo caused by the dose level, routes of introduction, and the time of application of each teratogenic agent.

All teratogenic agents, especially those listed in the summary tables, are arranged alphabetically irrespective of their pharmacological and other uses.

Terminology of the chemical teratogens, especially organic compounds and drugs, follows that given in *The Merck Index*, eighth edition, without referring to those terms which have been used in the original publications.

The *Appendix* provides additional data, particularly on the natural variability in various characteristics of the chicken eggs as an aid to their selection for uniformity in their use as an experimental material in medical and biological research.

For the convenience of the reader, a brief *Glossary* of Greek medical terms for malformations is enclosed.

The unified *Bibliography* to all publications cited in the text, tables, and illustrations, is arranged alphabetically according to the senior authors. References to the pages of citation thus provide an easy access to original sources of information for those who wish to investigate further the subjects of their interest.

A comprehensive, cross-referenced subject *Index* has multiple entries for all teratogens and all types of malformation, using both medical and vernacular terms. Identification is made of pertinent topics, key words, and the numbered tables and illustrations.

Both *Bibliography* and subject *Index* are important components of this monographic volume.

Acknowledgment

This analytical study and preparation of the manuscript were supported in part by grant LM 00495 from the National Library of Medicine, National Institute of Health, DHEW.

I am grateful for the cooperative efforts of various officers of Cornell University in bringing to fruition this task, long unfinished—the outgrowth of my busy academic life—as well as for the pleasant and considerate environment created by my colleagues and immediate associates.

ALEXIS L. ROMANOFF

Ithaca, New York
February 1972

Contents

List of Tables

* Summary tables, placed at the ends of chapters.

Pathogenesis of the Avian Embryo

Introductory Note

To save unborn—ennobled task—
Needs morbid causes to unmask.

Characteristics of Prenatal Development

The period of prenatal development of the avian embryo, like that of other living organisms, is the most delicate, hazardous, and highly consequential in the individual's existence.

On the one hand, the pattern of development is hereditarily predetermined: the embryo receives the constitutional endowment of its parents. On the other hand, during incubation the embryo is continuously subjected to the influence of numerous environmental forces, which may facilitate or interfere with its normal course of development, health, and survival before or after hatching. Although the heredity is beyond immediate control, the environmental conditions can, for the most part, be controlled or modified at will by experimental procedures, particularly in the avian embryo.

Natural Protective Mechanisms. The avian embryo—unlike the mammalian embryo—develops outside the mother's body. It is left on its own, surrounded not only by uncertain atmospheric conditions, but also by a hostile microbial environment. To meet the situation, the embryo is provided with a series of the most elaborate protective mechanisms. These mechanisms act against some of the microorganisms which can penetrate the eggshell from the outside, and especially protect the embryo from the microorganisms most commonly found in the organs of the mother.

Among the most prominent structures of the egg resisting outside infection are the cuticle of the eggshell (*Horowitz, 1903; Vadehra, Baker, and Naylor, 1970*), shell membranes (*Zagaevsky and Lutikova, 1944*), the egg albumen (*Fleming and Allison, 1924; Korotkova, 1955a; 1955b; 1957*), the amniotic fluid (*Dmitrenko, 1954*), and the allantoic fluid (*Dmitrenko, 1954*), among many others. For details on antibacterial properties of eggs, see *The Avian Egg*, pp. 250, 249–507, 653, 683, 684, by Romanoff and Romanoff (*1949*).

It has been observed (*Teriukhanov, Pogoniailo, and Tivelev, 1965*) that chicken eggs during the first 14 days of incubation are practically free from microbial, viral, and fungal infections. Examination of embryos dying at 4, 14, and 19 days of incubation revealed microbial infection only in the oldest group, and then in a small percentage (about 12 per cent).

Experiments with the Avian Embryo

The avian embryo, as an allantoic vertebrate, is confined within a closed system, the egg. It is isolated from maternal dependence and influences—unlike placental animals. A teratogen introduced into an egg system will remain there throughout the developmental period, and will not be lost externally through excretion. For that reason, chicken eggs especially have long been a standard material for studies of teratology. These eggs are abundant at all seasons, low in cost per unit, simple to handle in sufficient numbers, large enough to be examined with ease, and capable of giving results of the tests in a relatively short period.

In studies of congenital malformations, by using large numbers of experimental avian eggs, it is possible to conduct quantitative observations, which can reduce the effects of inherent variability (see Appendix) and provide for each teratogen the most diagnostically reliable values on the incidences of various pathogenic phenomena, and their specific syndromes.

Historic Use of the Avian Embryo. The avian embryo, particularly of the chicken, has long been a most popular experimental material. But the earliest workers were interested chiefly in the descriptive anatomy of the abnormalities, and no attempts were made to explain how the abnormalities appeared.

The nineteenth century saw the beginning of experimental teratogenesis on the avian embryo. The eggs were subjected to such varied treatments as temperature (*Prevost and Dumas, 1825; Dareste, 1891*), atmospheric pressure (*Giacomini, 1894*), poisonous gases and vapors (*Féré 1893–1899,* see Bibliography), mechanical shocks (*Marcacci, 1893*), electricity (*Lombardini, 1868*), and magnetism (*Windle, 1895*). On the whole, these different treatments seem to have produced similar types of abnormalities.

Since the beginning of the twentieth century, with the development of various techniques for the studies of the avian embryo, attention has been directed to the mode of occurrence of structural anomalies and their specificity with individual teratogenic agents.

Special Advantages of the Avian Embryo. During recent years, the developing chicken embryo has provided a valuable biological model for the study of toxicity and teratogenesis by numerous investigators.

Few vertebrates have proved as well suited for genetic study as the avian embryo, and especially the domestic chicken (Landauer and others, see Chapter III on hereditary factors). Adequate techniques are now available so that experimental interference with development may be achieved with relative ease.

Because the avian embryo relies solely on the yolk and albumen of the egg for its food supply until hatching, it has been used in studies of various nutritional deficiencies and excesses to great advantage. Moreover, the vitamin and mineral content can be altered by the controlled feeding of the hen laying the egg. These and many other food nutrients can also be added to the yolk and albumen through injection (see Chapter VI).

The avian embryo has been recognized as an excellent tool for the study of radiation effect on embryonic tissues, especially in local radiation (*Wolff, 1936b*). The separation of the embryos from many maternal contributions permits evaluation of responses attributed solely to the effects of ionizing radiation of the embryo. The strict environmental controls that can be maintained during chick embryogenesis minimize the number of variables that may alter the radiation response (see *Boland and Tranter, 1957; Clapp, 1964*).

The avian embryo is well adapted to a variety of bacteriological and viral experiments. Internally it is free from active microorganisms. It comes enclosed in the shell, which is readily opened for inoculation and other manipulations, and which can easily be sealed again for the period of reincubation. Also, the potential of the avian embryo as a tool for the study of infection and natural resistance has been emphasized (*Finkelstein and Ramm, 1962*).

The avian embryo's location outside the maternal organism has proved an advantage in chemical experimentation. In the production of malformations by means of chemical substances, chemoteratogens, the avian embryo is more suitable than the mammalian embryo (*Ancel, 1950*), because the direct effects of the chemical compounds and drugs can be observed.

The avian embryo is a good object for studying the effects of gaseous anesthetic agents because of its isolation from the maternal complications of anesthesia. In addition, secondary effects of anesthesia, found in the mammal, are eliminated from the study of the pharmacological effects on the embryo itself (*Smith, Gaub, and Moya, 1965*).

Application of Experimental Results

Most present knowledge about the possible causes of congenital anomalies and prenatal death has been derived from the already available extensive studies of the avian embryo. In spite of some basic differences in morphology and physiology between allantoic (avian) and placental (mammalian) vertebrates, the studies of toxicity and teratology of the avian embryo itself, as a whole or in part, are directly applicable to the mammalian and human embryos (*Dareste, 1891; Ancel, 1950*).

It is also obvious that the effective doses of different teratogens administered to the avian embryo could not be given directly to other species. Indirectly, the results are useful to the purposes of preventive medicine long before any clinical tests have been carried out on humans.

There is enough evidence to assume that the avian embryo still remains the most suitable experimental material among vertebrates, especially in preliminary quantitative evaluation for lethal and teratogenic effects of various harmful agents in prenatal development.

I

Dynamics of Development

Potentialities in life
Throughout the early tender age
Surpass environmental strife . . .
And expectations of the sage.

The developing avian embryo is structurally and functionally unlike an adult. It is a rapidly differentiating and growing system, possessing enormous potentialities for development. Under the influence of various internal and external morbid agents, the developmental processes can be disturbed or changed from normal to anomalous or even monstrous. The younger the embryo is, the greater the developmental divergencies that can be produced.

PERIODS OF EMBRYONIC SENSITIVITY

In the studies of congenital malformations, the embryonic period in Aves can be divided into several phases. Each phase is characterized by certain physiological responses of the embryo to various experimental procedures. These phases extend in sequence from fertilization (before laying of the egg) to the time of hatching.

Totipotentiality in Development

The most striking feature of the early phase is its totipotentiality for development. The blastoderm of the bird at the time of egg laying, that is, about 24 hours from the time of fertilization, consists of a two-layered structure containing close to 3000 cells (according to *Spratt and Haas, 1960,* there are approximately 60,000 cells, including those forming various extraembryonic structures).

It has been shown *in vivo* that if this cluster of cells in the blastoderm is mechanically cut into two equal parts, each part will form a normal embryo. If the blastoderm is cut into three parts, it will yield three embryos, and four parts may produce four normal embryos. This is outlined in the table on page 8.

The potentialities of different regions of unincubated duck blastoderm (*Lutz, Departout, Hubert, and Pieau, 1963*) are confirmed by the proportional development of incomplete parts of the embryo.

Fissuration in Relation to Presumptive Axis	Types of Duplicity	Reference
Two sections		Wolff (1948); Lutz (1949); Lutz, Beaudenon, and Mesure (1965)
Parallel	Complete twins	
Perpendicular		
Three sections		Lutz (1949)
Radially		
Parallel	Complete triplets	
Perpendicular		
Four sections		Wolff (1948); Lutz (1949)
Cross		
Parallel	Complete quadruplets	
Perpendicular		

This great regularity in the potentialities of the blastoderm was also observed in similar fragments cultivated *in vitro* (*Spratt and Haas, 1960*).

All these tests lead to the conclusion that the unincubated blastoderm in Aves is totipotential.

Localization of Potency

The blastoderm, after being incubated about 16 hours (that is, upon reaching the primitive streak stage), shifts its localized potency for development posteriorly (Fig. 1*B*). Then, at 20 hours or older, it acquires an elaborate developmental pattern (Fig. 1*C*), each section of which, in general, corresponds to a primordium of a definite organ or of a system (see also *Rawles, 1943; Romanoff, 1960, p. 181*).

From the first day on, damage by experimental procedures may not be general; instead, it is very likely to be limited to a local area of the developing embryo, such as the central nervous system, or the respiratory, the digestive, or the skeletal system (cf. effect of localized irradiation by X-rays, and Figs. 71, 72, and 73).

Now, it is also a well-established fact (*Lenicque, 1959*) that the above areas of histogenetic activity, being without boundaries, partly overlap each other. Thus during the process of differentiation, the specialized tissues can act on each other and modify the anticipated effects.

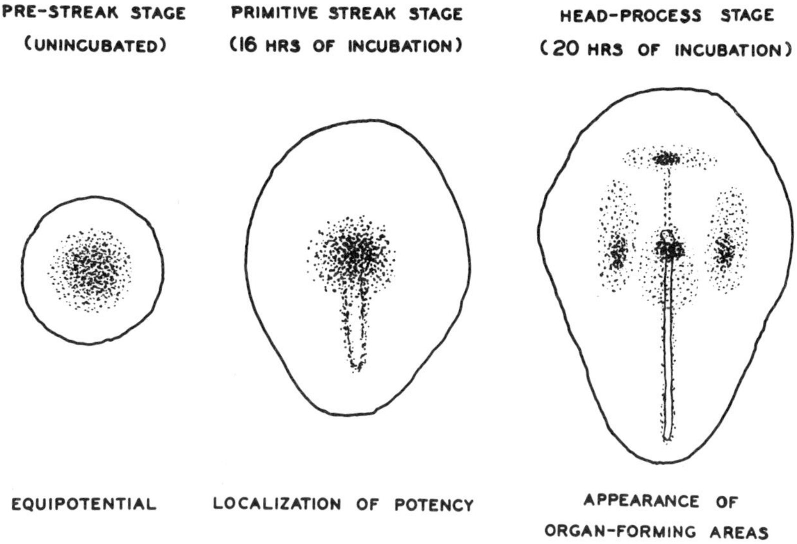

Fig. 1. Changes in the developmental capacity of the chicken blastoderm during the first 20 hours of incubation. (Romanoff, unpublished.) *A*, equipotentiality at the pre-streak stage (unincubated blastoderm); *B*, localization of potency at the primitive streak stage (about 16 hours of incubation); *C*, time of appearance of organ-forming areas at the head-process stage (about 20 hours of incubation).

Self-Differentiation

Between the second and fifth days of incubation of the chick embryo, as has been demonstrated (*Hamburger, 1938*), the limb primordia, transplanted either to the flank of a host embryo (of the same morphogenetic age) or into its intraembryonic coelom, often develop normally as extra limbs. These experiments also show that the controlling factor for the normal expression of the potencies of the limb primordium is not the axial orientation, but its location in the host with respect to neighboring structures possessing adequate food supply.

Similar self-differentiation of the skeleton (*Murray, 1926*) and musculature (*Hunt, 1932*) has been shown earlier, using the method of chorioallantoic grafting. More recently, a similar phenomenon was demonstrated with the lens primordium (*McKeehan, 1954*) grafted to the coelom of the host embryo.

From this it is apparent that when organ primordia are isolated from their original locale, they retain, in the new environment, the capacity for further differentiation along their specific course of development.

Functional Integration

During the midperiod of embryogenesis, between the eighth and twelfth days in the chicken embryo, development is marked by many adjustments to the initiation of functional activities by various organ systems. These include the transition from the mesonephric to the metanephric system, with the activity of the latter; the definitive expression of sex to the genetic male and the genetic female; and noticeable changes in the chemical composition of blood, among many others (*Willier, 1954*).

Growth and Maturation

The period from about the fourteenth day to the time of hatching is characterized by very rapid growth and by progressive increase in the amount of glandular tissues and their domination of functional activity. During this period, the embryo is prepared for independent existence outside the eggshell.

During each of these five stages of prenatal development in Aves, the embryo shows a highly characteristic response to external environmental changes and to various teratogenic agents.

EMBRYOLOGICAL BASIS FOR TERATOLOGY

The developing avian embryo, as was stated before, is a rapidly differentiating and growing system. In the course of this highly delicate process, a distortion of normal development may easily occur. Some aberrant manifestations are closely associated with several specific developmental phenomena in embryonic life, such as power of organization, inductive capacity, ability for self-regeneration, faculty of wound healing, and survival of tissues after death.

Power of Organization

The developing blastoderm has a regulative potentiality for lateral parts as well as for anterior and posterior parts.

When fragments of the blastoderm are cultured, the developmental ability differs according to which fragments are taken and the stage at which they are isolated (*Fargeix, 1969*). When duck blastoderms are separated in the eggs before incubation, lateral halves are more likely to develop than anterior or posterior halves. But in the 6- and 9-hour-old blastoderms, the incidence of development diminishes for the lateral parts and increases for the anterior and posterior halves. Some results are shown in the tabulation below.

Age at Fissuration (hours)	Incidence of Development (%)	
	Lateral Parts	Anterior and Posterior Parts
6	8.5	4.6
9	3.4	5.3

It would seem that the most important center of organization exists during the first stages of embryonic development (*Fargeix, 1970*).

Inductive Capacity

There are inductors and inhibitors of embryonic differentiation, that is, agents that promote specialization and those that inhibit specification in a position which is already occupied.

The whole development of the embryo may be explained by a chain of inductions in which each structure or cell group (anlage) acts successively as an inductor and as a reacting tissue.

The following are examples of such processes:

1. Induction of the vertebral column by the notochord and the spinal cord (*Strudel, 1953; Watterson, Fowler, and Fowler, 1954; Grabowski, 1957*).
2. The development of supernumerary limbs by means of grafting the mesodermal components of a primary limb bud under the ectoderm of an indifferent area (*Kieny, 1960*).
3. The formation of the metanephros by two components, namely, the ureteral bud and its related mesenchyme (*Grobstein, 1955; Calame, 1961; Wolff, 1966*).

Induction of Lens Formation. It has been shown (*Wedlock and McCallion, 1968*) that a new lens may appear in the explanted optic vesicle after excision of the original lens only if ectoderm is included with the explant. The frequency of lens formation in cultures of optic vesicles and ectoderm from embryos of various ages is shown in the accompanying table.

Age of Embryo (hours)	Frequency of Lens Formation (%)
48	70
60	33
68	20
72	0

Induction of Secondary Embryos. Regionally differentiated secondary embryos can be induced by extensive contact between host ectoderm and graft mesoderm 18 to 19 days of age (*Grabowski, 1957*). The secondary embryos, one-half to three-quarters normal size, are complete in every detail and are composed of tissues derived from the ectoderm of the host and the mesoderm of the graft.

Ability to Self-Regenerate

The regenerative capacity of embryonic organs has been recognized by experimental embryologists. Among these organs, the central nervous and skeletal systems are especially notable.

Central Nervous System. It has been observed (*Clearwaters, 1954*) that the younger the embryo at the time of operation (transection of the spinal cord at 2 to 5 days of incubation), the speedier and more complete is the repair.

It has also been noted (*Gutierrez, 1965*) that after removal of the diencephalic roof of the chicken embryo at 24 hours of age, normal epiphyseal regeneration consistently occurs. After a similar operation performed at the age of 72 hours, there is no regeneration.

The hindbrain of chicken embryos at 35 hours of age is capable of regeneration after extirpation of one half of one or two rhombomeres (*Källén, 1955*). The regenerating cell mass develops in a normal way in nearly half of the cases; in the rest, abnormalities of different kinds develop.

Skeletal System. It has been observed (*Zwilling, 1942*) that removal of the tail anlage from the chicken embryo is followed by a certain degree of restitution, or regeneration, of the tail. The capacity for regeneration is restricted by the age of the embryo and by the amount of material removed, as the accompanying data indicate.

Amount Removed	Incidence of Developed Tails (%)	
	1 Day of Age	2 Days of Age
Entire tail thickening	—	12
Entire streak	38	—
Anterior fraction	—	43
Posterior fraction	—	66
Posterior fraction of the streak	12	—

In embryos which failed to form a tail, the body usually ended abruptly at the posterior level of the hind limbs.

Capacity for Wound Healing

The faculty of chick embryos for healing wounds varies significantly with age (*Weiss and Matoltsy, 1959*). Wound healing *in vivo* does not occur until at least the tenth day of incubation.

When cultured on a plasma clot *in vitro,* both skin and cornea fragments are capable of migratory wound healing at all ages.

Survival of Tissues after Death

A certain amount of local growth in dead embryos, at 5 days or younger, has been observed in such tissues as liver and blood cells, as well as extraembryonic areas (*Byerly, 1926*).

In further studies, it was found (*Wilburg, 1937*) that various tissues can survive at 38° C from 3 hours to 9 days after death, as shown in the accompanying table.

Embryonic Tissue	Time of Survival
Heart	
Ventricle	3 hours
Atrium	$3\frac{1}{2}$ hours
Iris (of the eye)	5 hours
Brain	6 hours
Intestine	6 hours
Liver	6 hours
Skin	
Of the head	7 days
Of the thigh	8 days
Area pellucida (yolk sac)	9 days
Amnion	9 days

II

Spontaneous Malformations

Vast deviations from the norm
In crucial embryonic stage
May take a stunted phantom form
Which brings a new, most wondrous page.

Spontaneous malformations are considered to be only those abnormalities which occur independently of any experimental intervention. They include the deviations from normality due to the general influence of undetermined genetic constitution, or due to the response of the embryo to the accidental deviations of intrinsic and extrinsic environment during development, or both.

The present discussion is based primarily on the statistical evidence of egg infertility, of mode of embryonic mortality, and of malpositions and malformations of the embryo.

Most kinds of teratological development are fatal to the embryo before hatching.

INFERTILITY IN EGGS

The distribution of infertile eggs in a flock of 202 breeding hens is shown graphically in Fig. 2. The majority of these hens produced less than 10 per cent infertile eggs. Some produced up to 50 per cent infertile eggs, and in a few instances, the percentage was as high as 85 per cent. The average infertility for the entire flock was 11.2 per cent.

The above information is of some importance here, since the incidence of malformations is often calculated on the basis of fertile eggs examined.

(For detailed information on this subject, the reader is referred to Chapter II, on "Fertilization and Fertility," of Romanoff, 1960, *The Avian Embryo,* pp. 75–114.)

VARIANCE IN GROWTH

It is a well-established fact (*Romanoff and Romanoff, 1949, p. 230*) that the period between ovulation (or fertilization) and oviposition (or egg laying) extends from about 24 to 28 hours. Consequently the size

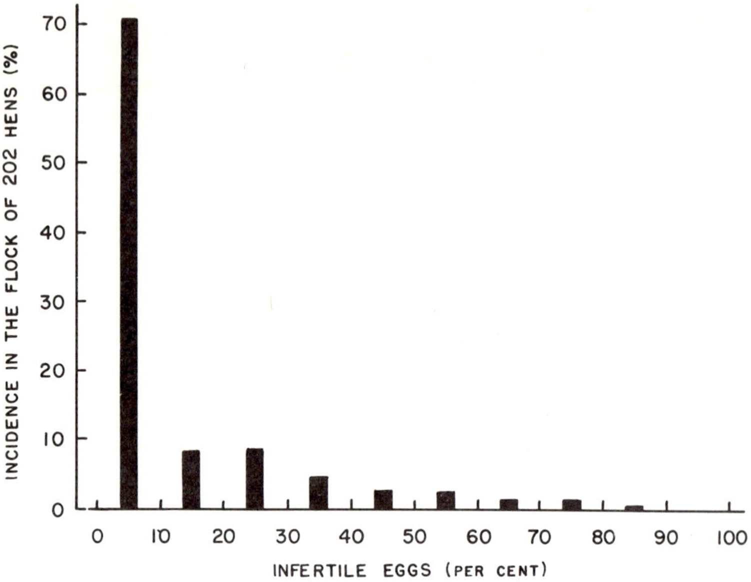

Fig. 2. Distribution of a flock of hens according to the fertility of their eggs. (After Romanoff, unpublished.)

of the blastoderms at the time of laying may vary (*Bernier, Taylor, and Gunns, 1951*), while they are still in the state of cellular differentiation (*Zusman, 1968*). The frequency distribution of the size of blastoderms is shown in Fig. 3.

Among 24-hour blastoderms of turkey embryos (*Olsen and Marsden, 1950*), those judged to be the largest in size and most clearly visible by candling had an average embryonic mortality at hatching of 12.4 per cent; the group of medium size, 35.9 per cent; and the group with the least-developed blastoderms, 71.6 per cent.

The difference in survival between the embryos of the three different groups was found to be due largely to mortality during the first week of incubation, as is shown below.

Size of Blastoderm at 24 Hours	Distribution of Mortality (%)		
	1–7 Days	8–20 Days	21–28 Days
Large	2.9	0.8	8.8
Medium	21.5	1.7	12.7
Small	61.9	0.8	8.9

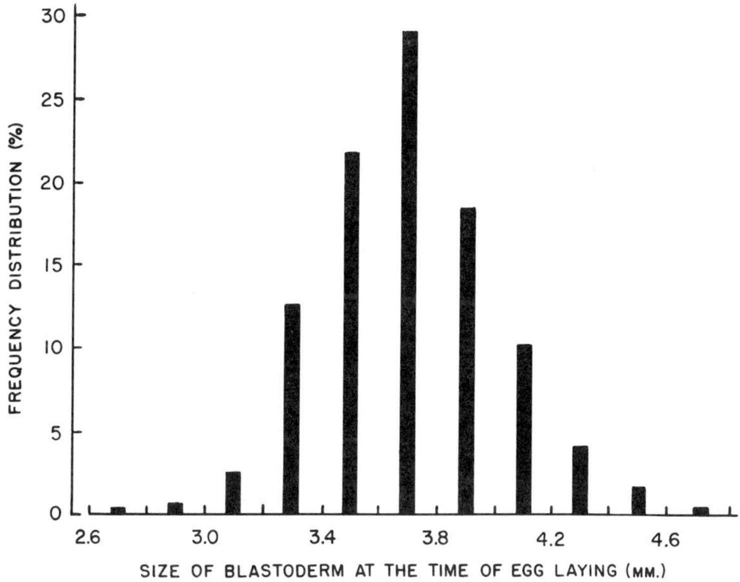

Fig. 3. Variation in the size of blastoderms at the time of egg laying resulting from the length of the period between ovulation, or fertilization, and oviposition, egg laying, as determined by the maximum diameter of 1677 blastoderms in unincubated eggs. (Plotted from the data of Bernier, Taylor, and Gunns, 1951.)

Slow-developing blastoderms are chiefly responsible for the so-called normal peak of mortality which occurs in the chicken on the fourth day of incubation (cf. Critical Periods, p. 25).

The relationship between the number of somites at 38 hours of incubation and the time interval between fertilization and egg laying (*Bernier, Taylor, and Gunns, 1951*) is illustrated in Fig. 4.

There is also correlation among the total time of development prior to laying, the size of the embryo, and the embryonic mortality. These relationships are shown in the accompanying table.

Time of Development in the Oviduct (hours)	Size of Embryo at 38 Hours (number of somites)	Embryonic Mortality at Hatching (%)
20–22	9.6	25.0
23–24	10.6	21.6
25–26	11.5	21.5
27–28	11.8	26.7
29–30	13.0	60.0

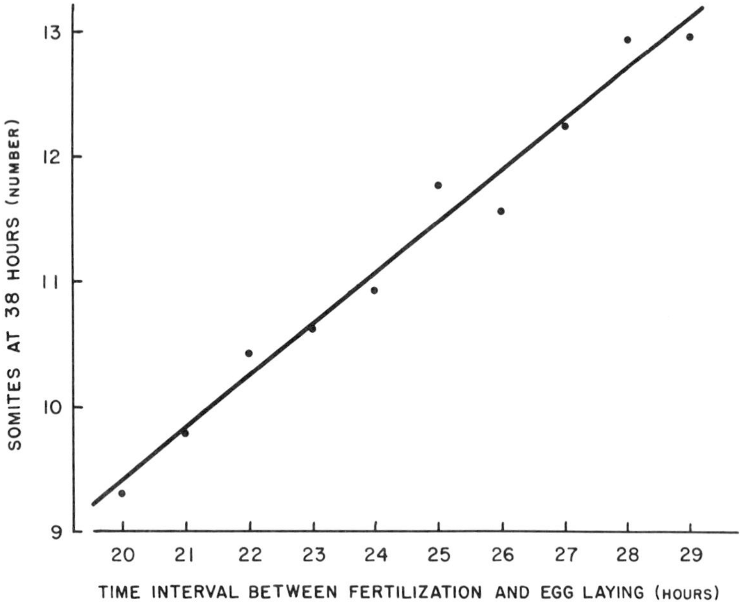

Fig. 4. Relationship between the number of somites at 38 hours of incubation and the time of development in the oviduct before egg laying. (Based on observations of 742 fertile eggs, by Bernier, Taylor, and Gunns, 1951.)

Eggs which are more advanced in development at the time of laying have a higher embryonic mortality than those with development times between 23 and 27 hours.

Variation in size of the blastoderm is considerably diminished with increase in the rate of egg production and shortening of the time interval between fertilization and egg laying to 23 to 25 hours—a normal time of retention of the egg yolk in the oviduct.

Variation in the growth of 20- and 48-hour embryos is shown in Fig. 5*A* and *B*. Here again the number of somites of the 48-hour embryos is obviously related to the total incidence of embryonic mortality at hatching (*McNally and Byerly, 1936*), as shown below.

Number of Somites	Embryonic Mortality (%)
14	60
16	54
18	52
20	42
22	64
23	78

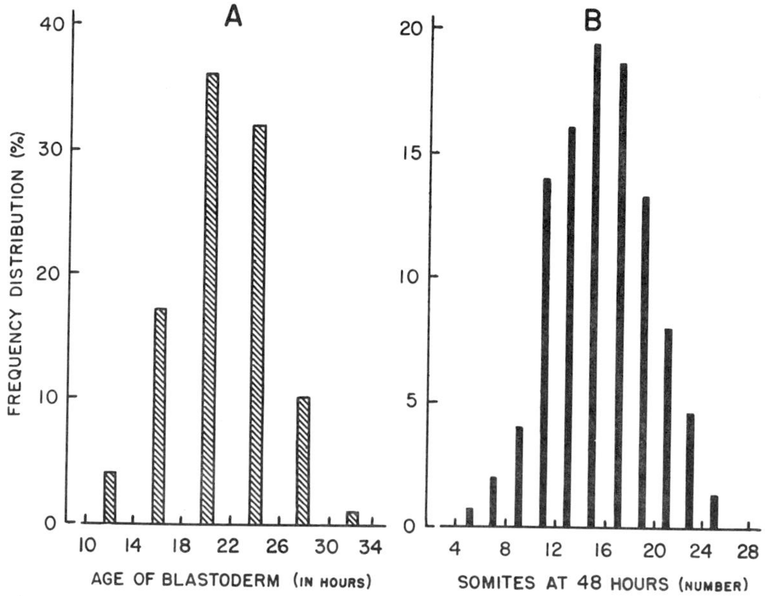

Fig. 5. Variance in the growth of 20- and 48-hour embryos. *A*, frequency dstribution of sizes of 424 embryos incubated for 20 hours, by corresponding ages (equivalent to about 10 to 34 hours) of average embryos (rearranged from the data of Lenicque, 1959); *B*, frequency distribution in somite numbers among 153 embryos about 48 hours of age (rearranged from data of Sadler, 1955).

The lowest embryonic mortality is observed in eggs with an average of about 20 somites, followed by a rapid increase in mortality in eggs averaging a smaller or greater number of somites.

Figure 6 shows dramatically the observed range in size of individual blastoderms at 40 and 66 hours of incubation. The great divergence in the morphological age of the embryo accounts for the variability in weight of embryos at 15 days of incubation (Fig. 7) and still more at 18 days (Fig. 8), just preceding the third critical period (see Fig. 13) when many undersized and some oversized morbid embryos begin to die.

Dwarfism

Study of the respiration of intact individual embryos (*Romanoff, 1940b*) reveals that the rate of oxygen consumption is directly proportional to the body weight of the living embryo.

Of 22 moribund embryos observed (*Romanoff, 1941*), three (14

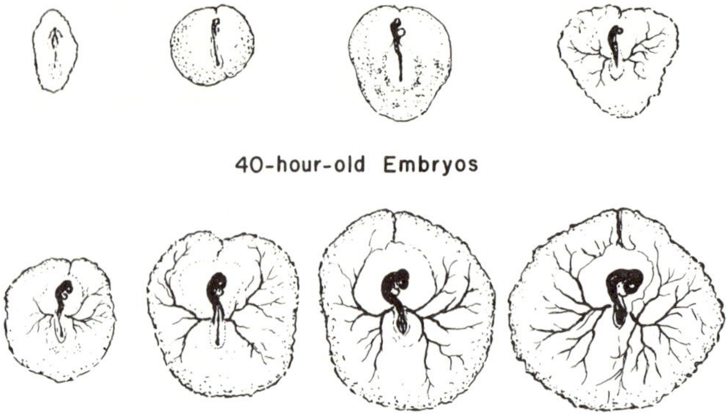

Fig. 6. Variability in the size and appearance of 40- and 66-hour blastoderms incubated under optimum conditions. (Sketched by the author from mounted specimens, after Romanoff, unpublished.)

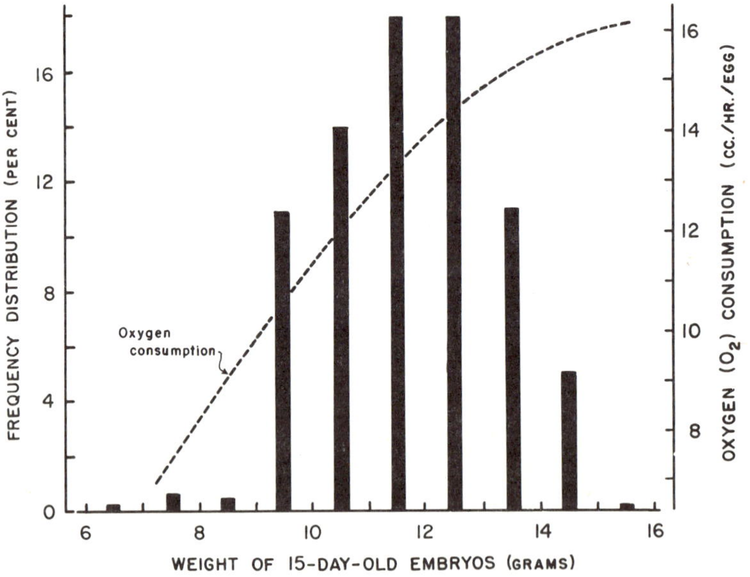

Fig. 7. Frequency distribution of weights of 15-day-old embryos and their rates of oxygen consumption. (Based on 87 observations of Romanoff, unpublished.)

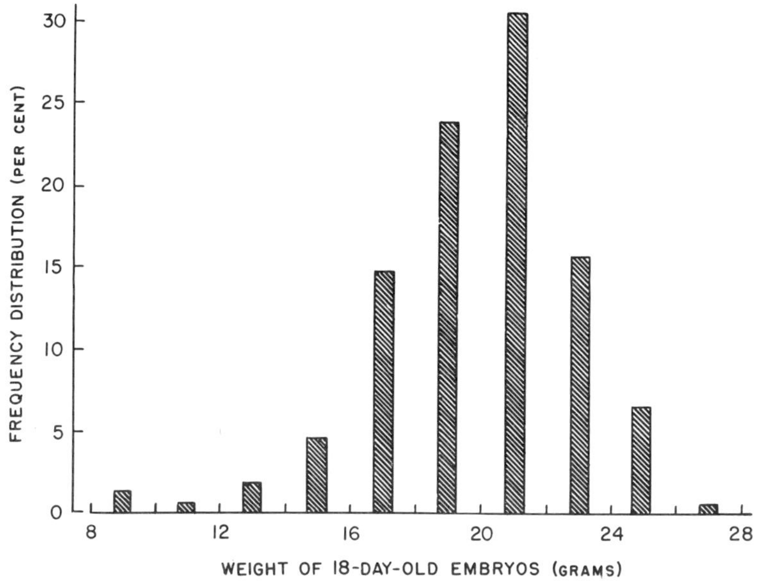

Fig. 8. Frequency distribution of the weights of 18-day-old chicken embryos, based on 446 observations. (Romanoff, unpublished.)

per cent) showed a marked decrease in total oxygen consumption during the 4 or 5 days before death. In the embryo *C* (Fig. 9), for example, the respiratory exchange on the day previous to death was about 60 per cent below the normal value for embryos of that age.

The above experiments illustrate the spontaneous appearance of dwarf embryos. This phenomenon could be caused by a disturbance to the developmental pattern by some general alteration of physiological functions. It may be initiated by factors such as thyroid abnormalities or nutritional deficiencies, or it may be genetic—nonhereditary or hereditary, a lethal mutation (*Landauer, 1940; Asmundson, 1945*).

DISTRIBUTION OF HATCHING

The distribution of hatching of several representative species of birds (Fig. 10) shows that this process usually occurs within very narrow time limits. Only a very few embryos hatch before or after the average time for a species.

Under optimum conditions of incubation (*Romanoff, 1967, p. viii*) any differences in weight of hatched chicks with an accelerated or re-

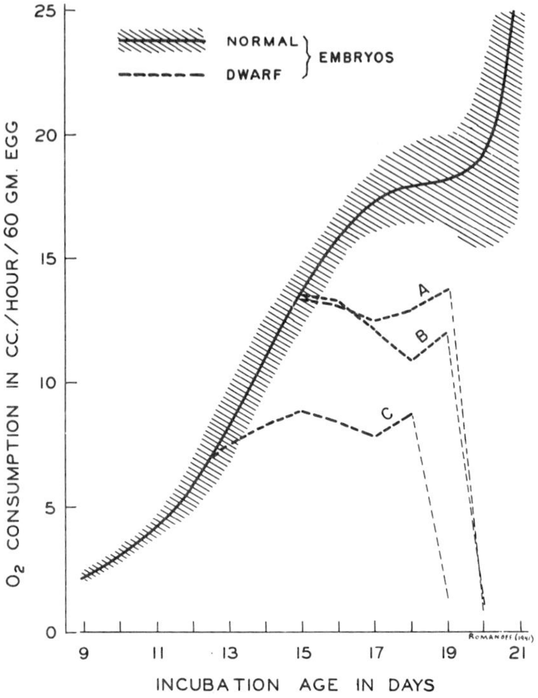

Fig. 9. Detection of spontaneous dwarfism in chicken embryos by the daily measurement of oxygen consumption of the intact incubating eggs. (After Romanoff, 1940b, and Romanoff 1941.)

tarded time of hatch result chiefly from variations in the original weight of the egg (*Romanoff, 1933a; 1960, p. 1148*), as shown below.

Original Weight of Egg (grams)	Weight of Embryo at Hatching (grams)
30	18.8
40	28.8
50	31.0
60	35.0
70	38.4

There is a positive correlation between the original egg weight and the weight of the newly hatched chick (without the reserve yolk sac), although the individual variability (Fig. 11) is very high.

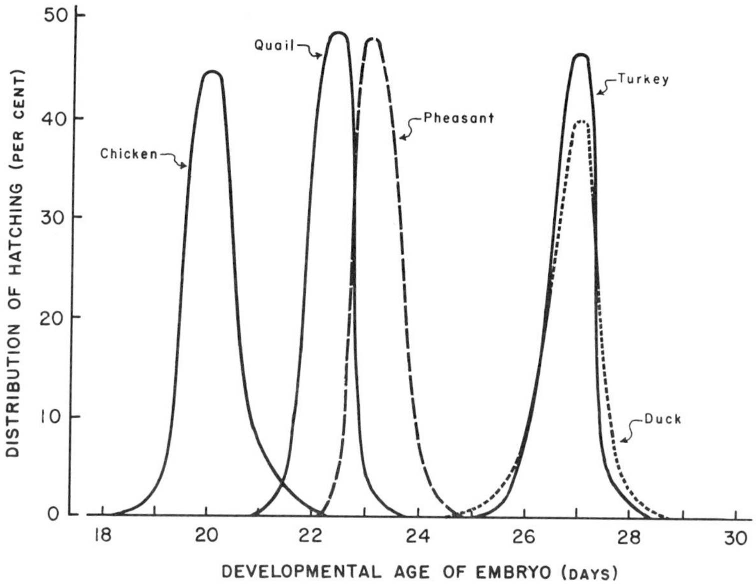

Fig. 10. Frequency distribution of hatching eggs within homogeneous populations of birds. (Observations were made on 1169 chicken eggs, *Gallus gallus*—after Romanoff, Smith, and Sullivan, 1938; on 976 quail eggs, *Colinus virginianus*—after Romanoff, 1934, 1938; on 2230 pheasant eggs, *Phasianus colchicus*—after Romanoff, 1934, 1938; on 351 turkey eggs, *Meleagris gallopavo*—after Romanoff, 1938; and on 998 duck eggs, *Anas platyrhynchos*—after Romanoff, 1943a, and Romanoff, unpublished.)

EMBRYONIC MORTALITY

Numerous observations confirm that the incidence of embryonic survival at hatching is exceedingly consistent for each individual bird. Figure 12 gives the frequency distribution of a flock of 475 pullets (the first year of laying) according to their averages of embryonic survival at hatching. From this chart it is obvious that a few birds have a very low embryonic survival, whereas others have nearly 100 per cent.

Critical Periods. It is well recognized by avian embryologists that the distribution of embryonic mortality throughout the incubation time falls into three critical periods. In the chicken, these periods of susceptibility occur at about the fourth, the eleventh, and the nineteenth days of incubation (Fig. 13*A* and *B*), though susceptibility at the eleventh day is pronounced only under certain unfavorable conditions.

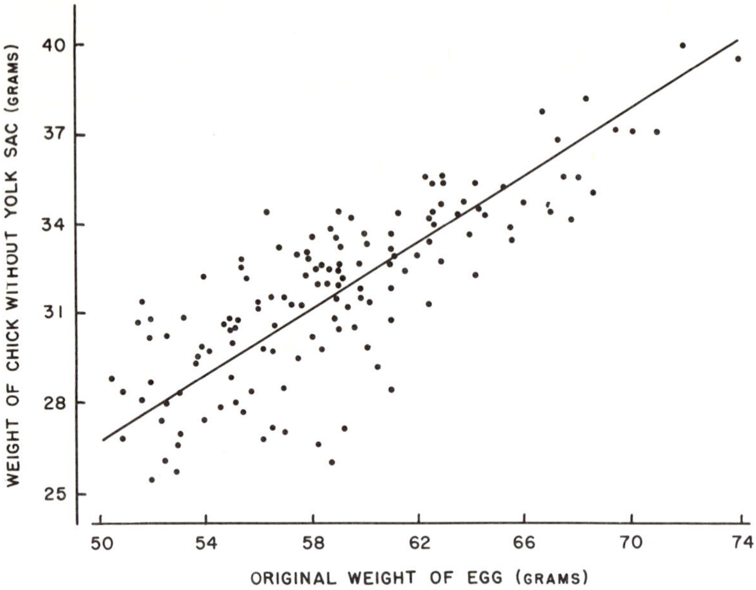

Fig. 11. **Relationship between the initial weight of eggs and the weight of newly hatched chicks (without their spare yolks).** (Based on experimental observations of Romanoff, 1933a.)

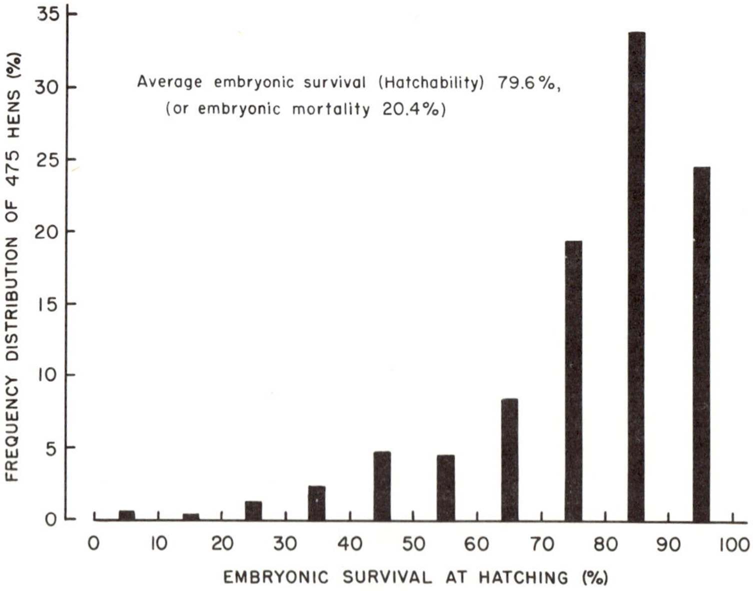

Fig. 12. **Frequency distribution of 475 hens (first year) according to the average embryonic survival at hatching of their fertile eggs.** (Plotted from the data of Titus, 1953.)

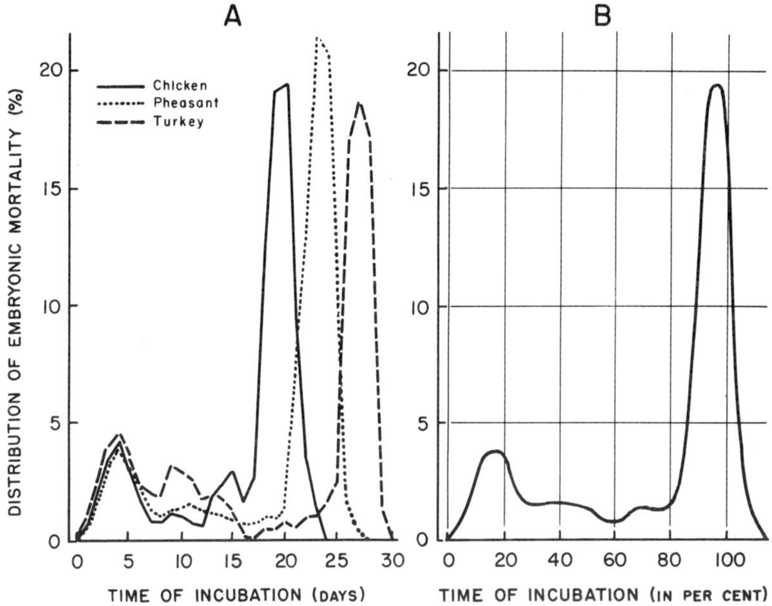

Fig. 13. Frequency distribution of embryonic mortality during the incubation period. (Plotted from the data of Romanoff, unpublished, and Romanoff, 1949; see also Martin and Insko, 1935, on turkey eggs.) *A*, distribution of 1452 dead chicken embryos (*Gallus gallus*), 2230 dead ring-necked pheasant embryos (*Phasianus colchicus*), and 168 dead turkey embryos (*Meleagris gallopavo*); *B*, distribution of mortality of the three species combined, plotted against the percentage of developmental time. *Note:* similar distribution curves of embryonic mortality have been observed in quail (Romanoff, 1934), ducks (Romanoff, 1943a), pigeons (Riddle, 1930), and others. See also Needham's (1942, p. 64) polygons showing normal distribution of embryonic mortality in several species of birds.

MALPOSITIONS

During incubation the chicken embryo usually orients itself so that the body is parallel to the long axis of the egg, with the head toward the blunt end. The normal position of the chick at hatching is therefore with the head at the blunt end of the egg, turned to the right and lying under the right wing (Fig. 14*A* and *B*). Since the beak is pointed out toward the air cell, this position is most natural and facilitates the establishment of pulmonary respiration with piping through a membrane (Fig. 14*B*) long before the actual breaking of the shell (Fig. 14*C*).

Types of Malpositions. There are several unusual positions which may hinder or definitely prevent the embryo from hatching. These positions are commonly referred to as malpositions.

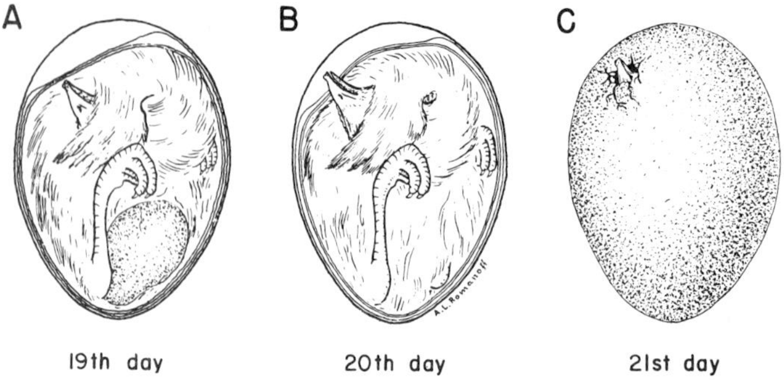

Fig. 14. The normal position of the chicken embryo at the time of hatching. (After Romanoff, original, unpublished.) *A*, prior to the breaking of membranes (amnion, allantois, and egg membrane); *B*, after the breaking of membranes; *C*, at the time of making a hole in the shell.

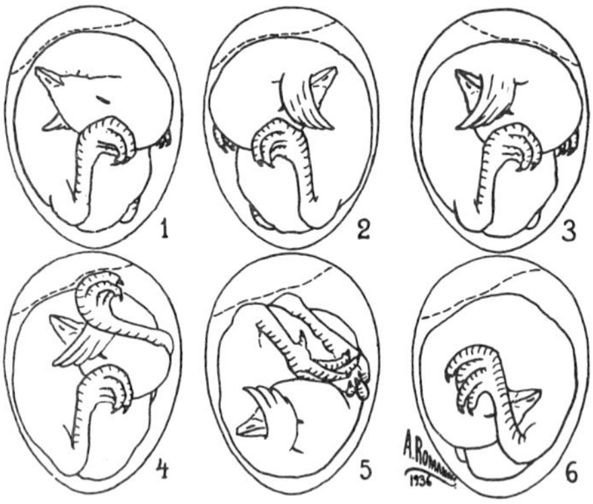

Fig. 15. Various forms of malposition of the chicken embryo observed prior to hatching. (After Romanoff, 1936b; see Sanctuary, 1925; Hutt, 1929; Byerly, 1930b.) Positions of the embryo: *1*, normal except that beak lies above instead of under the right wing; *2*, head turned to the left instead of under the right wing; *3*, head normal, but embryo rotated so that the beak is turned away from the air cell; *4*, leg over the head; *5*, head at the small end; *6*, head between the thighs.

The malpositions that have been observed in the chicken embryo (*Sanctuary, 1925; Hutt, 1929; Byerly, 1930b;* and others) are shown in Fig. 15. They may be classified as follows:

1. The head of the embryo is at the blunt end of the egg, turned to the right and positioned over instead of under the right wing.

2. The head of the embryo is again at the blunt end of the egg, but turned to the left instead of the right. Consequently, the beak is pointed in the opposite direction from the air cell.

3. The head position is normal, but the whole embryo is rotated away from the air cell.

4. The head of the embryo is at the blunt end of the egg, with one leg on top of the head or over the beak.

5. The embryo is upside down. That is, the head is at the sharp end of the egg and the feet at the blunt end.

6. The head of the chick is buried between the thighs.

Although there might be a few more odd types of malposition, they are either rare or are modifications of one of the forms described above.

Incidence of Malpositions. Under ordinary conditions various forms of malpositions occur naturally with a consistent frequency. It has been shown (*Byerly, 1930b*) that the most frequent malposition is number *5*, with the head in the small end of the egg, and the least frequent is number *1*, with the beak over the right wing (Fig. 15).

Observations of dead embryos at the time of hatching indicate that over one half of them are in malposition, as shown by the data below.

Dead Embryos Examined	Total in Malposition (%)	Reference
1490	51.7	Sanctuary (1925)
5050	55–85	Hutt (1929a)
395	56.4	Byerly (1930b)

Relation to Mortality. The frequency of malpositions is related to the total mortality. If the number of dead embryos, especially during the latter part of incubation, increases, the number among the dead in various malpositions also increases, but at a much greater rate. The failure of embryos to attain normal position for hatching might account for the peak of mortality at the last critical period (Fig. 13).

The 20-day embryos in conventional malpositions *6*—head between thighs, *5*—head in small end of the egg, and *2*—head to the left, as detected by X-rays (*Robertson, 1961b*), have the following embryonic mortality.

Malposition	Embryonic Mortality (%)
Head between thighs (*6*)	55.6
Head in small end of the egg (*5*)	55.1
Head to left (*2*)	20.0

MALFORMATIONS

Spontaneous malformations have been observed by several investigators in a large number of eggs with dead embryos. The table below gives the total incidence of terata in the chicken eggs.

Number of Eggs Examined	Incidence of Terata (%)	Reference
—	6.5	Alsop (1919)
3,000	5.0	Dunn (1923b)
11,797	8.1	Hutt and Greenwood (1929)
2,087	9.5	Byerly (1930b)
286	5.0	Blackwood (1960)

Types and Incidence of Malformations

The relative statistical occurrence of individual types of spontaneous malformations is presented in Table 1.

The most frequently occurring types of malformations are arranged in descending order below.

Anomaly	Incidence (%)
Monomicrophthalmia (smallness of one eye)	1.9
Anterior abnormalities	1.8
Microphthalmia (small eyes)	1.3
Posterior abnormalities	1.2
Short arched beak	1.1
Microphthalmia (unilateral or bilateral)	1.1
Chondrodystrophia (morbidity of cartilage)	1.0
Exencephalia (exposure of the brain)	0.9
Duplicitas (double monsters)	0.8
Cranioschisis (fissure of the skull)	0.7
Inversion of the heart	0.7
Ectopia (displacement of organs)	0.5

In many cases the malformed embryos show more than one of the malformations listed in Table 1, and therefore are recorded more than once. This is especially noticeable in the occurrence of experimentally induced malformations.

Table 1 Incidence of Spontaneous Malformations in Avian Embryos

Type of Malformation	Embryos Observed	Anomalies (%)	Reference
Acardia (absence of the heart)	11,797	0.010	2
Acleiencephalia (imperfectly closed brain)	773	0.039	6
Anencephalia (lack of the brain)	2,198	0.180	3
Anencephalia, unilateral or bilateral	1,448	0.120	7
Anencephalia	621	0.480	12
Anophthalmia (absence of the eye)	1,448	0.480	7
Anophthalmia	284	0.340	10
Anophthalmia	621	0.160	12
Anterior abnormalities	3,413	1.750	11
Anouria (taillessness)	621	0.320	12
Beak			
Crooked	3,933	0.483	13
Cross	1,448	0.280	7
Cross	621	0.480	12
Malformed	17,771	0.020	2, 4
Malformed	3,933	0.305	13
Mandibles, both absent	3,933	0.025	13
Upper, absent	2,198	0.140	3
Upper absent	284	0.340	10
Upper absent	3,933	0.280	13
Upper, arched	621	1.130	7, 12
Upper, short	1,448	0.760	7
Brain, mid-, folded	284	0.340	10
Celosomia (hernia of the viscera)	773	0.130	6
Cerebral hernia	1,448	0.070	7
Chondrodystrophia (morbid cartilage)	2,198	1.000	3
Cleft palate (facial cleft)	284	0.340	10
Cranioschisis (fissure of the skull)	284	0.680	10
Cyclopia (both eyes fused medially)	11,797	0.020	2
Duplicitas (double monsters)	11,797	0.060	2
Duplicatas	5,974	0.070	4
Duplicatas	773	0.065	6
Duplicatas	708	0.800	8
Duplicity			
Anterior	2,198	0.090	3
Posterior	2,198	0.360	3

Table 1 *(Continued)*

Type of Malformation	Embryos Observed	Anomalies (%)	Reference
Ectopia (displacement of organs) see also Heterotaxia (5)	11,797	0.050	2
Ectopia	2,198	0.016	3
Ectopia	5,974	0.013	4
Ectopia	708	0.510	8
Ectopia	621	0.480	12
Encephalocele (hernia of the brain)	284	0.340	10
see also Exencephalia (1, 4, 8)			
Exencephalia (exposure of the brain)	11,797	0.850	2
Exencephalia	5,974	0.160	4
Exencephalia	708	0.100	8
Eyes			
Both absent	3,933	0.178	13
One absent	3,933	0.153	13
Head, abnormal	3,933	0.661	13
Heart, inverted	2,449	0.680	11
Heterotaxia (transposition of viscera)	14,073	0.163	5
see also Ectopia (2, 3, 4, 8)			
Hyperencephalia (lacking sinciput)	11,797	0.520	2
Hyperencephalia	2,198	0.360	3
Hyperencephalia	5,974	0.130	4
Hyperencephalia	708	0.240	8
Limbs, malformed	621	0.113	12
Lordosis (abnormal curvature of the spine)	773	0.260	6
Malformations			
Anterior	3,413	1.750	11
Posterior	3,413	1.160	11
Microphthalmia (small eyes)	11,797	0.860	2
Microphthalmia	2,198	1.270	3
Microphthalmia	5,974	1.270	4
Unilateral	773	0.390	6
Unilateral or bilateral	1,448	0.280	7
Unilateral or bilateral	708	0.796	8
Unilateral or bilateral	284	1.020	10
Monomicrophthalmia (one eye small)	2,198	1.865	3
Octocephalia (ears united below the face)	11,797	0.010	2
	5,974	0.012	4

Table 1 (*Continued*)

Type of Malformation	Embryos Observed	Anomalies (%)	Reference
Posterior abnormalities	3,413	1.160	11
Prognathism (projection of jaws)	11,797	0.070	2
	2,198	0.090	3
	5,974	0.070	4
Prosophthalmia (forward projection of eyes)	2,198	0.140	3
Trunk, anomalies of posterior	773	0.390	6
Twins (pigeons and doves)	20,000	0.004	1
	708	0.240	8
	607	0.179	9

Note: In most cases the data on the incidence of various malformations are calculated either from the total fertile eggs incubated (*), or from the total dead embryos examined (†).

[1] Riddle (1923). [2†] Hutt and Greenwood (1929). [3†] Byerly (1930b). [4†] Hutt (1930). [5*] Taylor (1934). [6] Wolff (1936)—cited by Lutz-Ostertag and Senaud (1966). [7*] Landauer and Baumann (1943). [8*] Olsen and Haynes (1948). [9†] Ikota and Kondo (1953). [10†] Grabowski and Paar (1958). [11†] Goff (1959). [12] Duraiswami (1961). [13†] William, Gopalakrishnan, and Raman (1969).

Abnormality	Incidence (%)
Heart	
Variation in shape and size	4.3
Dextrocardia	2.0
Transposition of vessels	0.8
Auricles, faulted	0.3
Apex, anomalies	0.8
Defects of arteries	
Pulmonary	1.8
Aorta	3.0
Ducti arteriosi	13.2
Carotids	1.5
Defects of veins	
Allantoic and vitelline	0.5

Anomalies of Heart and Blood Vessels. Upon close observation of 400 dead embryos at 18 to 20 days of incubation (*Marshall, 1951*), it was found that 121 embryos (32 per cent) demonstrated one or more abnormalities of the heart itself or of the blood vessels. These are summarized in the table at the bottom of page 33.

These spontaneous anomalies—in fertile eggs with 14 per cent total embryonic mortality—are classed as major causative factors in producing dead-in-shell embryos.

III

Genetic Mutations

Heredity may play its role
Through mutants of each parent cell:
Transmit defects, with full control
of Nature's freaks . . . inside the shell.

The great majority of congenital malformations in birds are caused by genetic factors (see Chapter II, Spontaneous Malformations). A variety of these hereditary defects are determined by factors transmitted through the germ cells from generation to generation. Often they appear in a regular Mendelian fashion.

These hereditary abnormalities, however, are susceptible to modification by the influence of environment, both physical and chemical. In a broad sense, each avian embryo is a product of heredity and environmental influence.

HYBRIDIZATION

Crossing of pure breeds of fowls would be likely to decrease the embryonic mortality in inverse proportion to the viability of the parent breeds (*Warren, 1942*), as shown below.

Crossing	Embryonic Mortality (%)
Females to males of the same breed	27.8
Females to males of different breeds	16.6

It also has been demonstrated (*Byerly, 1930a*) that the incidence of teratism in hybrids is reduced to below that in either parent stock; percentages are given in the table below.

Crossings	Incidence of Malformations (%)
"A" male × "A" female	5.94
"B" male × "B" female	3.71
"A" male × "B" female	2.76
"B" male × "A" female	3.57

Crossbreeding in turkeys (*Clark, Runnels, and Livesay, 1944*) resulted in decreased average embryonic mortality, the difference being 8.2 per cent of fertile eggs in favor of the crossbreeding.

It appears that no significant reduction in embryonic mortality can be expected from a crossing of families of birds with naturally high viability (*Landauer, 1967*).

Intergeneric Crossings

Hybridization usually becomes highly detrimental between more distantly related species.

It has been demonstrated (*Romanoff, 1960, pp. 109–111*) that fertile eggs can be obtained from intergeneric matings. Such matings may occur naturally among wild birds (*Suchetet, 1897; Guyer, 1909*). Embryonic mortality in such eggs is usually high, and it may go as high as 100 per cent.

Pekin and Muscovy Hybrids. Ducklings produced by Pekin females (*Anas platyrhynchos*) mated with Muscovy males (*Cairina moschata*) are usually sterile. The male testes have no sperm, and females are hermaphrodites.

The embryonic malformations observed (*Rigdon, 1968*) are located in the distal half of the body and in both legs. There is considerable scoliosis involving the thoracic and lumbar areas of the spine, with presence of spina bifida.

Inbreeding

Mortality. Numerous observations show that with extended close inbreeding—brother and sister mating—embryonic mortality increases. This increase takes place at different rates in each family, as shown in the accompanying table (*Dunn, 1927b; Jull, 1929*).

Successive Generations	Total Embryonic Mortality (%)		
	Family A	Family B	Family C
P	30	4	15
F_1	44	40	49
F_2	55	43	50
F_3	—	54	81
F_4	50	82	100
F_5	58	100	

From these data it is obvious that embryonic mortality increases significantly at the beginning of inbreeding. Families in which the embryonic mortality reaches over 70 per cent tend to become extinct.

Observations also show (*Byerly, Knox, and Jull, 1934*) that inbreeding causes some increase in embryonic mortality during the first week of incubation and a much greater increase during the third week, as indicated below.

Time of Incubation (days)	Embryonic Mortality (%)	
	Control	Inbred
1–5	9.5	10.5
18–22	15.5	33.5

Malformations. The incidence of abnormalities observed (*Goff and Tumlin, 1950*) in 18-day embryos of an inbred stock of 295 hens was considerably higher during the second year than during the first year. The results are presented in the table below.

Abnormalities	Incidence in Inbred Stock (%)	
	1st Year	2nd Year
Total abnormalities	6.0	15.0
Micromelia	8.0	13.5
Chondrodystrophy	7.7	21.2

Among the malpositions, the one most frequently observed was that with the head in the small end of the egg, or type 5 (Fig. 15).

In extensive inbreeding of Polish greenfeet-fowl (*Czaja, 1939*) there were observed various forms of skeletal abnormalities, such as curvature of the skeleton, noticeably in the lumbar region, and bending of the limbs backward and outward.

It is believed (*Landauer, 1967*) that the deleterious consequences of inbreeding are brought about by genes with harmful effects of a specific nature, for which the stock had been heterozygous.

Successive full-sibling matings of Japanese quail lead to depression in embryonic survival (*Sittmann, Abplanalp, and Fraser, 1966*).

MUTATIONS

Most of the mutations which produce hereditary malformations occur and recur spontaneously (see Chapter II, Spontaneous Malformations). In addition there are many malformations induced by various mutagens.

Spontaneous Mutations

Chromosome aberrations have been observed in 5.2 per cent of 115 early-dead embryos examined (*Bloom, 1969*). The aberrant chromosomal components were all found in phenotypically abnormal embryos. Of the macroscopically abnormal embryos, 13.3 per cent had chromosome aberrations. These included two cases of haploidy, four cases of triploidy, and one case of trisomy, which had been described previously (*Zartman and Jaap, 1968*).

It also was found (*Doina, Onea, and Marcu, 1969*) that gamma radiation (^{60}Co) in doses of 970 to 1730 R, applied to 72- and 124-hour embryos, produces chromosomal alterations.

Observations (*Donner, Chýle, and Sainerová, 1969*) of three chicken embryos with anophthalmia, brachygnathia superior, hernia umbilicus, and retarded growth, and also of one hatched chick with left-sided anophthalmia and brachygnathia superior, showed that all were carriers of triploidy with a ZZZ sex chromosome.

There are a number of lethal (see Summary Table 2) and nonlethal hereditary mutations which have been recognized in birds. Some of the mutations which are nonlethal prior to hatching, but which cause distinct types of malformations, are listed in Summary Table 3.

Regional Syndromes. Some hereditary malformations may appear as predominant structural defects in different parts of the embryo.

- Defects of beak
  - Short upper beak (*Landauer, 1933*)
  - Crossed beak (*Landauer, 1938*)
  - "Donald duck" (*Broekhuizen and Albada, 1953*)
- Defects of toes
  - Diplopod embryo (*Landauer, 1956a*)
  - Ametapodia (*Cole, 1967*)
- Syndactyly (*Taylor and Gunns, 1947*)
- Achondroplasia (*Lamoreux, 1942*)
- Phocomelia (*Landauer and Dunn, 1930a*)
- Defects of skeleton (*Lamoreux, 1942; Hollander, 1945,* in pigeon; *Landauer, 1947b*)

These inherited anomalies are reproduced as effectively as normal traits. However, some anomalies can be entirely inhibited by breeding (*Landauer, 1946*), for example, traits such as an abnormally short beak.

Induced Mutations

There is enough evidence to show that hereditary mutations can be induced in birds experimentally. Among the mutagens potent in the chicken is irradiation by X-rays, which causes disturbance or dislocation of chromosomes (*Newcomer, 1959; Ohno, Kittrell, Christian, Stenius, and Witt, 1963*).

Early observations (*Dunn, 1925*) showed that hereditary rumplessness (anouria), which is transmitted as a dominant trait, may also appear as a nongenetic or "accidental" phenomenon. In this case, the spine ends abruptly posterior to the third or fourth synsacro-caudal vertebra (*Landauer, 1928*).

Now it is known that not only some physical treatments, but also various other mutagens, including chemical mutagens, can produce defects like hereditary mutations. The injection of insulin into chicken eggs prior to incubation or during the early hours of development (*Landauer, 1945; 1947b*) leads to the production of abnormalities which have many features in common with two recognized types of genetic anouria.

Experimental Phenocopies

It has been shown (*Landauer, 1954;* and others) that there is a relationship between phenocopies, or nonhereditary abnormalities, and hereditary mutants. That is, modifications of development with end effects morphologically similar to genetic syndromes can be produced by various physical and chemical agents.

Below are listed a few typical syndromes of malformations which closely resemble certain mutant conditions and are here referred to as phenocopies.

Anouria. Significant increases in the frequency of anourous embryos occur after the injection not only of insulin (*Landauer, 1945; Landauer and Bliss, 1946*), but also of cysteine hydrochloride, cystine, *dl*-glutamic acid hydrochloride, thioglycollic acid, and *l*-malic acid (*Landauer, 1945*), and of Trypan blue (*Kaplan and Grabowski, 1967*).

Anourous chicken embryos have also been produced by treatment with a variety of other teratogens, such as variation in the temperature of incubation (*Danforth, 1932*).

Micromelia and Beak Deformities. These anomalies have been produced by riboflavin deficiency (*Romanoff and Bauernfeind, 1942*), sulfanilamide (*Ancel, 1945*), insulin (*Landauer, 1947a*), the nicotinic acid antagonist methyl pyridyl ketone (*Ackermann and Taylor, 1948*), thallium (*Karnofsky, Ridgway, and Patterson, 1950*), triazene (*Dagg, Karnofsky, Stock, Lacon, and Roddy, 1955*), chilling (*Ancel, 1958*), and others.

Chondrodystrophy. Chondrodystrophy of a nutritional type has been produced by riboflavin deficiency (*Romanoff and Bauernfeind, 1942*), biotin deficiency (*Cravens, McGibbon, and Sebesta, 1944*), and others.

Achondroplasia. Some form of achondroplasia in the avian embryo can be produced by the following agents: (a) deficiencies in the egg and (b) injections.

Among deficiencies in the egg are biotin (*Couch, Cravens, Elvehjem, and Halpin, 1948b; Cravens, McGibbon, and Sebesta, 1944*), riboflavin (*Romanoff and Bauernfiend, 1942*), and manganese (*Lyons and Insko, 1937a; Caskey and Norris, 1940*).

Among injected chemical agents are nicotinic acid (*Ackermann and Taylor, 1948*) and thallium (*Karnofsky, Ridgway, and Patterson, 1950*).

Abnormal Vertebral Column. Abnormalities of the vertebral column have been induced by thallium (*Karnofsky, Ridgway, and Patterson, 1950*), quaternary ammonium and cholinergic drugs (*Karnofsky, Ross, and Leavitt, 1954*), and insecticides—parathion and Diazinon in the chicken (*Khera, 1966*) and in the duck (*Khera and Bedok, 1967*).

From these observations it is reasonable to visualize a common developmental cause of many tendencies, through which effects are imposed either by heredity or by forces of the external environment.

The developmental defects of experimental origin discussed above, similar to genetic defects, presumably are brought about by a common intervention in metabolic events on the cellular level.

HEREDITARY MALFORMATIONS

There are many congenital anomalies in birds which have been attributed to genetic factors.

In some instances, various manifestations of an abnormality appear at an early stage, when they can not be easily observed by morphological methods. This is the case in several lethal defects, such as the homozygous Creeper (*Landauer, 1932b;* and others), polydactyly in the pigeon (*Hollander and Levi, 1942*), synophthalmia (*Landauer, 1956c*), and ametapodia (*Cole, 1967*), in embryos which die in the earliest period of incubation. (See lethal factors, in Summary Table 2.)

Hamburger (*1942*) has also shown histologically that four different abnormalities may appear in the homozygous Creeper embryo which cause defects of the blood vessels, skeleton, and eyes.

One of the most pronounced characteristics of hereditary malformations is the appearance of complex syndromes. In addition to the basic abnormality, there is often a great variety of small aberrations (see Table 1) which can not be explained by a common basis of development, but which still may have a bearing on an individual's well-being and often on its survival.

Lethals

Many genetic abnormalities, both lethal and nonlethal, are now known in birds. Of these, more than half involve the skeleton, whereas few are known to affect the muscle. Some lethal factors are also noted for causing retardation in growth and characteristic deformities of the embryo, its organs, and the extraembryonic membranes.

Usually only homozygous embryos, the carriers of identical lethal factors from both parents, are likely to die at some stage of incubation before hatching. The detailed effects of lethal factors can be observed under laboratory conditions.

A list of some obvious embryonic lethal factors, observed in several species of birds, is given in summary Table 2.

These lethal defects are identified chiefly by a typical gross abnormality. However, many other more common anomalies evidence themselves in conjunction with each lethal character.

Growth. The most extensively studied hereditary "Creeper" (*Landauer, 1930; 1932b; 1933; 1939*), a homozygous mutant, weighed only about one-quarter normal (Fig. 16*A*, with *insert*) at the eighteenth day of incubation. The length of the long bones (tibia, tarsometatarsus, and femur) was about one quarter that in normal embryos of the corresponding age (Fig. 16*B*).

In the "crooked-neck dwarf" mutation, the homozygous embryos of both chicken and turkey (Fig. 17) weigh half as much as the normal embryos at hatching time (Fig. 18*A*).

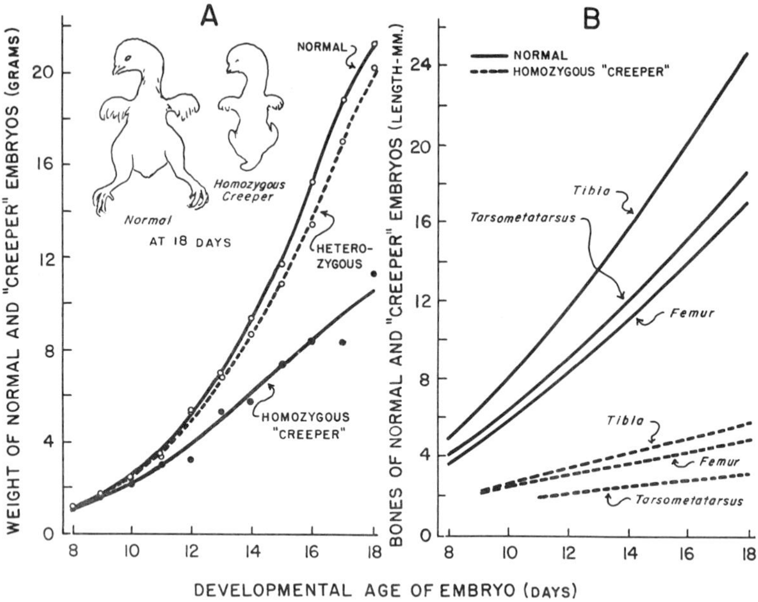

Fig. 16. Growth in body weight and in dimensions of long bones of a normal and a genetic mutant "Creeper" embryo. (Rearranged and plotted from the data of Landauer, 1939.) *A,* wet weight of normal and genetic mutants—homozygous and heterozygous "Creeper"—with an *insert* showing the appearance of normal and Creeper embryos; *B,* measurements of long bones of both normal and genetic mutants. *Note:* for verification of normal growth of organs see Chapter II in *Biochemistry of the Avian Embryo,* by Romanoff, 1967.

Weight of Organs. The organs of the homozygous "Creeper" weigh considerably less than those of normal embryos (Fig. 19). Proportionally to body weight, each organ changes with development according to its functional activities (insert of Fig. 19), although the absolute weights in the homozygous embryos are always below those in normals.

Changes in Extraembryonic Membranes. The amount of amniotic fluid is much greater in the turkey with the "crooked-neck dwarf" lethal than in normal turkeys (Fig. 18*B*), because the mutant embryo is unable to swallow the albumen entering the amniotic sac during the later part of incubation (*Romanoff, 1960, p. 1120*).

Lethal Syndromes. Many hereditary abnormalities manifest themselves by degenerative changes, which appear somewhat characteristic for each lethal gene involved. Several representative lethal characters are shown in Fig. 20.

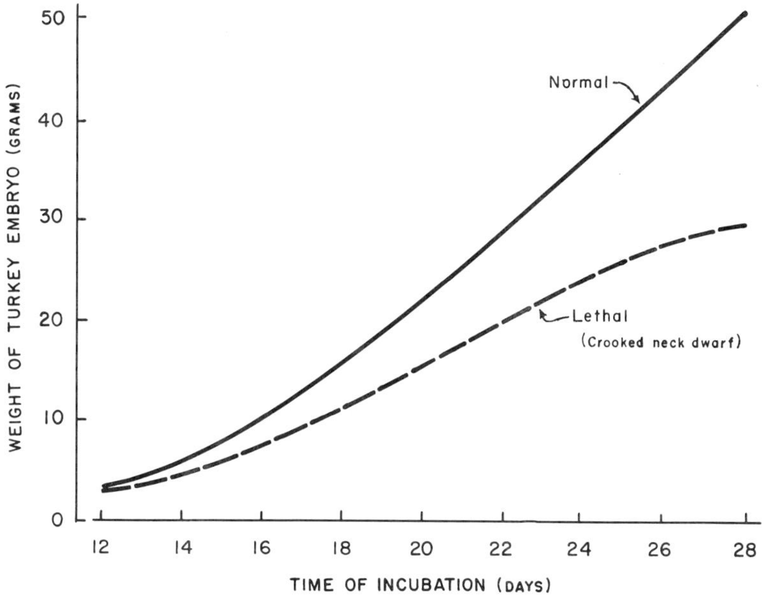

Fig. 17. Growth of normal turkey embryos and ones with lethal defect ("crooked-neck dwarf"). (Redrawn after Asmundson and Pun, 1956.)

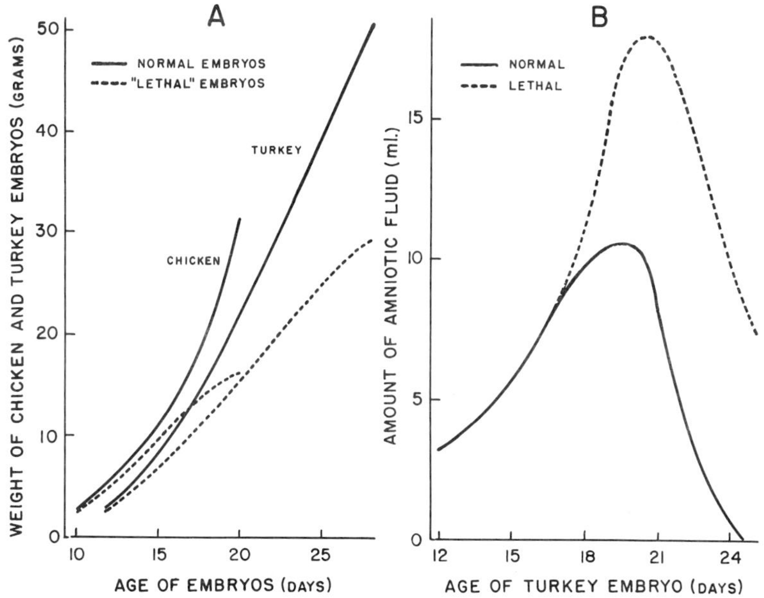

Fig. 18. Effect of a "crooked-neck dwarf lethal" character on the growth of chicken and turkey embryos. (Plotted from the data of Pun, 1954—on chicken, and of Asmundson and Pun, 1956—on turkey.) *A*, growth of normal and "crooked-neck lethal" chicken and turkey embryos; *B*, the ability of the normal, and the inability of the mutant embryos to swallow the amniotic fluid (from Asmundson and Pun, 1956).

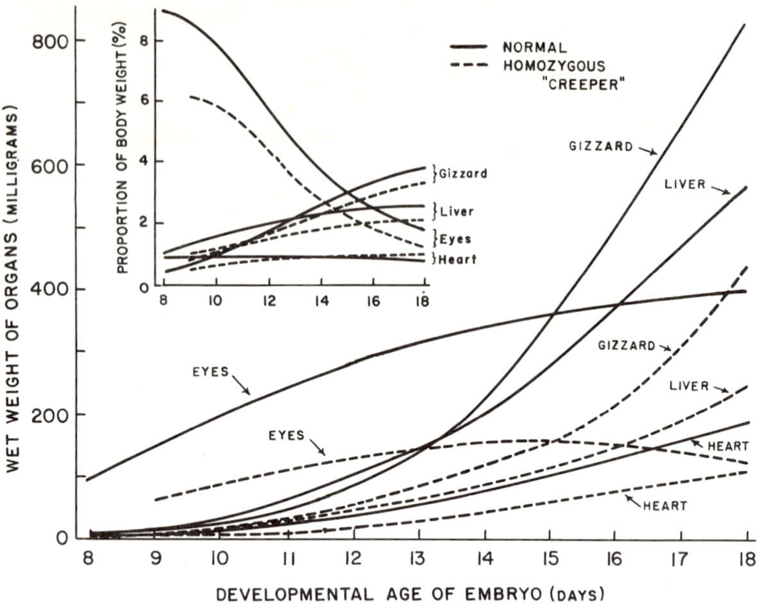

Fig. 19. The absolute growth in weight of eyes, heart, liver, and gizzard of the normal and the homozygous "Creeper" mutants. (After Landauer, 1939.) *Insert:* weight of organs (eyes, heart, liver, and gizzard) proportional to body weight.

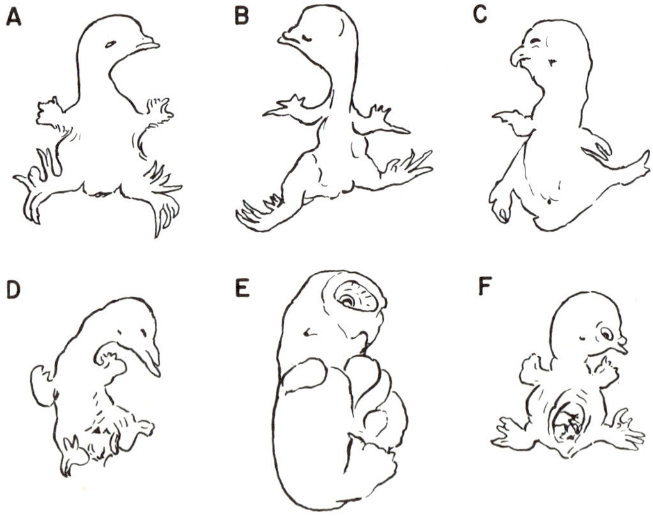

Fig. 20. Various forms of embryonic malformation induced by genetic lethal characters in birds. *A,* a diplopod mutation (after Landauer, 1956a); *B,* a form of polydactyly (after Taylor and Gunns, 1947); *C,* nanomelia, a recessive mutation (after Landauer, 1965b); *D,* polydactyly in the pigeon (Hollander and Levi, 1942); *E,* a talpid lethal character (Cole, 1942); *F,* a talpid mutation (after Abbott, Taylor, and Abplanalp, 1960).

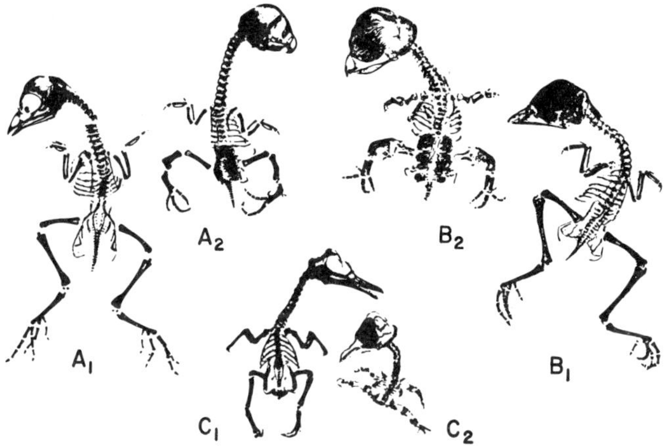

Fig. 21. Malformations of the embryonic skeleton caused by genetic factors. A_1, control, and A_2, chondrodystrophy of the 21-day-old chicken embryo—with a parrotlike beak, and reduced and bent long bones of the legs (after Lamoreux, 1942); B_1, control, and B_2, a lethal mutation, called nanomelia, of the 20-day-old chicken embryo—with malformations of the skull and of the long bones of the appendicular skeleton (after Landauer, 1965b); C_1, control, and C_2, achondroplasia of the pigeon embryo—with extraordinary shortening of all the bones, except the cranium (after Hollander, 1945). See Summary Table 2, on hereditary lethal factors.

The presence of complex syndromes indicates (see also Summary Table 2, for specific lethal factors) that the abnormalities affect various parts of the embryo and its organ systems.

The same phenomenon is observed in genetic malformations of the skeletal system of the embryo caused by lethal factors (Fig. 21). The three lethal mutations—*A*, chondrodystrophy, *B*, nanomelia, and *C*, achondroplasia (in the pigeon)—involve deformities of the skull, the vertebral column, and the limbs.

NATURAL SUSCEPTIBILITY

Individual differences among embryos in response to teratogens are controlled by the genetic background. In some families the hereditary defects follow a simple Mendelian pattern of transmission.

Such differences in susceptibility are present in certain inbred strains, and they manifest themselves in response to the contingencies of the environment—nutritional deficiency, pathogenic infection, and other factors.

Susceptibility to Atmospheric Environment. It was observed (*Romanoff, 1951*) that accidental exposure of chicken embryos to a temperature above 50° C for a short time at the midpoint of incubation resulted in the early death of all but a few embryos. The surviving embryos, incubated at the optimum temperature of 37.5° C for the remaining period, all hatched into strong and healthy chicks.

Susceptibility to Nutritional Deficiencies. It has been observed (*Davis, Norris, and Heuser, 1938*) that there is a wide variation in embryonic mortality in eggs of different families of hens on riboflavin-deficient diets. This is illustrated in the table below.

Families of Hens	Embryonic Mortality (%)
A	36.6
B	69.3
C	80.8
D	86.7
E	98.3

Susceptibility to Chemical Treatments. Experiments with progenies derived from mating of high- and low-susceptibility lines (*Landauer, 1965a*) show a difference in the incidence of micromelia after treatment with nicotinamide at three stages of development. The data are summarized in the accompanying table.

	Incidence of Anomalies (%)		
	Treated at 84 Hours	Treated at 96 Hours	Treated at 120 Hours
Low-susceptibility line			
Micromelia	10.0	39.6	12.7
High-susceptibility line			
Micromelia	34.6	64.5	12.9

Another case of reciprocal crosses between breeds of low and high resistance to insulin-induced anouria (rumplessness) is shown below (*Landauer, 1948*).

Breed Crosses	Incidence of Anouria (%)
Leghorn	25.7
Leghorn (female) × jungle fowl (male)	26.7
Jungle fowl (female) × Leghorn (male)	9.3
Jungle fowl	9.3

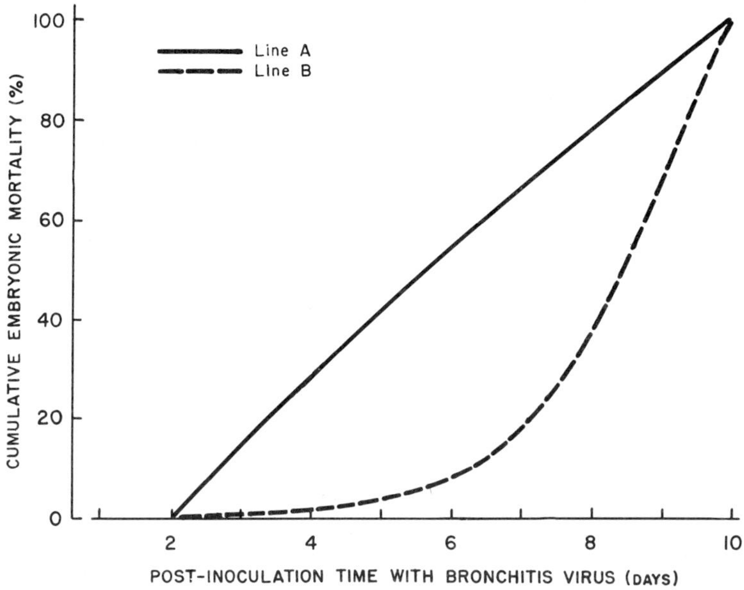

Fig. 22. Cumulative mortality of the embryos of two inbred lines of chickens in response to inoculation of bronchitis virus on the tenth day of development. (Plotted after Purchase, Cunningham, and Burmester, 1966.)

Variation in susceptibility to insulin exists within as well as between breeds, as was demonstrated in another experiment (*Landauer, 1945*) on the effect of insulin. An obvious difference in response of two breeds was found, as shown below.

	Insulin-Treated (0.05 cm³)		
	Leghorn	Creeper	Control
Mortality (%)	74.7	62.5	25.1
Anouria (%)	41.6	17.1	1.0

The difference between the breed lines is due in part to variation in response because of genetic dissimilarity.

Susceptibility to Pathogenic Infections. Study of the pathogenicity of bronchitis virus injected into the allantoic sac of embryonated eggs (*Purchase, Cunningham, and Burmester, 1966*) showed that in one of two inbred lines there was a significant delay in embryonic mortality (Fig. 22).

In general, the inter-strain differences in response to teratogens are controlled by the genetic background in the same way as the manifestations of the major mutant genes.

Summary Table 2 Hereditary Lethal Mutations Responsible for Malformations and Death of the Avian Embryo (Chicken Embryos, Unless Otherwise Stated)

Designated Characters	Types of Abnormality	Mortality Peak	Reference
Achondroplasia	*In Pigeon.*—Extraordinary shortening of all the bones (except the cranium) (Fig. 21*C*).	Very high (late)	Hollander (1945)
Albinism, sex-linked	*In turkey.*—Absence of melanin pigment in the eye, and blindness.	75% (females)	Hutt and Mueller (1942)
Ametapodia	Eliminated or reduced metapodial region of both fore and hind limbs.	High (early)	Cole (1967)
Beak, curled	Abnormality of the beak: upper beak curled upward, lower beak downward; embryonic death at hatching caused by inability to break the shell.	93% (late)	Broekhuizen and Albada (1953)
Beak, short upper	Reduction in length of the upper beak (48%), and shortening of the long bones of the extremities.	100%	Landauer (1941)
Cerebral hernia	Hernia of the brain, upper mandible much reduced in size and pointed upward, lower mandible almost entirely missing.	93% (late)	Marble, Harper, and Hammer (1944)
Chondrodystrophy	Extremely shortened and bent tibia and tarsometatarsus, particularly short curved beak (Fig. 21*A*).	30%	Lamoreux (1942)
Cornish lethal	Short lower extremities, especially tarsometatarsus (25%).	66%	Landauer (1935)
Creamy gier	*In carrier pigeon.*	100% (all females)	Lienhart (1937)
Creeper	Retardation in growth, extreme shortness of the long bones of the extremities, malformations of extremities, skull, and eyes (Figs. 16 and 19).	100% (early)	Cutler (1925); Landauer and Dunn (1930a); Landauer (1932b; 1933)
Crest	*In duck.*—Malformed cranial integuments and upper beak, exposed brain.	100%	Rüst (1932)

Crooked-neck dwarf	Marked dwarfism of the embryo (at 20 days, 17/35 grams), slight reduction in the upper beak, and crooked neck (Fig. 18*A*).	100% (20 to 21 days)	Asmundson (1945); Herrmann, Clark, and Landauer (1963)
	Greatly reduced growth, edema (10%); skeletal abnormalities: crooked neck, crooked spinal column, slightly shortened upper beak, spindly legs, curly toes, and reduced sternum without keel and lateral processes; esophageal atresia; hypertrophy of adrenals: hydramnios (Fig. 18*B*).	—	Pun (1954)
	In turkey.—Retarded growth (Fig. 17), decreased weight of the thyroid, testes, liver, and proventriculus, slightly shortened upper and lower beak, rigid crooked neck, reduced pectoral and leg muscles, decreased amount of amniotic fluid (Fig. 18*B*).	100% (late)	Asmundson and Pun (1956)
	In quail.—Reduction in body size, neck rigid, sharply bent to one side (usually to the right), beak positioned over the right wing (cf. malpositions in Fig. 15, Chapter II), the maxilla shorter than the mandible.	High (14 to 19 days)	Sittmann and Craig (1967)
	Degeneration of the trachea and thymus.		Allenspach (1970)
Diplopodia			
In Leghorn	Partial doubling (polydactylism) of metatarsals and digits of the feet and of the metacarpals and digits of the wing, shortening and curving of the long bones of the leg, and shortening of the maxillae (Fig. 20*B*).	64% (18 to 22 days)	Taylor and Gunns (1947); Landauer (1956a)
In Minorca	Polydactylism in both pairs of extremities, reduction in length of the long bones, and shortening of the upper beak (Fig. 20*A*).	99% (18 to 22 days)	Landauer (1956a)
Ectrodactyly	Reduced or completely absent legs, reduced and upturned beak.	100% (17 to 20 days)	Abbott and MacCabe (1966)

Summary Table 2 (*Continued*)

Designated Characters	Types of Abnormality	Mortality Peak	Reference
Foot, split	Second digit of the foot duplicated and hallux absent; also reduced premaxilla, defective eyes, sometimes accompanied by cataract.	95% (17 to 19 days)	Abbott and Goetinck (1960)
Long bones, short	*In turkey.*—Short embryo with reduced length of long bones, thickened femur, tibia, tarsometatarsus, and humerus.	80% (late)	Asmundson (1944)
Maxillae, abnormal	Absent or much reduced maxillae, frequently with nasal bones smaller than normal, upper beak bent to one side.	100% (late)	Asmundson (1936)
Micromelia	Much shortened and thickened extremities, short and curved upper beak, and defective lower beak.	100% (late)	Asmundson (1942b); Zwilling (1956a)
	Greatly shortened but straight long bones of legs and wings, short but proportional beak.	28%	Hays (1944)
	Extremely short and bent legs, significantly shortened wings, and microphthalmic eyes.	100%	Bernier (1951)
	Shortening of the upper beak, poor skeletal calcification—containing little calcium but abnormally high content of magnesium.	High (late)	Kawahara (1956)
	In quail.—Shortening of the long bones of legs and wings by about one-half, reduction of the lower mandible and shortening and curving of the upper beak (parrot beak), celosomia.	100% (11 to 16 days)	Hill, Lloyd, and Abplanalp (1963)
	In duck.—Reduced growth (at 25 days, 36/43 grams), conspicuous shortening of limbs, especially of the long bones, and reduction of upper mandible.	100% (late)	Ash (1966)
Microphthalmia	Reduction in size of the eyeballs (bilateral microphthalmia) to approximately one-half normal diameter.	66%	Jeffrey (1941)
	Minute cysts in the retina, signs of degeneration, absence of pectin, reduction of eyeball to one half of normal size.		Gruenwald (1944)
	In pigeon.—Production of small-eyed blind offspring, small eyelids.		Hollander (1948)

Nanomelia	Incompletely formed or entirely absent caudal vertebrae, hypoplasia of the extremities, brachycephaly, and a parrotlike beak (Fig. 20*C*).	100% (18 to 21 days)	Landauer (1965b)
Neck, short (In Durking)	Somewhat reduced body size, very short neck, abnormal beak, and, often, short tail.	100% (8 to 9 days)	Landauer (1959)
Polydactyly	*In pigeon.*—Polydactyly of feet and wings, shortened legs, undersized beak (Fig. 20*D*)	17% (early)	Hollander and Levi (1942)
Spine, short	*In turkey.*—Crowding together of the vertebrae, with consequent shortening of the neck and body.	70%	Asmundson (1942a)
"Stickiness"	Extreme softness of bones, bent tibia (33%), in malposition "head-under-left-wing" (8%) (see Fig. 15, various forms of malposition).	63% (late)	Byerly and Jull (1932)
Synophthalmia	Malformations ranging from synophthalmia to microcephaly—from rudimentary upper beak to cyclopia.	100% (early)	Landauer (1956c)
Talpid	Duplication of the digits fused together, shortening of the spinal column and the proximal bones of the appendages, and severe polydactyly (Fig. 20*E* and *F*).	100% (8 to 10 days)	Cole (1942); Abbott, Taylor, and Abplanalp (1960); Hinchliffe and Ede (1968)
Tremor, congenital	(Affects chicks after hatching.)	42%	Hutt and Child (1935); Sittmann (1967)
Winglessness	Absence of wings, partial or complete absence of the lungs, air sacs, and definitive kidney (metanephros).	High (very late)	Waters and Bywaters (1943); Zwilling (1949)
White, dominant	*In canary.*	100% (early)	Duncker (1924)
White, recessive	(Not observed.)	80%	Dunn (1923a)
	In quail.	92% (12 to 18 days)	Sittmann and Abplanalp (1965)

Summary Table 3 Hereditary Nonlethal Mutations Responsible for Malformations in the Avian Embryo (Chicken Embryos, Unless Otherwise Stated)

Designated Characters	Types of Abnormality	Reference
Albinism	*In quail.*—Sex-linked.	Sittmann, Wilson, and McFarland (1966)
Baldness	A featherless (supra-orbital) area on top of the head.	Jaap (1937)
	Possibly due to inherent defects in the epidermis.	Sturkie (1941)
Downlessness ("naked")	Complete downlessness.	Serebrovsky and Petroff (1929); Hutt and Sturkie (1938)
	A variable syndrome including polydactyly of feet and wings, undersized beak, short down.	Sturkie (1942)
"Frizzle"	A high degree of frizzling—upward turning of the body plumage.	Landauer and Dunn (1930b)
Leg, anomaly	A permanent crippling anomaly (18%).	Morgan (1963)
"Loco," congenital	Lack of control over the muscles of the neck (24%).	Knowlton (1929)
	In turkey.—Loss of balance caused by noise or vibration.	Cole (1957)
	In quail.—A behavior disorder expressed at hatching as pronounced lack of balance.	Sittmann, Richards, and Abplanalp (1965)
Muscle, abnormality	Muscular dystrophy.	Asmundson and Julian (1956)
Neck, arched	Defect of the head and neck.	Conner and Shaffner (1953)
Osteodystrophy, ungular	Loss of the toenails and terminal phalanges, and digit III.	Warren, Mueller, and Hutt (1944)
Polydactyly	A duplicate type of polydactyly in the foot and in the wing.	Warren (1941)
Polydactyly, preaxial	Preaxial polydactylism of wing and foot.	Landauer (1956b)
Rumplessness (anouria)	A suppression in development of the free caudal vertebrae and of the oil gland; modification of the rump region.	Dunn and Landauer (1934)
Wings, doubling of both	*In pheasant.*—Forking of the radii of both wings.	Bissonnette (1941)

IV

Atmospheric Changes

Environment has its effect
On living embryo, its doom . . .
Through changing forces, or neglect—
From heat to noxious gases' fume.

The development of the avian embryo is easily disturbed by the physical components of the atmospheric surroundings. These components, under appropriate conditions, may play a significant role in chains of teratological events.

The effect of atmospheric environment on the developing embryo can be observed in two distinct periods: the time before incubation (between egg laying and incubation) and the time during incubation.

It is important here to trace all the pathways by which atmospheric environment modifies avian embryonic development.

INCUBATING TEMPERATURE

Temperature during incubation is the single most significant environmental (physical) factor in the development of the avian embryo. All deviations from the optimum temperature of incubation (about 37.5° C), toward either lower or higher levels, greatly affect the rate of growth, the incidence of embryonic mortality, and the occurrence of some specific malformations of the whole body of the embryo and of its individual organs.

Growth of the Early Structures

Some of the most conspicuous gross anomalies in avian embryonic development are due to changes in temperature. Especially during the early stages of embryonic life, the deviations from the normal pattern of growth are very pronounced.

Growth of the Early Embryo. At the early stages, the measurements of the growth of the embryo most commonly employed include the following: (1) the size of the blastoderm, (2) the extent of the area vasculosa, (3) the time of appearance of specifically organized structures, and (4) the number of somites.

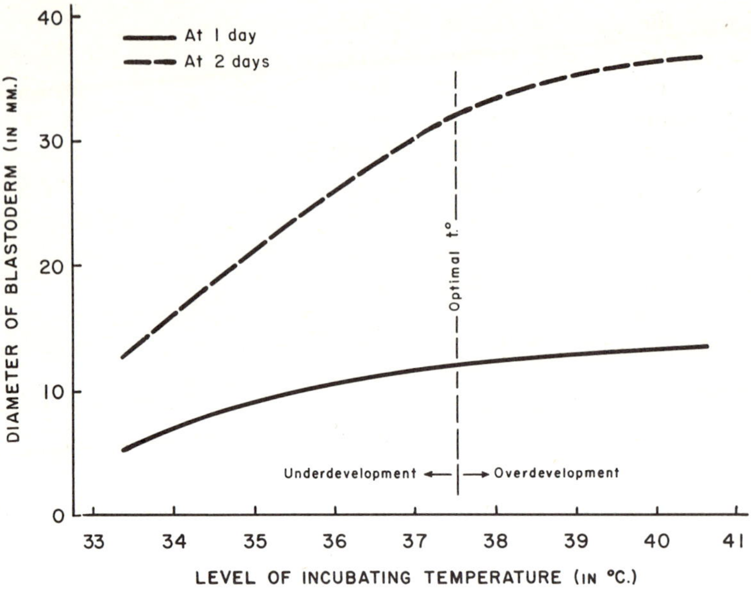

Fig. 23. Effect of different levels of incubating temperature on the size of the blastoderm. (Plotted from the data of Romanoff, Smith, and Sullivan, 1938.)

Size of Blastoderm. The size of the blastoderm at 1 and 2 days of development, at temperatures ranging from 33.5 to 40.5° C, is shown in Fig. 23. At the end of the first day of incubation, the increase in diameter of the blastoderm over the original size (4.4 mm) is about 13 per cent at low temperature and 54 per cent at high temperature. At the end of 2 days, the difference in size at extreme temperatures is still greater, showing that the retardation effect of low temperature is much greater than the acceleration effect of high temperature.

In studies of low temperatures down to 15° C (*Romanoff, Smith, and Sullivan, 1938; Funk and Biellier, 1944*) there is no visible growth of blastoderm, as shown below.

Incubating Temperature (°C)	Daily Size of Blastoderm (mm)			
	At 0 Days	At 2 Days	At 4 Days	At 7 Days
15	4.5	4.5	4.5	4.5
27	4.5	5.1	5.2	5.3
29	4.5	9	15	32
32	4.5	12	26	
37.5	4.5	32		

Size of the Area Vasculosa. The effect of a wide range of incubating temperatures during the first 4 days is shown in Fig. 24. There is an obvious regularity in the growth of this structure in relation to the temperature levels at the successive stages of development. The *insert* of Fig. 24 shows the thermal threshold for the possibility of survival at hatching (*Romanoff, 1936a*).

The changes in the size of the blastoderm and of its area vasculosa show very definitely that low temperature affects the embryo by slowing the processes of development, whereas high temperature accelerates development.

However, the extent of underdevelopment or overdevelopment should not always be taken as a true measure of embryonic growth. It has been observed (*Panum, 1863; Rabaud, 1899; Weber, 1906; Tur, 1908*) that blastoderms may develop without embryos (anideans), or even without the area vasculosa (*Grodzinski, 1931*). Such anomalies may occur from lengthy exposure to 25° C (see under Malformations, p. 80).

Growth of Early Structures. The time of differentiation and the appearance of the various early structures of the embryo are affected in a

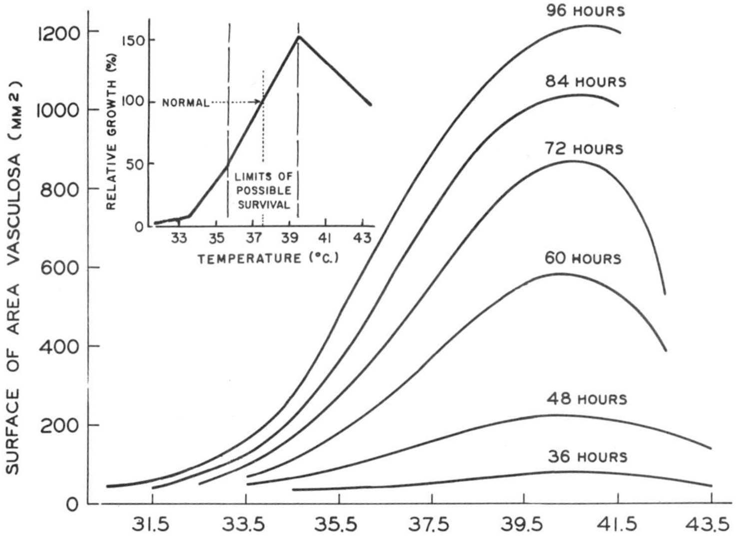

Fig. 24. Influence of incubating temperature on growth of the area vasculosa, indicated by increase in area. Area = $(s + h/2)\ \pi$, where s is the sagittal and h the horizontal measurement (in mm) of the area vasculosa. (After Romanoff, unpublished.)

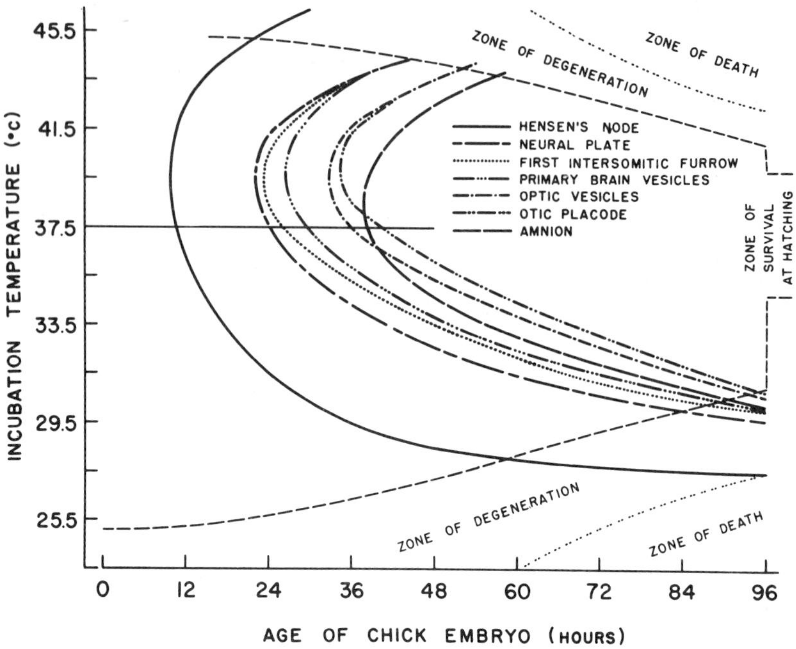

Fig. 25. Effect of various levels of incubating temperature on the time of appearance of early structures of the developing chicken embryo. (After Romanoff, *The Avian Embryo,* 1960, p. 196—based on a series of more than 350 mounted embryos.)

predictable manner by either depression or elevation of the incubating temperature (Fig. 25). In the presence of insufficient heat, differentiation is delayed; at temperatures above normal it is at first accelerated; then, with further increase in temperature, there is a delay in differentiation of all structures.

The average relative change in the growth rate of 31 observed structures during the first 4 days of development under different temperatures is shown in Fig. 26. However, the range of temperatures permitting survival of embryos is much narrower at the time of hatching (*insert* of Fig. 26).

When observations of some structures are separated by 12-hour intervals, the linear slopes of the effect of various temperatures become less steep as development progresses, and the period of the slopes is shifted from a high to a low temperature level (Fig. 27). All three linear slopes (heavy lines) have periods of the same length, within rather narrow limits.

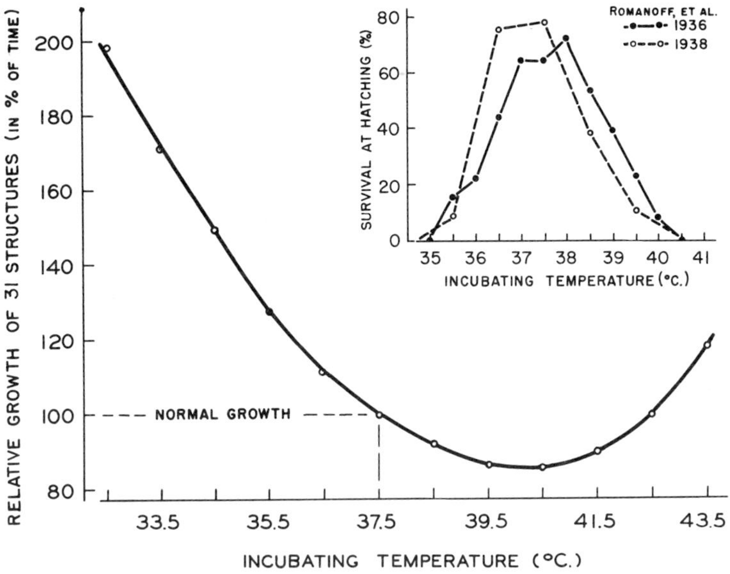

Fig. 26. Relative change in the growth rates of 31 structures during the first 4 days of development under the influence of various incubating temperatures. Data are expressed in terms of per cent of time required to reach the normal level of development. (After Romanoff, unpublished; see also Romanoff, 1960, p. 196, Fig. 76.) *Insert:* incidence of survival at hatching of embryos incubated at various temperatures. (After Romanoff, 1936a; Romanoff, Smith, and Sullivan, 1938.)

Number of Somites. The rate at which the first 40 somites are formed, at temperatures ranging from 29.5 to 41.5° C, is graphically shown in Fig. 28. It is obvious that below-normal temperatures cause a pronounced inhibition in the formation of somites, especially below 35.5° C.

Appearance of Embryonal Area. The appearance and relative size of the developing embryo at 12 to 96 hours, as influenced by various levels of temperature, is illustrated in Fig. 29. This figure dramatizes the profound effect of temperature on the early growth and differentiation of the avian embryo, which depends so much upon the heat provided by the environment.

Differential Development. Tazelaar (*1928*) demonstrated that simultaneously subjecting different main regions of the fertile egg to high and low temperatures would induce differential development of the two parts of the embryo. A good example of such development is given in Fig. 30, where in *A* the posterior end of the embryo was incubated for 3 days

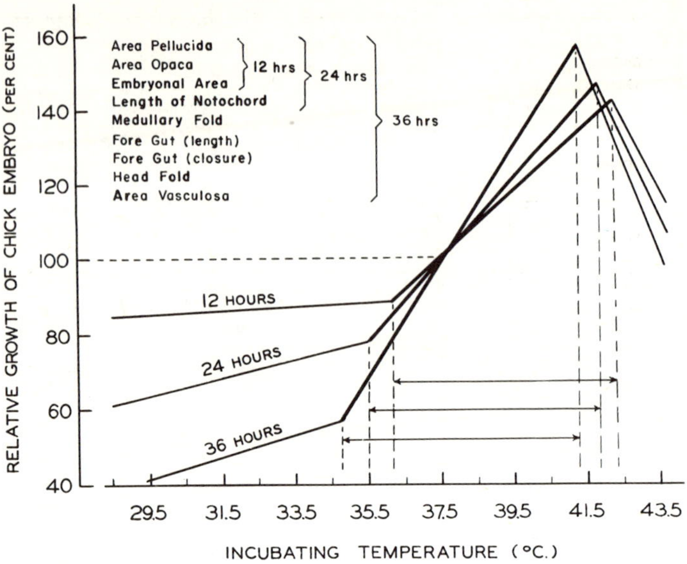

Fig. 27. Influence of temperature on the relative development of different embryonic structures of the chick at 12, 24, and 36 hours of incubation. (Romanoff, unpublished.)

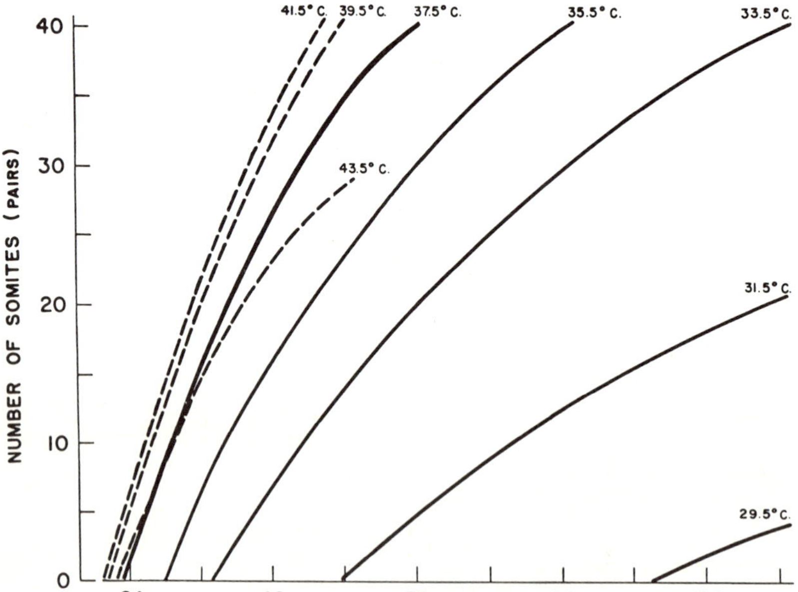

Fig. 28. The effect of temperature on the time of appearance of somites. (After Romanoff, 1960, p. 203.)

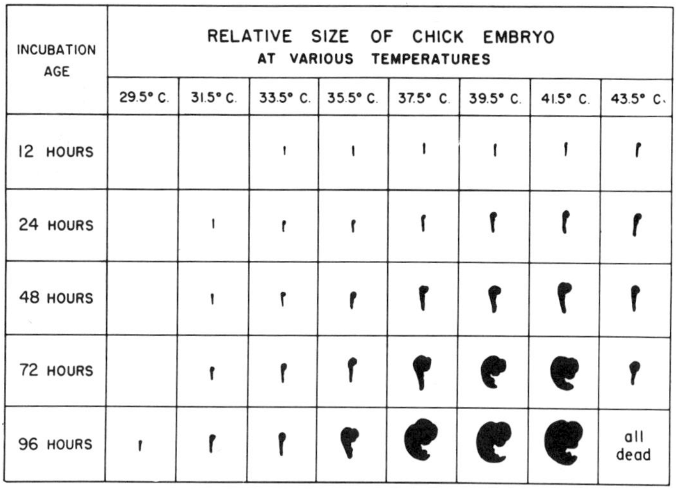

Fig. 29. Effect of temperature on the appearance and relative size of the embryonal area during the first 4 days of development. (After Romanoff, 1960, p. 1145; measured from mounted embryos.)

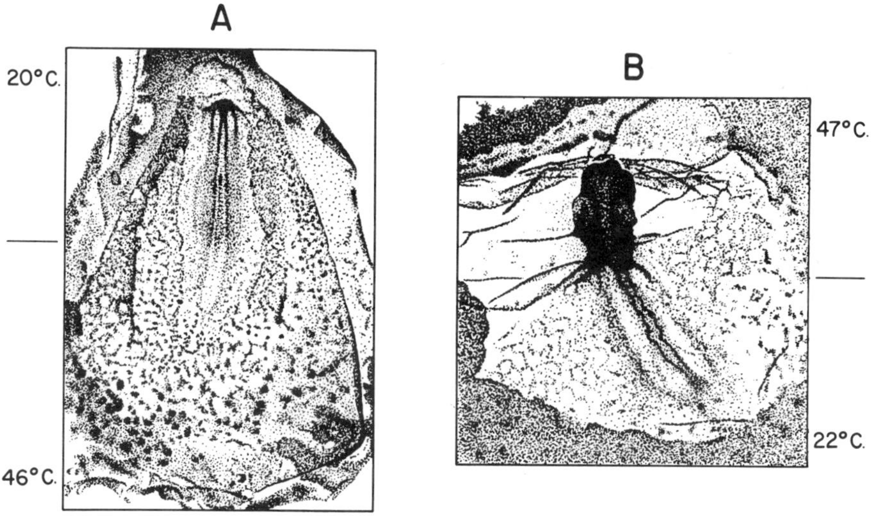

Fig. 30. Effect of low and high temperatures on the differential development of the two regions of the incubated egg. (Tazelaar, 1928; after Romanoff, 1960, p. 204.)

at 46° C, and the anterior end at 20° C; in *B*, the treatments of posterior and anterior parts were reversed.

In some embryos treated with a lateral temperature gradient, the somites shifted up to the heated side.

Growth of the Whole Embryo

The growth of the whole embryo under experimental conditions is most frequently measured by its weight. This method is the simplest and, for large numbers, can be statistically accurate in studies of the effects of temperature.

Continuous Exposure. Selected data on growth of chicken embryos incubated under representative temperatures which permit some survival at the time of hatching (*Romanoff, Smith, and Sullivan, 1938*) are given in the accompanying table.

Incubation Age (days)	Weight of Embryo (grams) At 35.5° C	At 37.5° C (Control)	At 39.5° C
4	0.032	0.059	0.098
8	0.63	1.15	1.62
12	2.3	5.2	7.2
16	9.1	16.6	14.9
18	13.8	22.2	20.5
19	17.4	27.0	22.1
20	19.3	31.8	Hatched
21	21.9	Hatched	
22	23.5		
23	27.5		
24	33.6		
25	Hatched		

The cumulative growth of the embryo in the early stages is much delayed at 35.5° C and accelerated at 39.5° C. In the later stages, growth at the high temperature is also considerably retarded. The hatch comes 1 day earlier than in controls, or 5 days earlier than at the low temperature.

The effects of different levels of temperature, especially between 33.5 and 37.5° C, are elaborated further in Fig. 31. In addition, *insert b* of this figure shows a close relationship between the weight of a

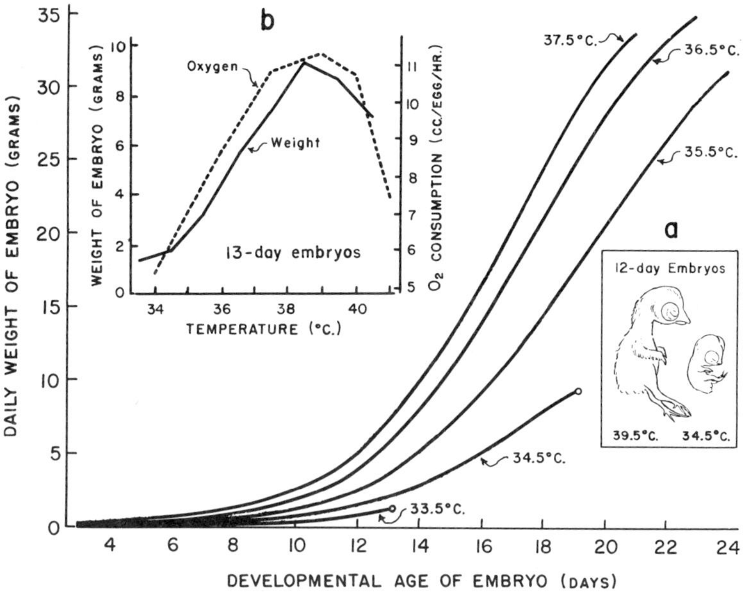

Fig. 31. Effect of different levels of temperature on the daily growth of the embryo. (Plotted from the data of Romanoff, Smith, and Sullivan, 1938). *Inserts: a,* relative size of the chick embryo at the age of 12 days, weighing 7.1 grams at 39.5° C, and 1.8 grams at 34.5° C; *b,* weight and oxygen (O_2) consumption of the 13-day embryo at various levels of incubating temperature. (The data on oxygen consumption after Greiff, 1952.)

13-day embryo and the rate of oxygen consumption at a wide range of incubating temperatures.

The rate of oxygen consumption per gram of dry tissue of the embryo falls noticeably at high temperatures, as shown by the accompanying table (*Greiff, 1952*).

Incubating Temperature (° C)	Oxygen Consumption of Embryo (cm^3 O_2/mg dry weight)
34	1.09
36	1.08
37.5	1.08
39	1.07
40	0.98
41	0.66

The drop in oxygen consumption at 41° C is due to a decrease in metabolic activity. Oxygen consumption stops completely with the onset of morbid state of the embryo (*Romanoff, Smith, and Sullivan, 1938*).

The precise changes in the growth of the embryo during continuous exposure to various levels of incubating temperature are revealed by plotting the relative daily weight of the embryo, expressed in percentage of weight of the control group (Fig. 32*A*). The weight of the 3-day-old embryo at 40.5° C is as high as 240 per cent of the weight at 37.5° C, whereas at 33.5° C it is only about 10 per cent of the normal weight.

In the later stages of incubation, both high and low temperatures cause the weight of the embryo to fall below normal, as is shown in the table below, in which the ratio of the embryonic weight at hatching to the original egg weight is expressed as a percentage (*Romanoff, 1936a*).

Incubating Temperature (° C)	$\frac{\text{Weight of Embryo}}{\text{Egg Weight}} \times 100$	
	Continuous Exposure	Exposure from 16th Day
30.5		67.1
32.5		71.1
34.5		72.6
35.5	68.5	72.6
36.5	69.2	72.6
37.5	72.4	72.7
38.5	69.1	72.5
39.5	68.9	71.7
40.5		69.7

High- and Low-Temperature Shocks. The relative susceptibility of the embryo to 24 hours of exposure to extremely high and low temperatures is shown in Fig. 32*B*. The plotted data on the percentage deviation of the wet weight of the embryo show that the immediate effect of a temperature shock is to accelerate the growth, by a high temperature, or retard it, by a low temperature. The 3-day-old embryo weighs about 40 per cent more than normal after exposure to a high temperature, and about 80 per cent less than normal after exposure to a low temperature. Later in development, the thermal effects are lessened, and from about the tenth day both high and low temperatures slightly inhibit growth.

The effects of 24 hours of exposure to low or high temperatures

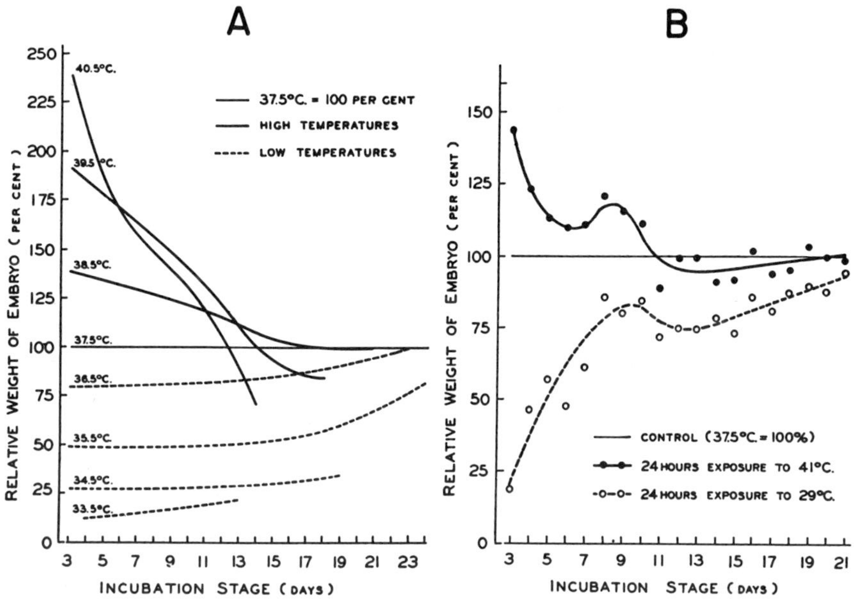

Fig. 32. Effect of various levels of temperature on the relative growth of the embryo. *A*, changes in the relative daily weight of the embryo under continuous exposure to various incubating temperatures (after Romanoff, Smith, and Sullivan, 1938); *B*, the immediate effect of 24 hours' exposure to high and low temperatures on the relative weight of the embryo at different stages of incubation (after Romanoff, 1939).

on the post-exposure development are shown in Fig. 33. These experimental data also indicate that the effect of temperature is greatest in the early stages, the growth of the embryo being accelerated by high temperatures and retarded by low temperatures. Again, the effects decrease as incubation progresses, and both high and low temperatures produce only slight inhibition after about the tenth day.

Time of Hatching. Fig. 34 shows that low temperatures delay the time of hatching, whereas high temperatures may slightly accelerate or, at extreme levels, also retard it. The earliest hatching of individual chicken eggs was observed on the eighteenth day of incubation, the latest on the twenty-eighth day (*Romanoff, Smith, and Sullivan, 1938*).

A similar phenomenon of delay in hatching has been observed with the eggs of pheasants and quail (*Romanoff, 1934*), of turkeys (*Romanoff, 1935a*), and of ducks (*Romanoff, 1943a*). (See also Fig. 10 for the *distribution of hatching* in several species of birds.)

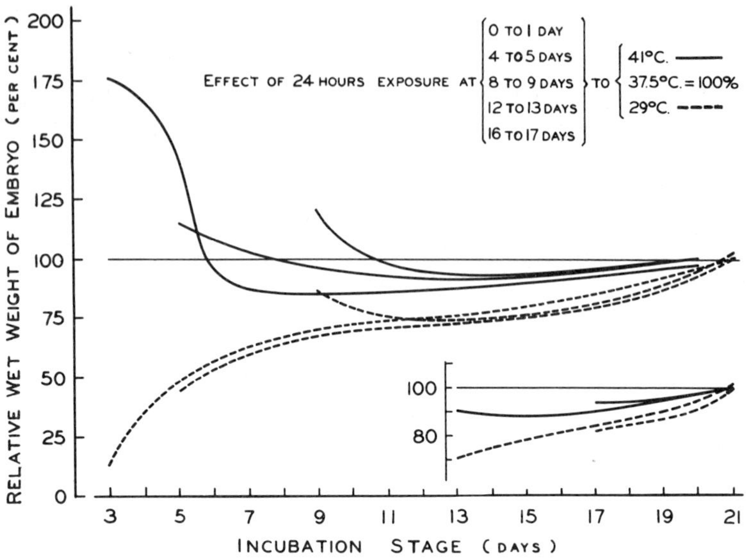

Fig. 33. Effect of 24 hours' exposure to either 29 or 41° C at 0, 4, 8, 12, and 16 days of incubation on the post-exposure relative growth of the embryo. (After Romanoff, 1939.) *Note:* the curves for exposure at 12 and 16 days are shown separately at the right lower portion of the figure.

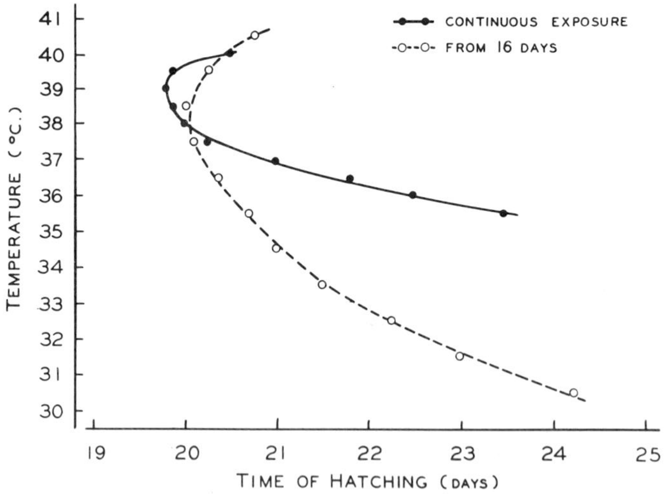

Fig. 34. The time of hatching of chicks after continuous exposure to different levels of incubating temperature, and after exposure from the 16th day—with preceding incubation at 37.5° C. (Plotted from the data of Romanoff, 1936a.)

Mortality

Temperature has a marked effect on mortality in the avian embryo. Some of the most severely affected embryos die early, whereas others reach the end of the incubation period. Beyond certain limits of high and low temperature, none of the embryos can survive.

Continuous Exposure. Under continuous exposure to various levels of temperature, the total embryonic mortality at hatching (Fig. 35) changes very rapidly with decrease or increase of temperature from the optimum of about 37.5° C. This is true with the eggs of chickens, ducks, pheasants, and quail. Study of turkey eggs (*Romanoff, 1935a*) indicates that they respond to temperatures in a way similar to the other species shown in Fig. 35.

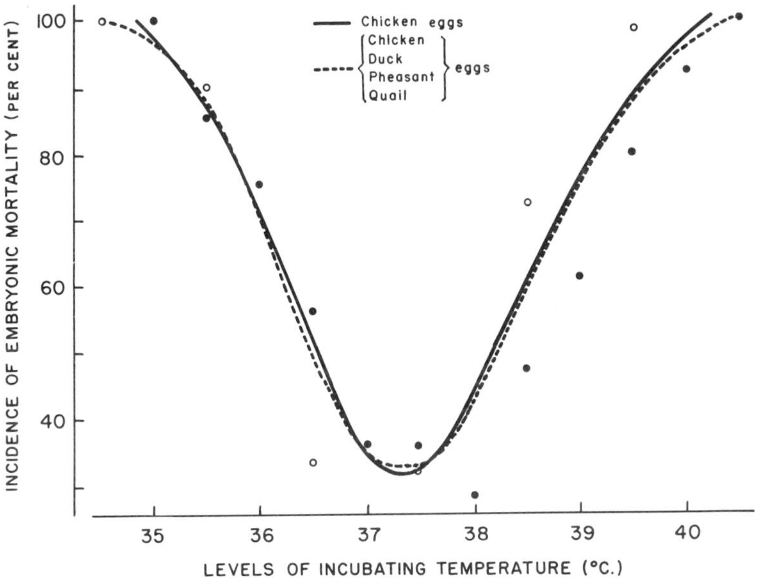

Fig. 35. Effect of various levels of incubating temperature on the incidence of embryonic mortality in chicken eggs. Series I—closed circles—is based on about 112 fertile eggs in each lot (after Romanoff, 1936a); Series II—open circles—is based on about 200 fertile eggs in each lot (after Romanoff, Smith, and Sullivan, 1938; see also Philips and Brooks, 1923.) The superimposed dotted curve shows combined averages of the incidence of embryonic mortality in the chicken and duck (Romanoff, 1943a), and the pheasant and quail (Romanoff, 1934). The total numbers of fertile eggs used for each species in these studies were as follows: 3680, 425, 960, and 1440, respectively.

Distribution of Embryonic Mortality. High and low temperatures cause irregularities in the distribution of dead embryos during the incubation periods. Normally, three critical periods are observed (see Chapter II on critical periods, Fig. 13). With abnormal temperatures, the curves of mortality are distorted; the last peak of mortality is shifted to an earlier date by high temperature and delayed by low temperature.

The actual mortality charts (Fig. 36) for both extremes of temperatures show four peaks of embryonic mortality. The first three are distinctly delineated for low-temperature (35.5° C) incubation but are barely distinguishable for high-temperature (39.5° C) incubation. It is also noticeable that excessive mortality occurs early at the high temperature, and late at the low temperature (see also *Cruz, Campillo-Sainz, and Munoz-Armas, 1966*).

Late Exposure. Exposure of 16-day-old chicken and 21-day-old turkey embryos to a wide range of temperatures showed that the time and incidence of embryonic mortality at hatching were affected more by an increase in temperatures than by an equivalent decrease (*Romanoff*

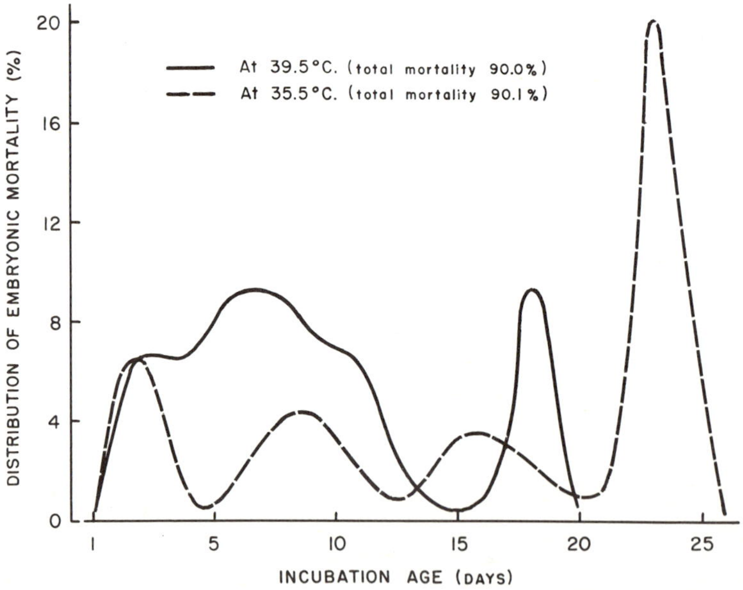

Fig. 36. Effect of continuous exposure to high and low temperatures on the distribution of dead embryos throughout the incubation period. (Plotted from the data of Romanoff, Smith, and Sullivan, 1938.) (See also the distribution of the dead avian embryos incubated under the optimum temperature of 37.5° C, in Chapter II, Fig. 13.)

and Faber, 1933; Romanoff, 1936a, on chicken; and *Romanoff, 1935a,* on turkey), as is evidenced by the data in the accompanying table.

Incubating Temperature (° C)	Incidence of Mortality at Hatching (%)	
	Chicken	Turkey
29.5	100.0	100.0
31.5	83.8	93.6
33.5	45.9	65.3
35.5	37.8	30.0
37.5 (control)	29.7	20.0
39.5	51.3	83.0
41.5	100.0	100.0

The above phenomenon is well illustrated by the survival of chicken embryos at hatching (Fig. 37), when one contrasts the effects of continuous and late (from 16 days) exposure to various levels of incubating temperature (with preceding incubation at optimum temperature 37.5° C). With exposure after the sixteenth day of incubation, the peak of survival occurs at a much lower temperature (about 36° C).

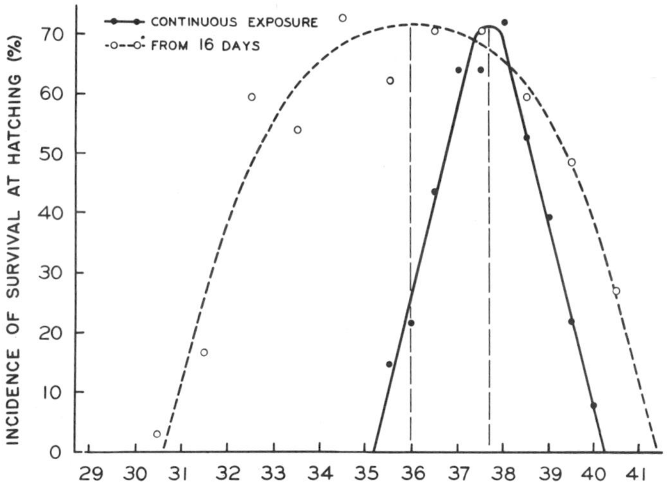

Fig. 37. Influence of various levels of incubating temperatures on the incidence of survival of the chicken embryo at hatching time—with exposure continuous or from the 16th day. (After Romanoff, 1936a.)

One- and Two-Week Exposures. The effects of 1- and 2-week exposures to abnormal temperature on the embryonic mortality (*Romanoff, Smith, and Sullivan, 1938*) are shown in the tabulation below.

Time of Exposure (weeks)	Embryonic Mortality at Hatching (%) At 34.5° C	At 40.5° C
First two	100.0	100.0
Last two	97.8	100.0
First	83.7	96.2
Middle	61.6	96.3
Last	32.0	100.0

From these data it is obvious that low temperature in the middle and, especially, in the last week has the least effect on the incidence of embryonic mortality. This is also well demonstrated in Fig. 38. Exposure during the first week could be considered the most destructive at both extremes of temperature; exposure in the last week is the least harmful at low temperatures.

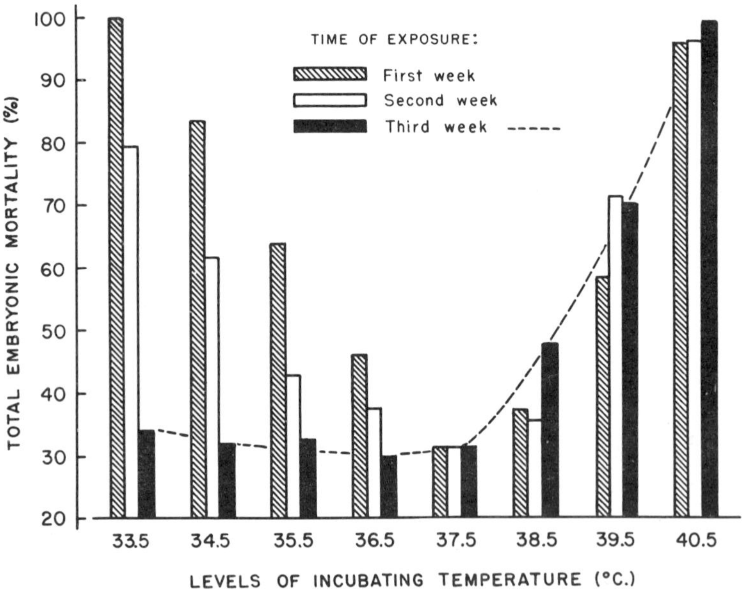

Fig. 38. Comparison of the effects of time of weekly exposure to various levels of incubating temperature on the total embryonic mortality at hatching. (Plotted from the data of Romanoff, Smith, and Sullivan, 1938.)

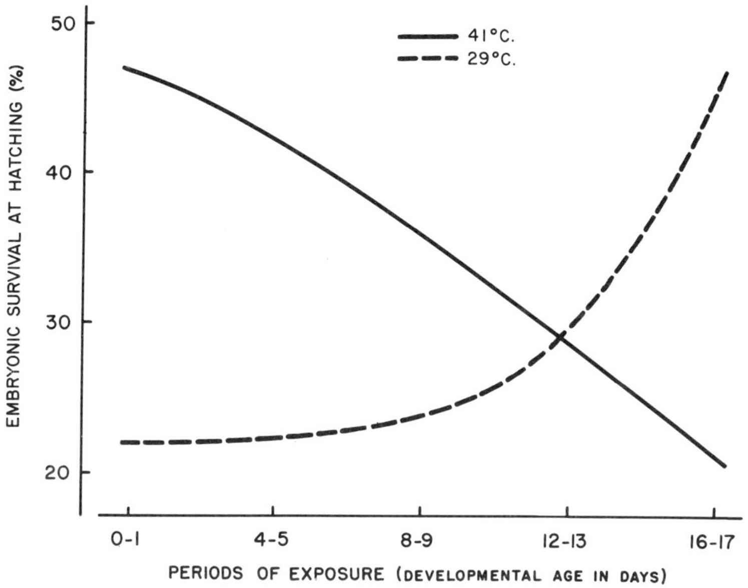

Fig. 39. Effect of a 24-hour exposure to 29° C and 41° C at different stages of development on the total embryonic survival at hatching. (After Romanoff, Smith, and Sullivan, 1938; based on normally developing eggs, from 100 to 200 in each lot. At the control temperature of 37.5° C, the embryonic mortality was 18.8 per cent.)

The Twenty-Four-Hour Exposure. The immediate effect on embryonic survival of 24 hours' exposure to 29 and 41° C is of special interest. The total survival of the embryo during the post-exposure period is noticeably pronounced with certain exposures (Fig. 39). Short exposures to the high temperature are the most injurious to the developing embryo at the late period of incubation; exposures to the low temperature are most injurious to the embryo at the early stages of incubation.

Lethal Exposures (High and Low Temperatures). When 13-day-old embryos were exposed to 55° C, the effect on the incidence of mortality (*Romanoff, 1951*) was found to be as follows.

Time of Exposure to 55° C (min)	Incidence of Mortality (%)
30	0.0
40	16.7
50	62.5
60	100.0

Repeated trials with ringneck pheasant (*Phasianus torquatus*) eggs (*Romanoff, 1951*), heated to 50° C in groups of 96, showed that all embryos survived this temperature for 70 min. Death occurred only after treatment for 80 min, and mortality rate increased irregularly to 100 per cent as the period of treatment was lengthened to 120 min.

It has been demonstrated (*Szarski, 1948*) that the death of the 10-day-old embryo, heated to a temperature of 45.5° C, is caused by cessation of the heartbeat, which ensues after a short period of intensified treatment.

The exposure of incubating chicken eggs to 20° C for 2 days at various stages of development produces a progressively heavy embryonic mortality (*Ancel, 1959*), as is shown in the accompanying table.

Periods of Exposure to 20° C (days)	Embryonic Mortality (%)			
	24 Hours After Exposure	On the 10th Day	On the 20th Day	At Hatching
3–5	38.7	43.7	—	—
4–6	12.0	28.4	50.7	86.5
5–7	35.7	50.0	70.6	89.5
6–8	39.2	57.2	84.0	91.2
7–9	30.1	30.2	79.8	96.3
8–10	46.4	—	82.2	96.7
Mean	33.7	41.9	73.5	92.0

There is almost a linear increase in mortality with increasing embryonic age at the time of exposure (see also Fig. 39).

Weight of the Dead Embryo. The weights of the individual embryos on the day of death, as the result of the exposure of eggs to various temperatures, are shown in Fig. 40. Up to the eighteenth day of incubation, the weights of the dead embryos were far below the weights of the control living embryos. But from the eighteenth day on, the weights become extremely variable, ranging from about 2 to over 30 grams.

When the dead embryos are selectively separated into two extreme temperature groups (low, 35° C, and high, 40° C) then the average weights for these groups become widely separated (Fig. 41). These curves, in general, describe the proportional weights of the dead embryos

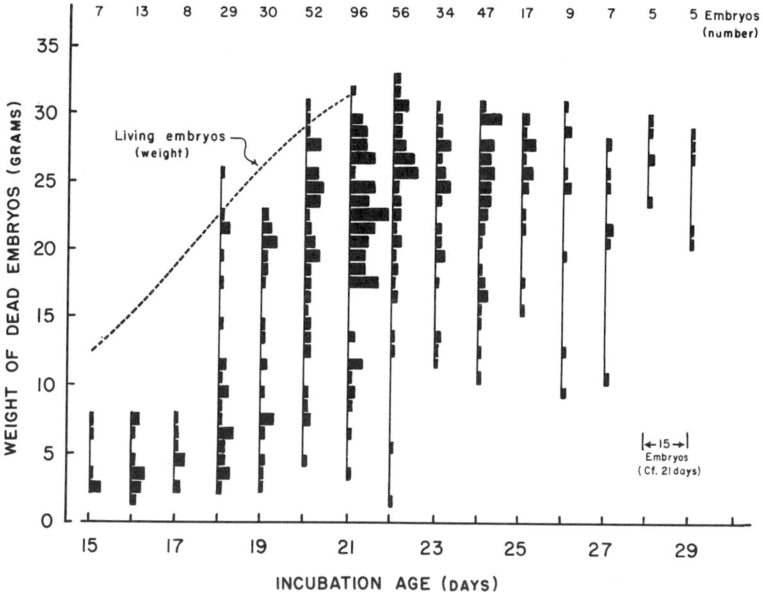

Fig. 40. Distribution of the weights of 415 dead embryos—from eggs incubated under temperatures within a range of 35 to 40° C, with intervals of 0.5° C. (After the data of Romanoff, 1936a, and Romanoff, unpublished.)

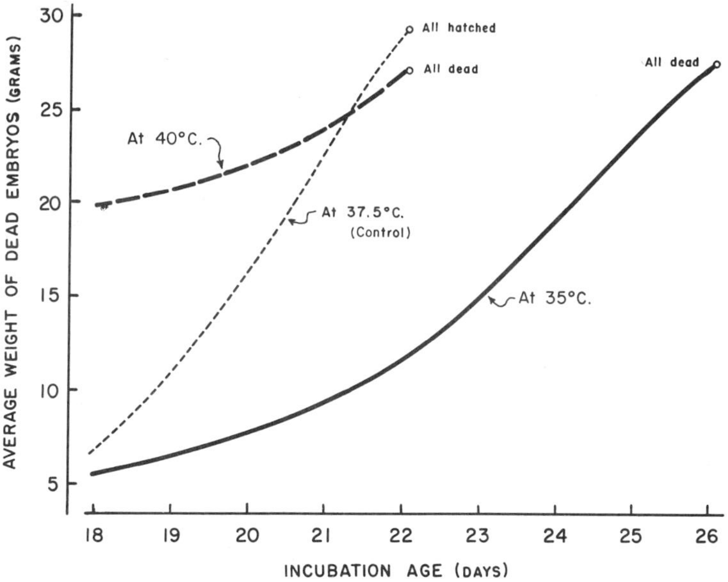

Fig. 41. Average weight of dead embryos at the two extreme temperature conditions sustained throughout the incubation period. (After the data of Romanoff, Smith, and Sullivan, 1938, and Romanoff, unpublished.)

to weights of living embryos of the corresponding age, as shown below (*Romanoff*, unpublished).

Incubation Age (days)	Proportional Weight of Dead Embryos (%)		
	At 35° C	At 37.5° C	At 40° C
18	36.5	33.1	94.7
19	54.7	50.6	96.8
20	73.8	52.5	100.7
21	95.8	75.2	103.5

With continued incubation, there is a tendency for the embryo to die suddenly, especially at high temperatures, and thus the weight of dead embryos approaches the weight of the living ones.

Malformations

It has been recognized since Dareste (*1891*) that the application of abnormal temperatures to the avian embryo is one of the simplest experimental procedures for obtaining prenatal malformations, under well-controlled conditions. Now it is known that even a small deviation in temperature from the optimum 37.5° C produces a noticeable percentage of abnormal embryos. The incidence of abnormalities usually increases with further departure from the optimum temperature. Some of the most abnormal embryos die early, but the less severely affected embryos reach their full development and may hatch.

Gross Abnormalities. Upon examination of 2740 living embryos aged 4 to 21 days and under exposure to various temperatures (33.5 to 40.5° C), 212 were found to have one or more structural defects (*Romanoff, Smith, and Sullivan, 1938*), as shown in the table below.

Abnormality	Incidence in 212 Embryos (%)
Liver, off-color	68.7
Deformed feet and toes	13.2
Celosomia (of liver, heart, and gizzard)	10.8
Edema (abdominal)	4.7
Shortened upper beak	1.4
Other abnormalities	1.2

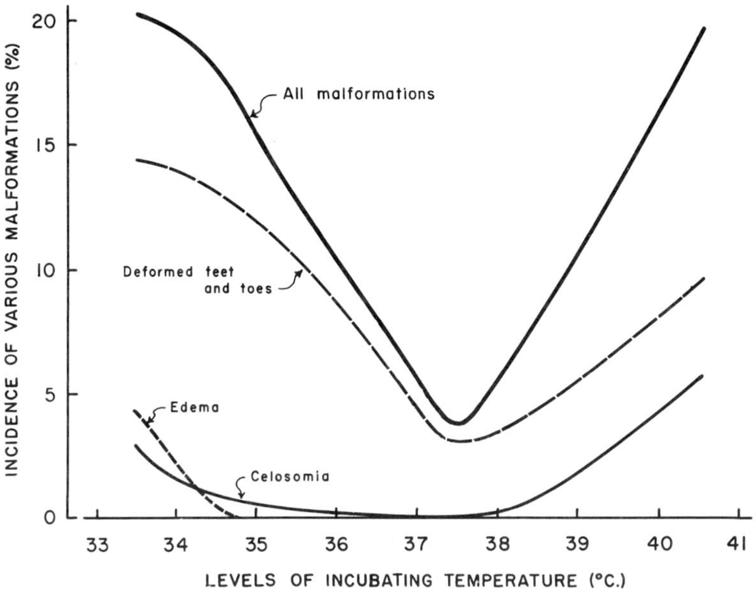

Fig. 42. Effect of different levels of incubating temperature on the frequency of malformations. A total of 212 malformed embryos were found among 2740 surviving embryos examined between the fourth and twenty-first days of development. (Based on the data of Romanoff, Smith, and Sullivan, 1938.)

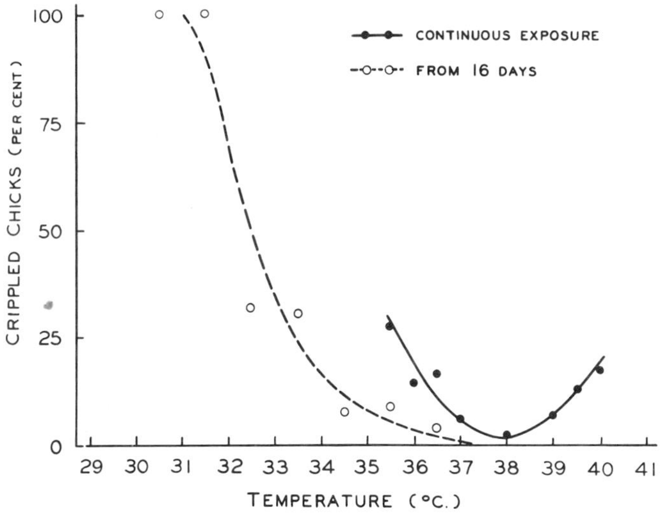

Fig. 43. Effect of various levels of incubating temperature on the occurrence of crippled chicks at hatching. (After Romanoff, 1936a.)

The largest number of defective embryos appeared under extremely low and extremely high temperatures.

The incidence of some typical malformations over the entire range of temperature is demonstrated in Fig. 42, and the incidence of crippling in the newly hatched chicks in Fig. 43. The frequency of crippled chicks evidently increases with high and low extremes of incubating temperature, but high temperatures from the sixteenth day onward do not produce such an effect (see also Fig. 42). It appears that crippling is a highly specific anomaly occurring only at low temperatures, both in the chicken (*Romanoff, 1936a*) and in the turkey (*Romanoff, 1935a*).

It has been demonstrated (*Alsop, 1919*) that abnormalities produced by excessive heat are located in somewhat different parts of the embryo than those produced by a limited amount of heat. This is clearly shown in the accompanying table.

	Three Days of Exposure	
Abnormalities	At 36.5° C	At 41.5° C
Total abnormalities (%)	67	90
In the head region (%)	17	46
In the neural tube (%)	83	54

Eggs incubated at normal temperature (about 37.5° C) showed nearly 6.5 per cent abnormal embryos. Many of these spontaneous abnormalities were different from the deformities produced under abnormal temperatures.

Specific studies of the effects of high temperature show that the occurrence of celosomia (*Ancel and Lallemand, 1941;* also *Delphia and Eveleth, 1961*) is very high, as is shown in the table below.

Temperature (° C)	Incidence of Celosomia (%)
38.5	0.0
39.5	49.2
40.5	62.7
41.5	50.0

Also, the effect of temperature shock at 45.5° C for 5 hours at five different stages between 0 and 19 hours of development (*Deuchar, 1952*) manifests itself as follows:

Age (hours)	Abnormal Gastrulation	Incidence of Abnormalities (%)				
		Necrosis	No Axial Tissues	No Head	No Trunk	Total
0	—	—	—	—	—	0.0
7	9.6	31.0	9.6	—	—	56.8
12	26.4	48.8	4.9	5.0	9.6	80.1
16	9.7	48.4	—	—	—	61.0
19	—	20.0	—	—	—	59.6

These anomalies, observed about 1 to 3 days after treatment, exemplify structural changes caused by high temperature. The embryos of unincubated eggs are practically unaffected by the high-temperature treatment.

The effect of low temperature on the morphology of the blastoderm in eggs incubated at 25° C has been observed by Harrison and Klein (*1954*) and is presented in the accompanying table.

Time of Exposure to 25° C (days)	Gross Morphology of Blastoderm	
	Anidean (formless) (%)	Degenerate Streak (%)
2	72	0
4	44	0
6	27	13
8	8	23
10	13	87

The abnormal and degenerate morphology of the embryo is also shown in the following tabulation.

Time of Exposure to 25° C (days)	Morphology of Embryo		
	Abnormal (%)	Degenerate (%)	Whole Anidean (Without Embryo) (%)
8	0	0	0
10	10	0	0
12	92	42	0
14	92	42	9
20	100	50	50

These studies (*Harrison and Klein, 1954*) show that fertile eggs kept for 10 days or longer at 25° C have a marked increase in the incidence of distinct abnormalities.

The effect of exposure at 25° C *in ovo* upon subsequent development of blastoderms *in vitro,* cultured on a yolk–albumen medium, was studied (*Harrison, 1957*). Following 10 to 14 days of incubation at the lowered temperature, the loss of potency of the blastoderm is correlated with the degree of degeneration of the primitive streak. This is illustrated by the blastoderm types in Fig. 44.

The drop in potency of the blastoderm occurs between the eleventh and twelfth days of exposure, as is shown in the accompanying table.

Blastoderm Type	Incidence at Two Exposures (%)	
	10 Days	13 Days
Streak	39	3
Diffuse streak	44	19
Node	8	16
Diffuse node	3	38
Anidean	8	25

With the increase in time of exposure there is a considerable increase in the incidence of anideans, that is, blastoderms without embryos—the highest degree of degeneration.

It was observed that a still lower temperature (20° C) produces malformations of the eyelid (ablepharia), of the limbs (ectromelia, hemimelia, syndactyly, ectrodactyly, polydactyly, atrophy of the fingers and toes), of the beak (harelip, atrophy), and hemorrhage and localized edema, which cause agenesis of down and buphthalmia (*Ancel, 1958*). The incidence of the most pronounced malformations is plotted in Fig. 45.

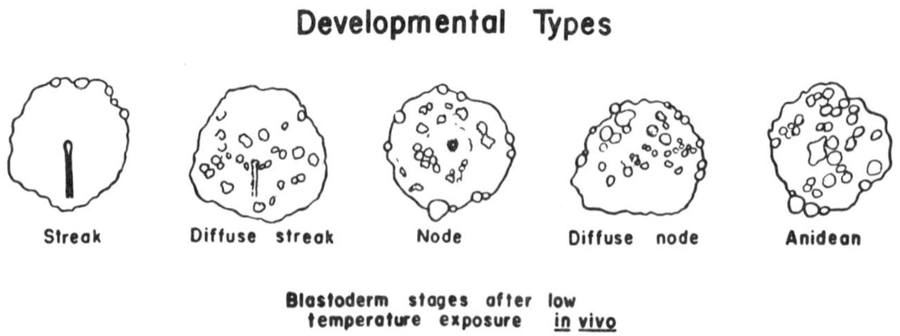

Fig. 44. Effect of exposure of chicken eggs to 25° C for 10 to 14 days on the development of blastoderms. (Rearranged after Harrison, 1957.)

From the plotted data it is obvious that with increasing age of the embryo the treatment has less effect, especially with respect to the variety and the localization of the malformations.

Growth and Malformations of Organs

The effects of temperature on growth and malformations have been observed.

Growth of Organs. Quantitative studies of the effects of various levels of incubating temperature on the growth of some organs reveal that the weight may deviate considerably under the influence of subnormal temperatures. The table below (*Romanoff,* unpublished) illustrates this point.

	Weight of Organ at Hatching (grams)		
Organ	At 36° C	At 37.5° C	At 38° C
Heart	0.38	0.32	0.28
Kidneys	0.35	0.26	0.34
Liver	0.90	1.10	1.11
Lungs	0.29	0.25	0.25
Gizzard	1.58	2.01	1.99

Under extreme temperatures, some deviations become more pronounced, as will be seen from the studies of individual organs.

Heart. The effect of various levels of incubating temperature on the absolute and proportional weight of the heart is shown in Fig. 46*A* and *B*. Although there is little change in the size of the heart at high temperatures, there is a noticeable increase at low temperatures.

The hearts from 11-day-old embryos exposed for 7 days to 32.5° C were the largest (*Leighton, Merkow, and Locker, 1964; Merkow and Leighton, 1966*).

Temperature (° C)	Weight of Hearts at 18 Days (mg)
32.5	203.8
37.5	153.3
42.5	81.1

The increase in size consists primarily of enlargement of the chambers of the heart, as well as thickening of their walls. It has been ex-

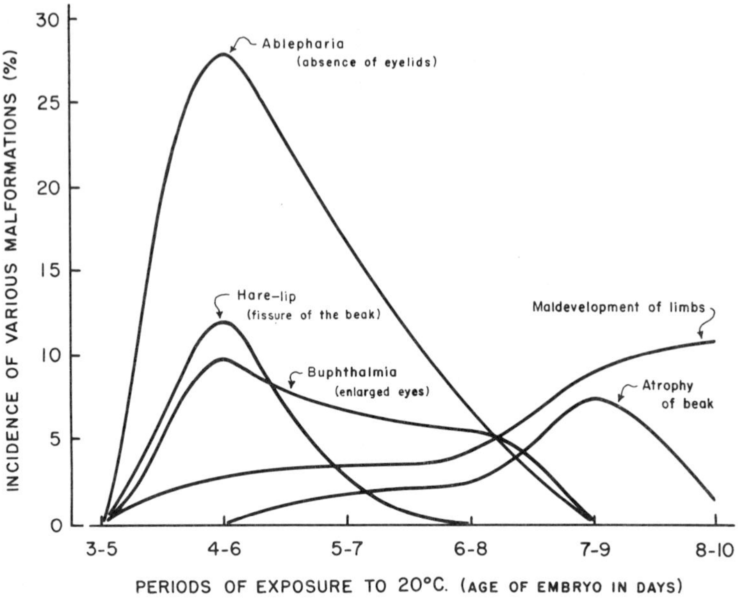

Fig. 45. Effect of 2 hours' cooling of eggs at various stages of incubation on the incidence of various malformations. (Based on observations of 1097 eggs by Ancel, 1958, 1959.)

plained that the environmental temperature influences the size of the intact heart not through a direct reaction of heart tissue to the changes in temperature, but through a hemodynamic response to congestive failure (*Leighton, Merkow, and Locker, 1964*).

Kidneys. High and low temperatures cause an increase in the absolute and proportional weights of the kidneys (Fig. 46*A* and *B*). It also has been shown that with exposure for 1 week (days 11 to 18) to very extreme temperatures (32.5 and 42.5° C), the kidneys, both mesonephric and metanephric, become enlarged (cf. accumulation of allantoic fluid in the allantois under influence of extreme temperatures, Fig. 48).

Liver. From Fig. 47*A* and *B*, it is evident that by exposure to high or low temperatures the liver is immensely reduced in size, especially by low temperature (below 35° C). The weight of the liver, whether absolute or in proportion to size of the body, follows very much the same curve.

Bones. It has been demonstrated (*Lerner and Gunns, 1938*) that with acceleration or retardation in growth of the embryo by a high or a low temperature (see Fig. 31), a corresponding change is noticeable

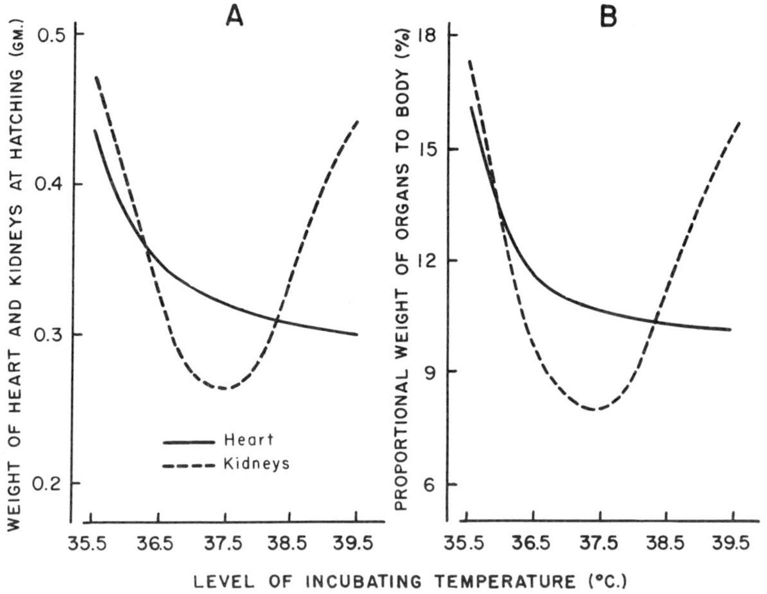

Fig. 46. Effect of different levels of temperature on the weight of the heart and of the kidneys at the time of hatching. (Based on the data of Romanoff, unpublished.) *A*, absoluate weight of the embryonic heart and kidneys; *B*, proportional weight of the heart and kidneys to the weight of the body. (See also Merkow and Leighton, 1967, on the embryonic heart in eggs incubated at 32, 38, and 42° C from the eighth to the eleventh day.)

in the length of bones, such as the femur, tibiotarsus, and tarsometatarsus. The growth of femur is shown in the table below.

Age of Embryo (days)	Length of Femur (cm)		
	At 36° C	At 38° C	At 40° C
11	0.48	0.60	0.66
13	0.63	0.77	1.09
15	0.91	1.10	1.42
17	1.24	1.47	1.75
19	1.52	1.76	2.07

It has been observed (*Romanoff, Smith, and Sullivan, 1938*) that with a high temperature, the limbs and beak at about 12 days of incubation are long and slender.

Allantois. Both high and low temperatures cause a pronounced reduction in size of the allantois (Fig. 48), with the amount of allantoic fluid not exceeding 2 cm^3 at its maximum.

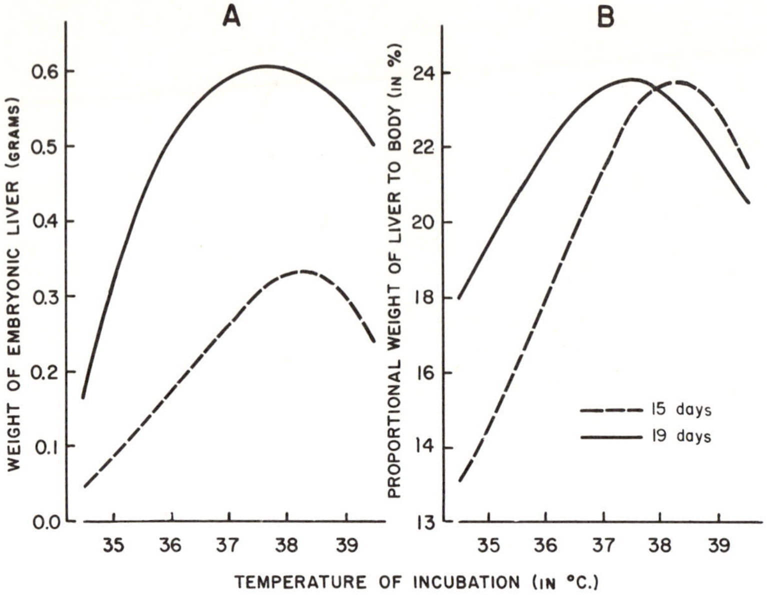

Fig. 47. Effect of different levels of incubating temperature on growth of the embryonic liver. (Plotted from the recalculated data of Romanoff, Smith, and Sullivan, 1938.) *A*, absolute weight of the embryonic liver; *B*, weight of the liver in proportion to the body weight.

The accompanying table (*Romanoff, Smith, and Sullivan, 1938*) shows that changes in the accumulation of allantoic fluid are characteristic of each temperature level.

Temperature (° C)	Amount of Allantoic Fluid (cm^3)					
	At 8 Days	At 10 Days	At 12 Days	At 14 Days	At 16 Days	At 18 Days
34.5	0.0	0.0	0.5	1.5	0.4	0.0
35.5	0.3	0.8	1.9	6.3	4.7	4.5
36.5	0.6	1.9	4.4	5.1	3.0	1.6
37.5	0.8	2.2	5.2	5.5	2.2	0.6
38.5	0.9	2.3	4.5	2.8	2.0	0.0
39.5	0.8	1.6	2.0	0.9	0.0	0.0
40.5	0.7	1.3	1.2	0.0	0.0	0.0

With maldevelopment of the allantoic sac, noticeable modifications occur in the arrangement and extent of the extraembryonic blood system, or in the extent of development of the area vasculosa (see Fig. 24).

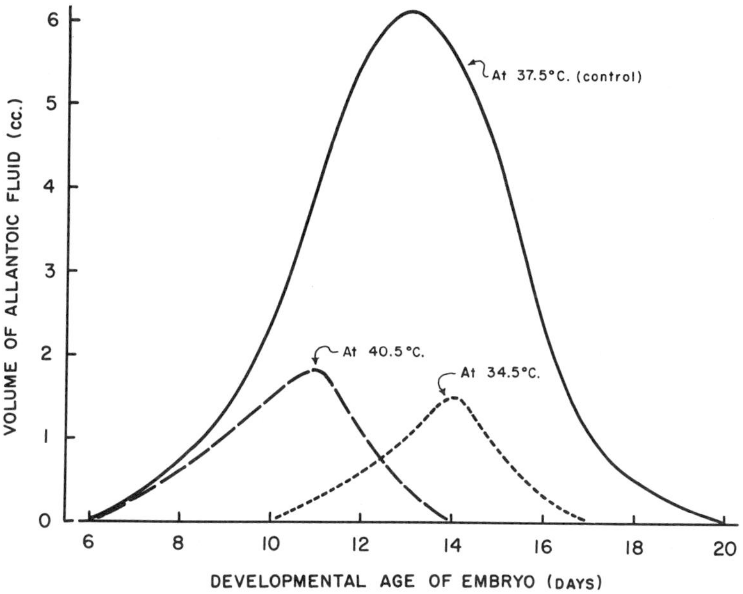

Fig. 48. Effect of extreme temperatures of incubation on the development of the allantois. (After Romanoff, Smith, and Sullivan, 1938.)

Similar observations were made by Tazelaar (*1928*) in a study of the effect of temperature during the first week of development.

With the reduction in the amount of allantoic fluid, there is an almost corresponding reduction in the amount of secretory uric acid (*Romanoff, Smith, and Sullivan, 1938*), especially under exposure to extreme temperatures (34.5 or 40.5° C). This is unquestionably related to the enlargement of the kidneys under these incubating temperatures.

Malformations of Organs. It has been observed (*Cruz, Campillo-Sainz, and Munoz-Armas, 1966*) that high temperature does not induce cardiac abnormalities in the chicken embryo, but that low temperature adversely affects the circulatory system, as is shown below.

	Malformations of Embryonic Heart		
Temperature (° C)	Ventricular (%)	Other Defects (%)	Total (%)
35.8	24.5	10.5	35.0
37.7	0.0	3.8	3.8
39.4	0.0	2.8	2.8

The same authors also observed that elevated temperature (39.4° C) led to malformations of the eyes and nervous system.

Heartbeat. It was observed (*Andersen, 1932*) that the heartbeat accelerates with increase in temperature. With exposure to 45° C, the heart ceases to beat during the second hour.

Rates of beat per minute of isolated 3-day embryonic chicken hearts, exposed to various levels of temperature, are shown in Fig. 49. The plotted data give a curve similar to that obtained by studies of hearts *in situ* (see *insert* of Fig. 49).

Abrupt exposure of the embryo to low (16.5° C) and high (58.5° C) temperatures results in linear changes in the rate of heartbeat (*Romanoff and Sochen, 1936*) with lower and upper breaks, followed by secondary linear changes (see *insert* of Fig. 49).

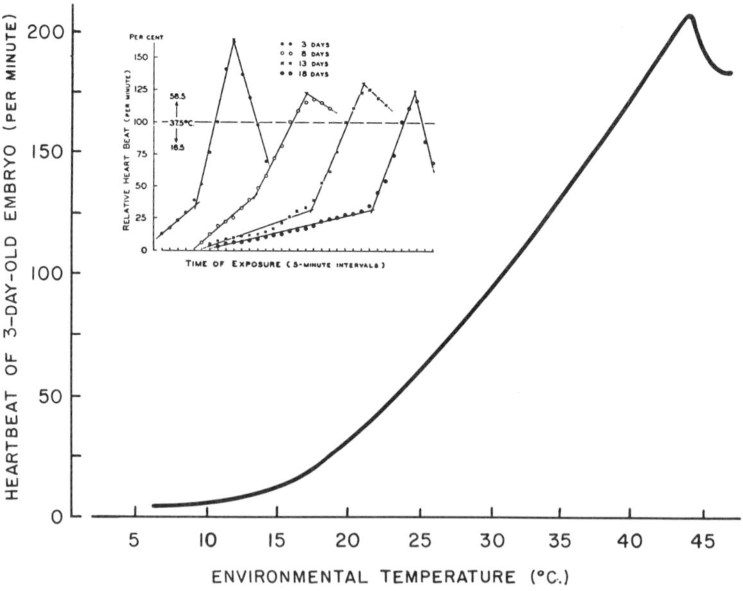

Fig. 49. Effect of different levels of temperature on the average rate of beat of 208 hearts, isolated from 3-day-old embryos, and suspended in chicken blood plasma. (Rearranged after Paff, Boucek, Nieman, and Deichmann, 1963.) *Insert:* relative changes in the rate and duration of the heartbeat of 3-, 8-, 13-, and 18-day-old chicken embryos during continuous exposure to low (16.5° C) and high (58.5° C) temperatures. (Romanoff and Sochen, 1936.)

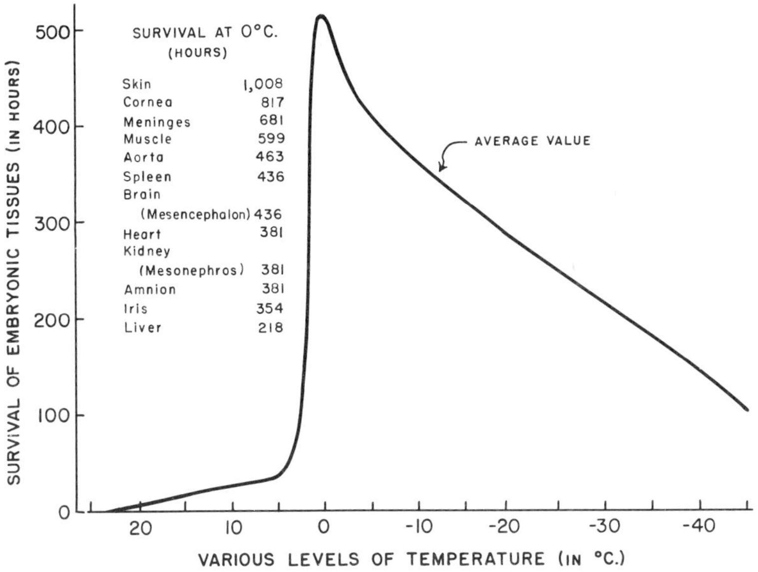

Fig. 50. Average survival of embryonic tissues of the chicken at various levels of temperature. (Recalculated from the data of Bucciante, 1931, 1933.)

Survival of Tissues. The effects of different levels of temperature on survival of individual tissues of the chicken embryo are shown in Fig. 50. The average curve demonstrates that all tissues retain viability for the longest time at 0° C. With decrease in temperature comes, first, a rapid decrease in viability, then a considerably slower decline, with some survival even beyond —40° C. With increase in temperature, there is a very rapid drop in viability of tissues up to 5° C, continuing to complete extinction at about 20° C.

Viability of individual tissues of 10-day-old chicken embryos heated to 45.5° C (*Szarski, 1948*) is shown below.

Organ Tissues	Viability at 45.5° C (hours)
Brain	1–2
Heart	1–3
Liver	1–3
Intestines	1–3
Iris (of the eye)	1–3
Skin	1–4
Area pellucida	7–13
Amnion	11–15
Allantois	11–15

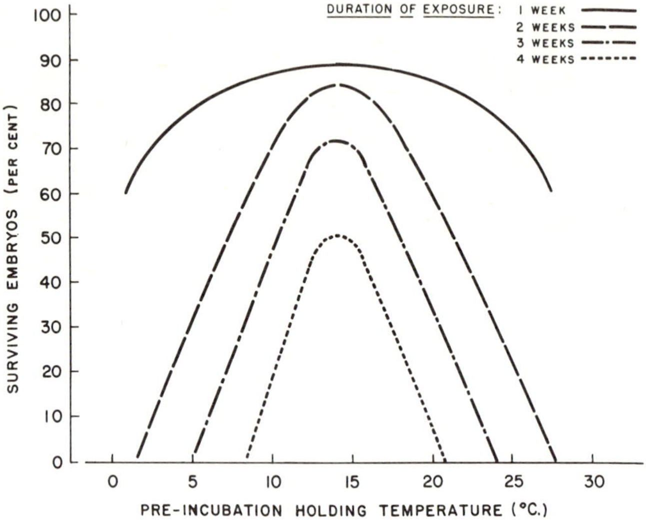

Fig. 51. Effect of preincubation temperature of fertile eggs on the survival of the chicken embryo. (After Romanoff, 1960, p. 200.)

At 45.5° C (*Szarski, 1948*), the contractility of muscle tissue and the conductivity of nervous elements are conserved longer than is the ability to regenerate *in vitro*.

Survival of Fibroblasts. Fibroblasts from the heart muscle of the chicken embryo survive for different periods at various temperatures (*Nemoto, 1929*) as is shown in the table below.

Temperature (° C)	Survival Time of Fibroblasts (hours)
5	72
12	144
23	264
30	288
33	168
45	24

Table 4 Effect of Exposure of Fertile Eggs to Low Temperature Before Incubation on Consequent Maldevelopment and Death of Embryos*

Temperature (° C)	Period of Exposure	Effect on Maldevelopment and Death	Reference
−18.0	½ hour†	No immediate death‡	Rabaud (1899)
−15.0	½ hour	No immediate death	Kumanov (1948)
	1 hour	50% dead	
	2 hours	100% dead	
−9.4	1½ hours†	No injury	Colasanti (1875), Elford (1921), Mussehl and Bancroft (1925)
−3.0	1 day	0% dead	Grodzinski (1934)
	2 days	4% dead	
	3 days	14% dead	
	5 days	80% dead	
−2.8	Several days	All dead	Pictet (1893), Mancini (1908)
−1.1	4 nights in succession	Normal hatch	Dougherty (1926)
±0.0	12 hours	Normal hatch	Mussehl and Bancroft (1925)
	24 hours	10% reduction in hatch	
	36 hours	25% reduction in hatch	
	48 hours	50% reduction in hatch	Mauro (1923)
	72 hours	Destroyed hatch	

* This is an expanded table originally published in *The Avian Embryo* (Romanoff, 1960, p. 198).

† The eggs were practically frozen.

‡ The hatched chicks showed numerous monstrosities.

Preincubation Temperature of Eggs

During the preincubation period of fertile eggs, the surrounding temperature plays a major role in the survival of the embryo (see *insert* of Fig. 183). At 14° C, the optimum storage temperature, only a few chicken eggs can survive over 4 weeks. The incidence of embryonic survival rapidly decreases at higher or lower levels of temperature (Fig. 51).

Resistance of Eggs to Cold. Table 4 shows that fertilized eggs can withstand a very low temperature (−18° C) for a short time. However, the subzero temperatures either destroy or reduce the hatch, with the occurrence of various abnormalities.

ATMOSPHERIC GASES

Gaseous environment in the incubation of the avian egg consists of respiratory gases, such as oxygen and carbon dioxide, and some volatile substances—by-products of metabolism.

Low Level of Oxygen (Hypoxia)

Growth. The exposure of fertile eggs to air free of oxygen (less than 1.2 per cent) resulted in about 20 per cent proportional dwarfing among anomalous embryos (*Grabowski and Paar, 1958*). The incidence of dwarfism varies with the age of the embryo at the time of exposure, as follows.

Age of Embryo at Exposure	Incidence of Dwarfism (%)
18 hours	0.8
1 day	5.4
2 days	7.7
3 days	3.9
4 days	2.3

In addition, there are 24 per cent midgets among the embryos which show no apparent anomalies.

Mortality. There is general agreement (*Barott, 1937; Cruz and Romanoff, 1944; Taylor, Sjodin, and Gunns, 1956; Grabowski and Paar, 1958; Nelsen, 1958, 1960; Grabowski, 1961; Yamamoto and Noguchi, 1964*), that continuous exposure to an oxygen (O_2) level below 21 per cent tends to increase mortality, or reduce viability of the embryo, proportionally to the decrease in oxygen concentration. This is partially illustrated by the effect of continuous exposure to various levels of oxygen on total embryonic survival (Fig. 52 with *insert*).

It was observed (*Naujoks, 1953*) that the total incidence of embryonic death caused by exposure to an oxygen concentration of 3 to

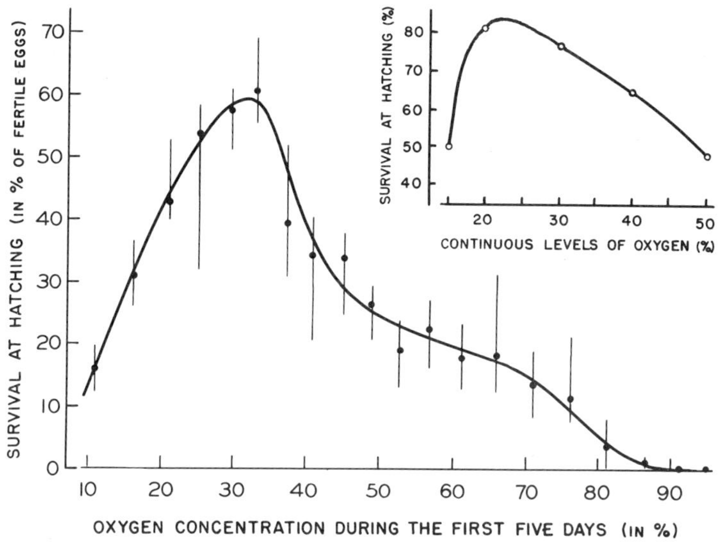

Fig. 52. Effect of different levels of oxygen concentration during the first 5 days of incubation on embryonic survival at hatching. (After Cruz and Romanoff, 1944.) *Insert:* effect of continuous exposure to different levels of oxygen on embryonic survival at hatching. (Barott, 1937.)

5 per cent for 3 to 5 hours varies with the age of the embryo at the time. This is shown in the table below.

Age of Embryo (days)	Incidence of Embryonic Mortality (%)
1	8.4
2	10.3
3	72.6
4–5	74.8

It is obvious that the younger embryos—during the first 2 days—are less susceptible to oxygen deficiency than the older embryos.

There are two categories of embryonic death: immediate and delayed (*Grabowski and Paar, 1958*). Immediate death occurs during or within a day after treatment and is directly attributable to the hypoxic condition. Delayed death occurs in embryos which survive the treatment by at least 24 hours.

Immediate embryonic death following 6 hours of exposure to various levels of oxygen pressure at different developmental stages is shown in Fig. 53.

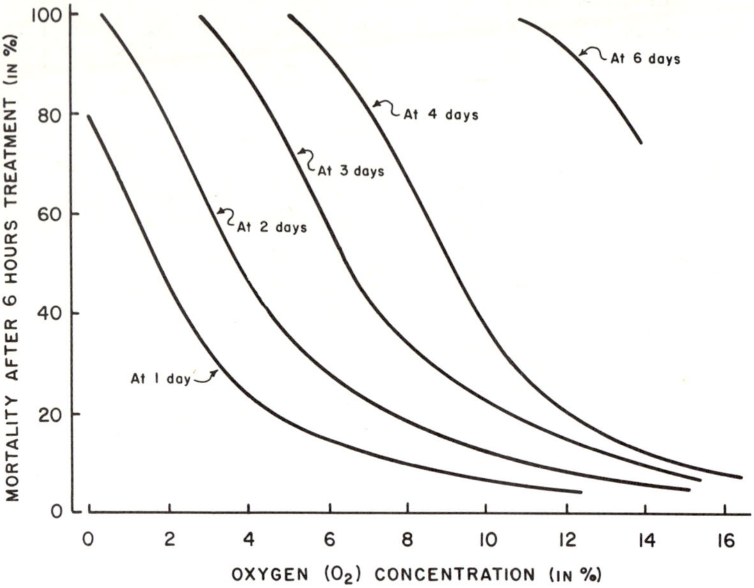

Fig. 53. Effect of oxygen deficiency for 6 hours on the rate of embryonic mortality. (Rearranged from the data of Grabowski and Paar, 1958.)

When fertilized eggs (warmed for 2 hours) were exposed for various lengths of time to air mixtures containing 6 or 8 per cent oxygen, 0.5 per cent carbon dioxide, and the remainder nitrogen (*Nelsen, 1960*), there were varying increases in embryonic mortality on the tenth day, as shown in the accompanying table.

Length of Exposure (hours)	Embryonic Mortality at 10 Days (%)	
	With 6% O_2	With 8% O_2
22	12	—
24	—	19
26	18	—
32	84	26
40	—	52

The lethal effect on the embryo of exposure to reduced oxygen pressure increases with the length of exposure, but decreases with the increase in level of oxygen in the air mixture.

Malformations. Chicken eggs incubated in pure oxygen during the first 2 days show no developmental abnormalities (*Gallera, 1951*).

Further observations on low-oxygen conditions (*Naujoks, 1953; Grabowski and Paar, 1958; Yamamoto and Noguchi, 1964; Noguchi, 1969*) yielded an extensive list of anomalies in experimental dead embryos. These included defects of most of the organ systems and numerous types of minor abnormalities. Among the embryos observed there had been many multiple anomalies.

These observations are presented in combined form in Table 5.

The frequency of occurrence of anomalous embryos is proportional to the degree of oxygen deficiency, but only within a definite narrow range. This is clear from the data (*Grabowski and Paar, 1958*) plotted

Table 5 Effect of Hypoxia on the Incidence of Various Malformations in the Chicken Embryo

	Incidence of Abnormalities (%)			
Anomalies	Naujoks (1953)	Grabowski and Paar (1958)	Yamamoto and Noguchi (1964)	Noguchi (1969)
Eye defects	28.8		23.7	33.7
Microphthalmia		35.7		26.0 (Microphthalmia and Anophthalmia combined)
Anophthalmia		7.8		
Nervous system	2.5		13.7	
Microcephaly		10.9		6.4
Omphalocephaly		9.3		
Midbrain anomaly		4.7		
Cranioschisis		3.1		
Anencephaly		2.3		8.7
Encephalocele		2.3		
Defective spinal cord	6.9		12.2	
Body	6.8			
Axis	6.8			
Rumplessness	35.6	9.3	1.4	10.0
Beak, short upper		10.9		26.9
Maxillary process		8.6		
Cleft lip and palate		8.6		22.0
Extremities	10.2			
Leg defects		17.9	10.1	11.7
Wing defects		6.3	5.1	27.4
Blood vessels		5.4		
Heart		3.1		
Subcutaneuos blebs		1.6	26.6	4.2
Umbilical hernia		3.9		

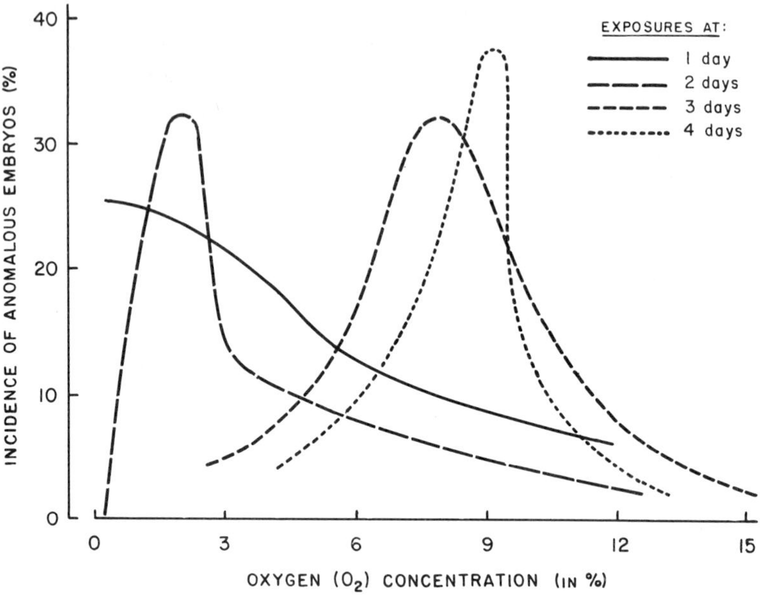

Fig. 54. Effect of graded levels of oxygen deficiency for a period of 6 hours on the incidence of anomalies in the early stages of embryonic development. (With modifications from Grabowski and Paar, 1958.)

in Fig. 54. The curves for embryos subjected to low oxygen concentration when 3 and 4 days old have very sharp peaks. Those for 1 and 2 days tend to be more extended.

It is also apparent from the accompanying table that older embryos are more sensitive to hypoxia than young ones.

Age of Embryo at Exposure (days)	Incidence of Malformations (%)		
	Grabowski et al. (1958)	Yamamoto et al. (1964)	Noguchi (1969)
1	1.7	7.1	10.0
2	24.0	48.2	43.5
3	32.6	44.7	68.9
4	21.1		60.9

The relative sensitivity of different structures of 3-day-old embryos to different degrees of hypoxia was further elaborated by Grabowski (*1961; 1964*). The frequency with which various structures are affected by 6 hours of exposure is shown below.

Structure	Incidence of Anomalies (%)	
	At 2–9% O_2	At 10–15% O_2
Entire head	0.5	0.6
Brain	4.3	2.5
Beak	7.0	2.5
Eyes	15.5	4.3
Extremities	6.4	1.2
Rump	11.2	8.6

The data show that the relative frequencies of anomalies in individual structures of the same age vary with the change of oxygen concentration.

Comparison of Vacuum and Air–Nitrogen Mixtures. It is notable that a comparison was made of vacuum and air–nitrogen mixtures as sources of oxygen-deficient atmosphere (*Grabowski, 1961*). The relative effects of exposures for 6 hours on the 3-day-old embryos under these two methods is shown below.

Oxygen Level (%)	Mortality (%)		Anomalies (%)	
	Vacuum	Air–N_2	Vacuum	Air–N_2
3.4	70	60	28	21
6.3	56	32	30	38
9.3	21	12	21	23
12.2	17	10	9	10
15.3	8	10	13	5

High Level of Oxygen

Growth. During the first 5 days of exposure of fresh fertile eggs to various levels of oxygen concentration (*Cruz and Romanoff, 1944*) the growth of the embryo was most rapid with 41 per cent oxygen, as is shown in the table below.

O_2 Concentration (%)	Weight of Embryo at 5 Days (mg)
21	51
31	71
41	89
51	71
61	68
71	56

After 60 hours of preliminary incubation under normal conditions, exposure of the embryos to 21, 31, or 41 per cent oxygen for 2 to 6 days still produces much greater growth at the higher oxygen levels than at 21 per cent, as is shown below.

Exposure (days)	Absolute Weight of Embryo (mg)		
	At 21% O_2	At 31% O_2	At 41% O_2
2	74	150	133
4	420	683	550
6	1148	1645	1593

Mortality. Embryonic mortality during 4 and 5 days' exposure to high oxygen pressure (*Cruz and Romanoff, 1944; Taylor, Sjodin, and Gunns, 1956,* respectively) rapidly increases with oxygen concentrations above 65 per cent. For embryos at these early stages of development, high concentrations of oxygen are so toxic that as the level approaches 90 per cent, none of the embryos can survive (Fig. 52).

With the increase in age, the embryos become less sensitive to elevated oxygen concentrations (*Taylor and Kreutziger, 1965; 1966*). At the age of 13 to 16 days (*Taylor and Kreutziger, 1969*) there is only about 15 per cent reduction in viability of embryos at hatching.

In view of the enormous consumption of oxygen by the developing avian embryo at hatching time (*Romanoff, 1941; Romanoff, 1967, p. 287*), short exposures to high concentrations of oxygen should have an insignificant effect on embryonic mortality.

Carbon Dioxide

Growth. The effect of continuous exposure to various levels of carbon dioxide on the absolute and relative growth of the developing embryo is shown in Fig. 55, with *insert.*

During the beginning of incubation, a low level of carbon dioxide (about 0.4 per cent) resulted in a slight increase in the size of the embryo, compared to normal. Under all other conditions, with large amounts of carbon dioxide, the size of the embryos rapidly diminished. The maximum amount of carbon dioxide permitting growth of the embryo, at least for a short time, is about 22 per cent (*Romanoff, 1930b; Romanoff and Romanoff, 1933*).

Mortality. With continuous exposure to various levels of carbon dioxide in the incubator, there is an inverse relationship between the carbon

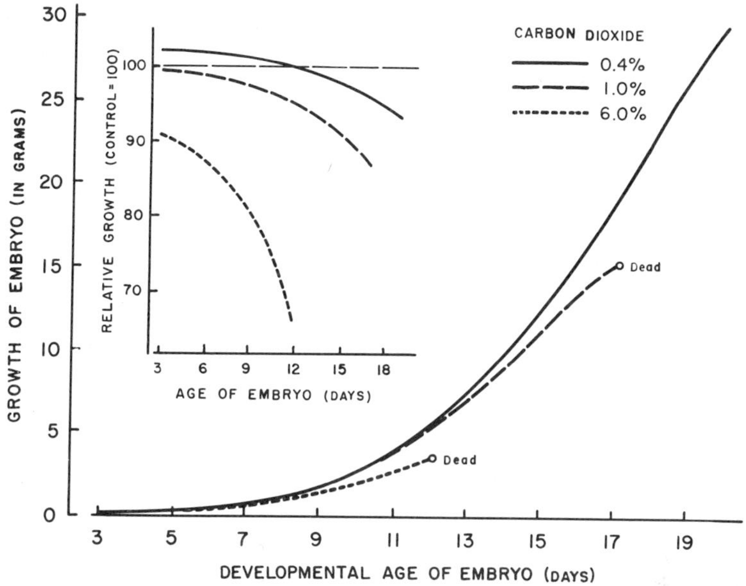

Fig. 55. Effect of carbon dioxide concentration on the growth of the developing embryo. (Plotted from the data of Romanoff, 1930b.) *Insert:* effect of carbon dioxide on relative growth.

dioxide content and the embryonic survival at hatching (*Lamson and Edmond, 1914; Romanoff, 1930b; Romanoff and Romanoff, 1933; Barott, 1937*).

The relative survival of embryos exposed at several developmental periods to various levels of carbon dioxide is demonstrated in Fig. 56. The plotted data show that the embryo's greatest susceptibility to carbon dioxide is from 0 to 4 days of development. The susceptibility declines with each developmental stage, from 5 to 8, 9 to 12, and 13 to 16 days to hatching time.

Malpositions Within the Eggs. At the end of hatching time, many embryos incubated under a continuously high level of carbon dioxide were found in abnormal positions within the eggs—usually with the leg twisted around the neck (*Romanoff, 1930a;* see Fig. 15, no. 4).

Malformations. During the first 2 days of incubation, a carbon dioxide concentration of 20 to 30 per cent, even in a gas mixture containing more oxygen than normal air, was found (*Gallera, 1951*) to cause some pronounced disturbances. The incidence of some malformations is shown in the table below.

Malformation	Incidence (%)
Platyneuria	73.6
Omphalocephaly	8.6
Diplocardia	1.9

Exposure of eggs to a high concentration of carbon dioxide for 1 to 3 days at the beginning of incubation (*Lutz and Lepy, 1958*) resulted in several malformations, the incidence of which is shown in the accompanying tabulation.

Malformation	Incidence (%)
Spina bifida	63.6
Encephaly	52.3
Anophthalmia and microphthalmia	27.2
Double monsters	4.1
Celosomia	3.1

A severe form of acleiencephaly is illustrated in Fig. 57.

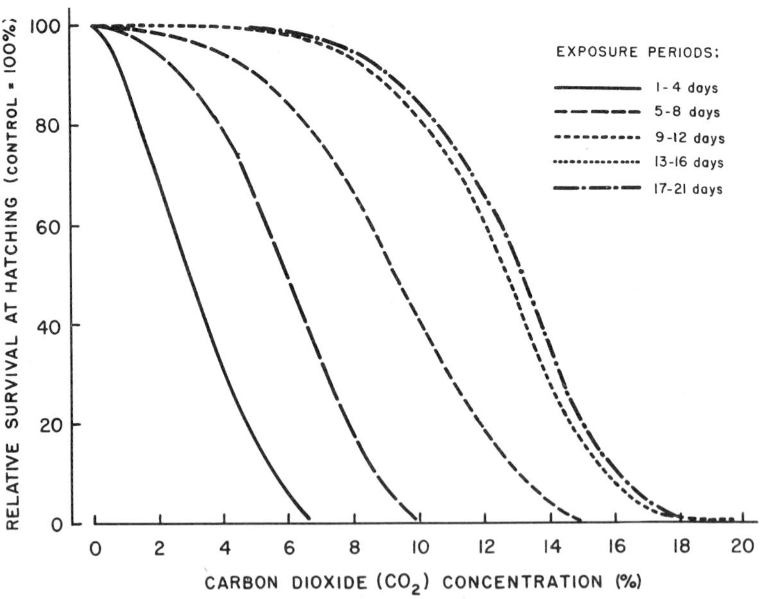

Fig. 56. Effect of carbon dioxide levels at various periods of incubation on the relative survival of the chicken embryo. (Composed from the data of Taylor, Sjodin, and Gunns, 1956; Taylor and Kreutziger, 1965, 1966, 1969; Taylor, Kreutziger, and Abercrombie, 1971.)

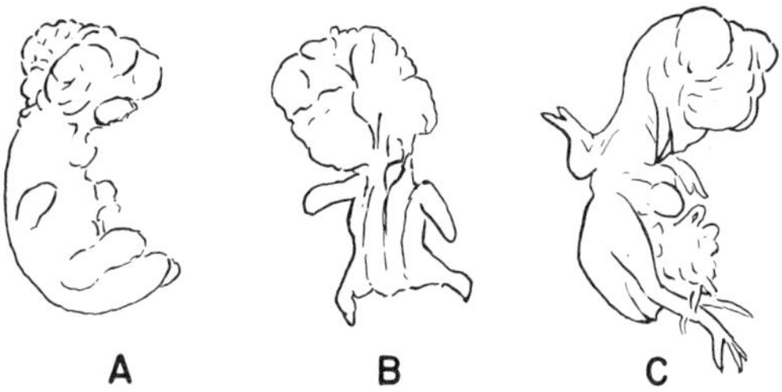

Fig. 57. Effects of exposure of the early embryo (during the first 3 days) to a high level of carbon dioxide. (Sketched from the photographs of Lutz and Lepy, 1958.) *A, B,* and *C* display mainly acleiencephaly; (anomaly of head and brain); *B* also shows spina bifida.

Nitrogen

Fertilized chicken eggs, incubated in an atmosphere of pure nitrogen for 6 hours (*Grabowski and Paar, 1958*), showed no apparent anomalies among developing embryos. Under similar conditions (*Nelsen, 1958*), exposure for 5 hours also resulted in normal development, but with some retardation in growth, and exposures from 8 to 12 hours produced general breakdown of the blastoderm.

Helium

It has been observed that substitution of helium (He) for nitrogen (N_2) in the atmosphere in which fertile eggs are incubated causes various developmental disturbances and increases embryonic mortality. In such a nitrogen-low atmosphere, the embryos die by the ninth day (*Volskii, 1959*), and 93 per cent of embryos show some abnormalities by the fourth day (*Boriskin, Oblapenko, Rol'nik, and Savin, 1962; Allen, 1963*). Experimentally, it was found that significantly fewer normal chicks can be hatched from eggs incubated in He–O_2 (*Weiss, Wright, and Hiatt, 1965*)—about 31 per cent, compared with a 73 per cent hatch under normal atmosphere. The detrimental effect of helium was expressed primarily in higher mortality during the second week of incubation, although early and late mortality, in the first and third weeks, was also higher.

Fertile chicken eggs incubated in 75 per cent helium and 21 per cent oxygen showed a higher metabolism (*Wright, Lessler, Weiss, and Hiatt, 1965*). The metabolic rate of 8-day-embryo homogenate in 3 hours was 15.4 μl O_2/gram dry tissue, as compared with a value of 11.9 μl in the control group.

Noxious Gases

Illuminating Gas. Chicken eggs were exposed for up to 20 hours, during their initial incubation, to natural gas (consisting of methane 93 to 95 per cent, ethane–propane–butane 4 per cent, and nitrogen 1 to 3 per cent).

The distribution of malformations observed (*Dugnat, 1965*) on the fourth day of development was as shown below.

Malformation	Incidence (%)
Spina bifida	36
Acleiencephaly	33
Anidean	17
Omphalocephaly	12
Hemimelia	2

Nitrous Oxide. Nitrous oxide (N_2O) has long been considered the least toxic and most inert of all anesthetic agents. Recent observations (*Rector*

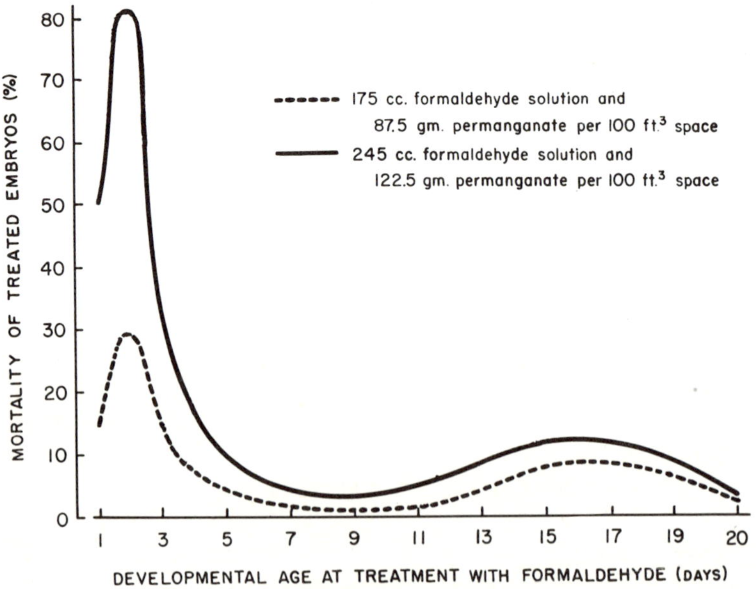

Fig. 58. Effect of formaldehyde gas on mortality of chicken embryos treated for 10 min at various stages of development. (Plotted from the data of Insko, Steele, and Hinton, 1941.) *Note:* all developing eggs were examined about 5 to 7 days following the treatment, except in later stages.

and Eastwood, 1964) indicate, however, that exposure to 80 per cent nitrous oxide (with remaining 20 per cent oxygen) during the entire period of incubation produced a delayed and reduced hatch. Among the 15 live experimental chicks, three were affected with spastic paralysis.

Further experiments (*Smith, Gaub, and Moya, 1965*) show that exposure for 6 hours of 3-, 4-, or 5-day-old embryos to 80 per cent nitrous oxide gas can result in both lethal and teratogenic effects, as is shown in the tabulation below.

Age of Embryo (days)	Mortality (%)	Anomalies (%)
3	22.9	5.0
4	8.8	0.0
5	0.0	0.0

Polluted Air. A study was made in France (*Dugnat, 1965*) on the detrimental effects of polluted air (gaz de Lacq) of the following compositions:

Components of Air	Amounts (%)
Hydrogen	31–51
Nitrogen	11–44
Carbon monoxide	8–33
Methane	9–23
Hydrocarbons	2–7
Carbon dioxide	3–4
Oxygen	0.7–0.8

Chicken eggs were exposed to the above gas for 10 hours, at developmental stages from 0 to 20 hours. The treated embryos were examined on the fourth day of incubation. The distribution of malformed embryos is shown in the accompanying table.

Malformation	Incidence (%)
Anidean	52
Spina bifida	39
Acleiencephaly	8
Omphalocephaly	1

Fumigant (Formaldehyde Gas). As an effective fumigant, formaldehyde (35 cm^3 formaldehyde solution with 17.5 grams potassium permanganate per 100 ft^3 of space) has been used in the disinfection of incubating eggs. In concentrations exceeding this "normal" concentration per space, the toxic effect of formaldehyde becomes very obvious (*Insko, Steele, and Hinton, 1941*).

Figure 58 shows that with the use of 175 and 245 cm^3 of formaldehyde, there is a positive relationship between the concentration of the fumigant and the mortality of embryos. The stage of incubation at which the embryos are most susceptible to injury from fumigation is the first critical period, about the second and third days of development (see critical periods, Fig. 13).

RELATIVE HUMIDITY

Since the early days of artificial incubation (*Reamur, 1749*), humidity has been considered a deleterious factor in the development of the avian embryo.

Growth. Growth of the embryo is somewhat hastened by low and retarded by high humidity (*Romanoff, 1928; 1929b; 1930a*). At hatching time, the weight of the embryo is lighter at low and heavier at high humidity. This has been observed in the incubation of pheasants and quails (*Romanoff, 1934*), ducks (*Romanoff, 1943a*), and turkeys (*Romanoff*, unpublished). The actual weight of newly hatched birds incubated at 40, 60, and 80 per cent relative humidity is shown in the accompanying table.

	Weight of Hatched Birds (grams)		
Species	At 40% R.H.	At 60% R.H.	At 80% R.H.
Pheasant	19.6	20.7	21.7
Quail	5.6	5.8	6.2
Duck	46.2	48.4	49.7
Turkey	60.6	62.5	64.5

The difference in weight of the chick at various levels of relative humidity has been partially accounted for by the relative amount of water in the tissue of the embryo (*Romanoff, 1929b*).

Mortality. With continuous exposure of the developing eggs to various levels of relative humidity, the most obvious effect is a change in embryonic mortality, as is shown in Fig. 59.

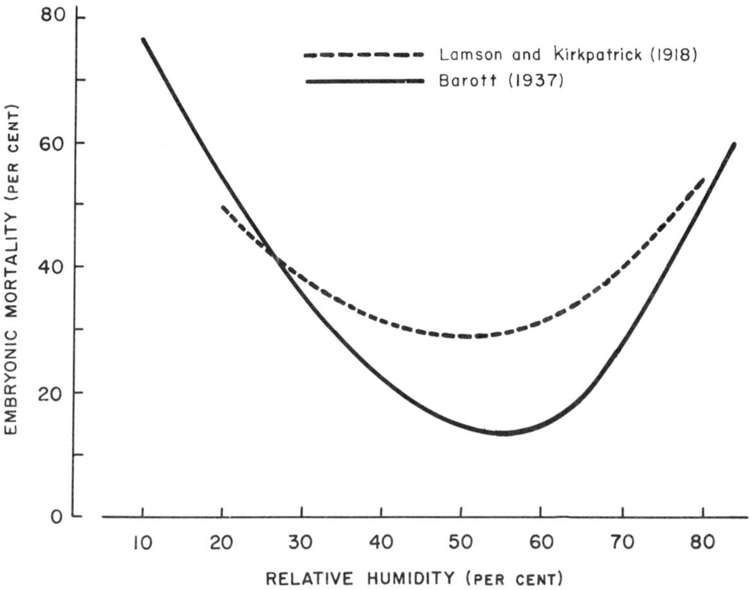

Fig. 59. Effect of various levels of humidity on mortality of the chicken embryo. (Plotted from the data of Lamson and Kirkpatrick, 1918; and Barott, 1937.)

This phenomenon was also observed with species of birds other than chickens, the data on which is shown below (on turkeys—*Romanoff*, unpublished; on pheasants and quail—*Romanoff, 1934;* on duck—*Romanoff, 1943a*).

Relative Humidity (%)	Total Embryonic Mortality (%)			
	Turkey	Pheasant	Quail	Duck
40	28	40	66	49
50	25	33	64	—
60	8	32	48	29
70	17	26	47	—
80	18	28	62	24

Observations show (*Romanoff*, unpublished) that surviving embryos incubated under high humidity become highly susceptible, and many die within 1 week after hatching.

Malformations. It has been shown (*Fischer, 1969*) that high and low relative humidities have different effects on the incidence of malformations of the aortic arches. The results are shown below.

Relative Humidity (%)	Malformations on 13th Day (%)
50	15.2
70	6.1

ATMOSPHERIC PRESSURE

Low Pressure

Growth. The relative growth of the embryo, as measured by total heat production, during exposure to reduced atmospheric pressure, is shown in Fig. 60.

(See *Romanoff, 1941; 1967,* Chapter VIII, on the relationship between the growth of the embryo and the metabolic rate.)

The reduced barometric pressure slows the metabolic activity and retards the growth of the embryo. Below 507 mm Hg, the growth is more retarded and is often interrupted by the death of the embryo.

During 36 hours' exposure to low atmospheric pressure, there is

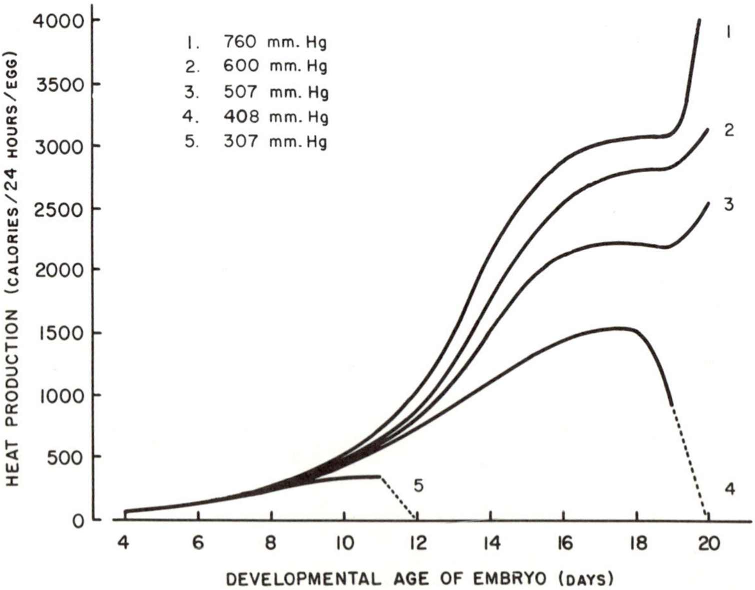

Fig. 60. Effect of reduced atmospheric pressure on the growth and development of the embryo, measured by heat production. (With some modifications after Romijn, 1967.)

an excessive daily loss of water in the egg (*Fraps, 1945*) in comparison with controls (*Romanoff, 1930a*), as is shown below.

Pressure (mm Hg)	Daily Loss of Water in the Egg (grams)
760	0.38
22	0.71
15	2.63
12	7.57

Mortality. With continuous exposure at a pressure of 307 mm Hg, the embryos become extremely susceptible and mortality occurs at all developmental stages (*Romijn, 1967*). In experimental observations (*Tengerdy, Fitch, and Moreng, 1969*) on the 16-day-old embryo, exposure to 460 mm Hg for different periods of time resulted in mortality as shown below.

Duration of Exposure (hours)	Incidence of Mortality (%)
24	40
48	60
96	100

In fertilized eggs exposed to a greatly reduced atmospheric pressure (*Fraps, 1945*) from the beginning of incubation, 12 hours daily for 3 days (for a total of 36 hours), there is a pronounced effect on the incidence of embryonic mortality.

Atmospheric Pressure (mm Hg)	Incidence of Mortality (%)
760	8
22	10
15	20
12	55

Malformations. Low atmospheric pressure (down to 280 mm Hg) causes various abnormalities in 3-day-old embryos (*Becher, 1939*), such as retardation in development and injuries to the central nervous system, as well as to the heart and circulatory system.

High Pressure

Growth. Observations show that under high pressure (about 2 atm) there is an acceleration of growth (*Cunningham, 1936; 1938*).

Age of Embryo (days)		Wet Weight of Embryo (grams)	
		Experimental	Control
9		2.07	1.46
9	Dry weight:	(1.16)	(0.71)
11		6.07	3.69

It also has been noted (*Cunningham, 1936*) that in the experimental eggs there is somewhat less water loss—about 0.23 gram daily, as compared to 0.34 gram in controls.

Mortality. It was observed (*Flemister and Cunningham, 1940*) that increased atmospheric pressure produces a high rate of embryonic mortality.

Twenty-four-hour exposures at the beginning of incubation to pressures up to 6 atm (*Nelsen, 1955*) immediately produce the following rates of embryonic mortality.

Pressure (atm)		Embryonic Mortality (%)
4		0
5		8
6a		35
6b	(72 hours after)	57

Malformations. The effect of various increases in atmospheric pressure during the second day of incubation (*Nelsen, 1955*) is shown in the table below.

Atmospheric Pressure (atm)	Incidence of Malformations (%)
3	0.0
4	16.8
5	91.7
6	100.0

It has also been seen (*Flemister and Cunningham, 1940*) that at a pressure over 1 atm, poorly developed extraembryonic vessels are coupled with a low number of red cells and a low hemoglobin content.

The chief deterrent effects of high pressure (i.e., 4 to 5 atm of added air) are due to the oxygen density (*Nelsen, 1955*).

V

Ionizing Radiation

Emitted rays have hidden bend:
To stimulate, then rearrange . . .
To break completely nature's trend
And leave behind a marked change.

Ionizing radiation can exist anywhere in the natural background, from the sources in the ground below (X-rays and gamma rays) and the sky above (galactic cosmic rays), and from naturally occurring and artifically made isotopes.

Since the early days of the discovery of ionizing radiation, it has been realized that irradiation is capable of interrupting embryonic development and causing malformations and death. Also, there is no known treatment capable of reducing the teratogenic effect of ionizing irradiation once irradiation has taken place.

X-IRRADIATION

It is common knowledge that irradiation with X-rays may severely damage the developing avian embryo and result in gross malformations of various organs.

Growth

Early Stages of Development. The effect of exposure of fresh fertile eggs to various doses of X-rays on the size of the 24-hour blastoderm is graphed in Fig. 61. Moderate doses of X-rays stimulate early embryonic development, a dose of about 250 R units being most effective, with an increase in the potential difference between the blastoderm and albumen (see *insert* of Fig. 61).

Large doses of X-rays inhibit embryonic development and diminish the potential difference. The delay in the time of hatching of eggs subjected to moderate doses of X-rays (up to 500 R) is about 12 hours.

Mitotic Activity. It has been observed (*Essenberg and Zikmund, 1938*) that X-rays in any dosage affect the mitotic activity (see under Malformations of Gonads). Recovery is rapid with dosages of 80 R or less, slow with dosages above 80 R, and rare with a dosage of 600 R.

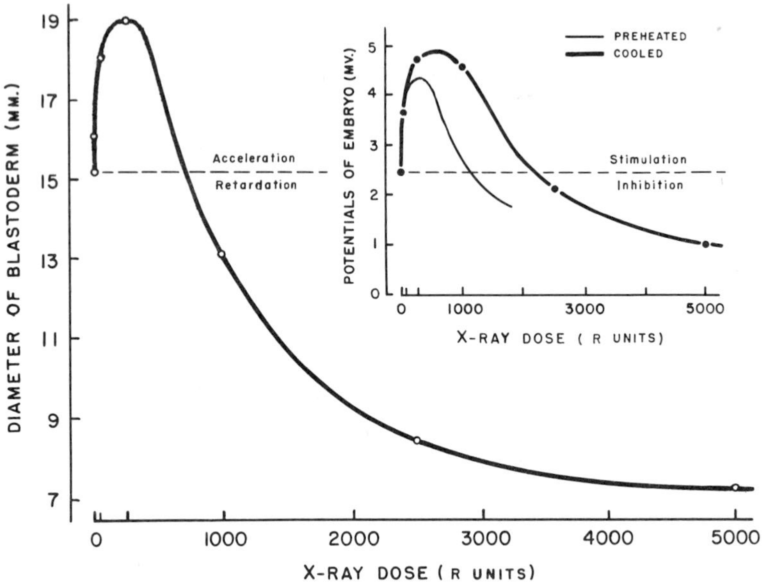

Fig. 61. Changes in diameter of the 24-hour blastoderm of the chicken embryo as a function of X-ray dosage. (After Bless and Romanoff, 1943.) *Insert:* effects of various doses of X-rays on the bioelectric potential of the 24-hour chicken embryo. (After Romanoff and Bless, 1942. See also Romanoff and Cottrell, 1939, and Romanoff, 1944.) Preheating eggs prior to irradiation shortens and decreases the extent of stimulation. (See Boland, 1954, on delay of the effect of X-rays by cooling eggs prior to radiation.)

Recently (*Diwan and Batra, 1969*), studies were made of the indices of the individual mitotic phases in erythroblasts from embryos treated with 250 R of X-rays at the definitive primitive streak age (about 18 hours), and examined at the age of 4 days. A rise in indices of all phases was found, as is shown in the table below.

	Mean Mitotic Index	
Mitotic Phase	X-Rayed	Control
Prophase	2.5	2.1
Metaphase	1.9	1.3
Ana-telophase	3.3	1.1
Mean	2.6	1.5

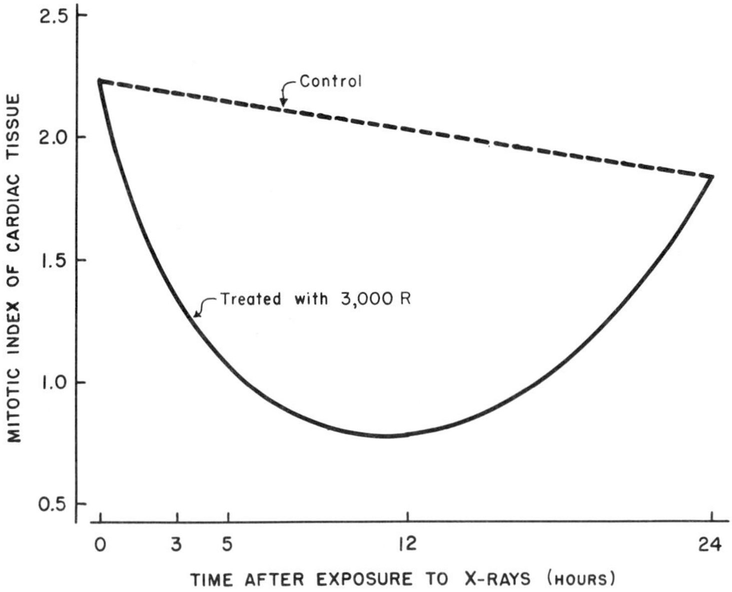

Fig. 62. Effect of local irradiation of the heart of the 3-day-old embryo on the mitotic index of the cardiac tissue. (Redrawn after Le Douarin, 1963.)

However, some developmental abnormalities were observed (see under Malformations).

Local radiation *in situ* of the 3-day-old embryo's heart with 3000 R units of X-rays causes only temporary decrease in the mitotic index of the cardiac muscle (*Le Douarin, 1963*). About 12 hours after radiation, the index begins to rise, and by the twenty-fourth hour it reaches the level found in unexposed embryos of a comparable age (Fig. 62).

Mortality

Since early observations (*Gilman and Baetjer, 1904*), it has been known that X-rays are toxic to the chicken embryo. At that time, it was observed that the effect of X-rays on embryonic mortality varies directly with the dosage and inversely with the age.

Direct and Indirect Effects on Mortality. It has been shown (*Schneller, 1951*) that X-rays act directly, destroying the cells and the embryo itself. The indirect effects of X-irradiation are mediated through the vascular

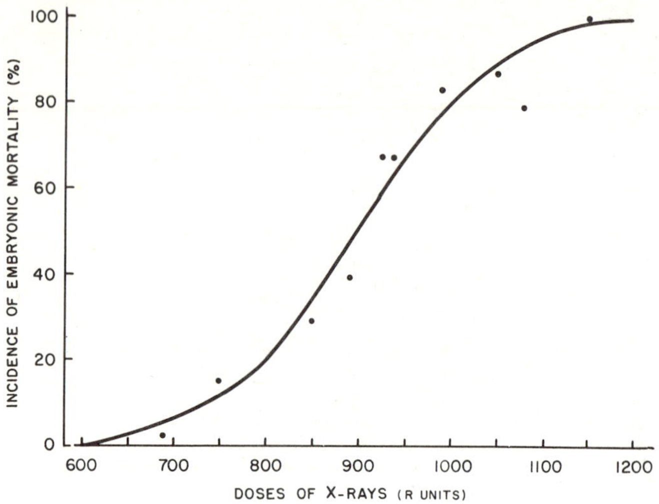

Fig. 63. Effect of various doses of X-rays on the incidence of mortality of 4-day embryos examined 6 hours after irradiation. (With some modification after the data of Boland, 1954.)

system of the yolk sac or of the extraembryonic membrane, the allantois (see The Effect of Shielding of the embryo, p. 117).

The direct effects, depending on the dose, produce early death of the embryo, within 24 hours after irradiation (*Karnofsky, Patterson, and Ridgway, 1950*). The indirect effects of X-rays cause a delayed mortality.

Dose–Mortality Relationship. The effect of various doses of X-rays on embryonic mortality has been frequently observed (*Essenberg, 1935; Krayevoy, 1938; Romanoff and Bless, 1942; Bless and Romanoff, 1943; Boland, 1954; Boland and Tranter, 1957; Oftedal, 1956; Le Douarin, 1963; 1968; Stearner and Christian, 1968*). A general curve derived from typical observations is shown in Fig. 63. The curve of the mortality rate plotted against the dose is S-shaped. At about 50 per cent mortality, the curve is almost linear, and it is possible to determine the lethal dose 50 per cent (LD_{50}) more accurately.

Age–Dose–Mortality Relationship. The age of the embryo has a definite influence on the incidence of mortality after exposure to X-irradiation. It has been demonstrated (*Le Douarin, 1968*) that with age the embryo becomes more resistant. When groups of 3-day-old embryos are treated with gradually decreasing doses of X-rays, the time for approaching the LD_{50} becomes progressively longer, as is shown in the table below.

Dose of X-rays (R units)	Time Required for LD_{50} (days)
1550	1
1400	2
1200	3
1050	4
1000	5
900	6

These data, graphically presented in detail in Fig. 64, also indicate that the curves of mortality become less widely separated with development, indicating a decreasing effect of X-rays as embryonic development progresses (see Fig. 65, on the acute lethal response of the embryo with age).

Causes of Acute Death. The cause of acute death is not grossly evident. However, it becomes obvious when lethal conditions are observed in

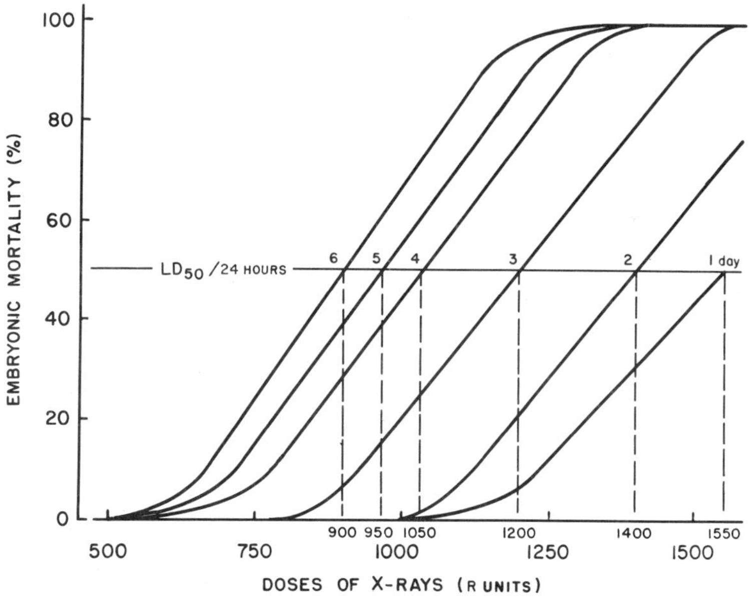

Fig. 64. Cumulative curves of early embryonic mortality showing the effect of various doses of X-rays, during the following 1, 2, 3, 4, 5, and 6 days of development. (After Le Douarin, 1963, 1968; see also Tyler and Stearner, 1960; and Fig. 65.)

dead embryos irradiated at various stages of development (*Goff, 1959*). These conditions are as follows:

Age at Irradiation (days)	Symptoms at Death
2–3	General cell death
4–5	Hemorrhage of brain
6	Malformed eyes and legs
7–16	No specific symptoms

Microscopically, defects are visible in brain at 4 to 5 days, in mandible at 5 days, in eyes and legs at 6 days, and in down feathers at 12 days.

Measurement of the LD_{50} at various stages of development has been used to great advantage in evaluating the relative susceptibility of the embryo to X-rays (*Essenberg, 1935; Wilson, 1935; Lea, 1941; Karnofsky, Patterson, and Ridgway, 1950; Schneller, 1951; Boland, 1954*).

Boland (*1954*) made special tests for the reliability of LD_{50} determinations. These include the effect of embryo age, dose rate, and quality of radiation.

The table below shows that there is a small but consistent increase in radiosensitivity from the third to the fifth day of incubation.

	Radiosensitivity (LD_{50})	
Age of Embryo (days)	6 Hours After (R)	10 Days After (R)
3	946	799
4	887	729
5	872	648

The results at 6 hours and 10 days after irradiation confirm each other.

Increasing the target-object distance causes only small alterations in LD_{50}, as is shown below.

Dose Rate (R/min)	Radiosensitivity (LD_{50}) (R)
151	834
66	847
36	907

There is a small but consistent decrease in effectiveness when the dose rate is lowered from 151 to 36 R/min. However, the effect of variations in intensity of radiation has been studied by Lea (*1941*) in detail; the lethal dose was found to be fairly constant at high intensities and unstable (with an increase) at a very low intensity. (See Fig. 81, under Relative Biological Effectiveness.)

The difference in the effect of hard and soft X-ray radiation is insignificant, as the data below indicate.

	Radiosensitivity (LD_{50})	
Dose Rate (R/min)	6 Hours After (R)	10 Days After (R)
250	891	724
400	854	738

The results at 6 hours and at 10 days after irradiation do not show any real difference, although one might be detectable by use of a very large number of eggs. (See under Relative Biological Effectiveness, p. 143.)

Changes in Radiosensitivity with Age. The most detailed observations of the LD_{50} under well-controlled conditions, at nearly all ages of the developing embryo, have been carried out by Goff (*1959*). He was able to establish a continuous curve for the acute lethal response to X-rays (Fig. 65). From this curve it is apparent that the most sensitive period of the embryo occurs on the eighth to ninth day. The next most sensitive period appears on the sixteenth day with, possibly, a less conspicuous one on the fourth day. (See also *insert* of Fig. 67, by *Reyss-Brion, 1956.*)

The periods of sensitivity to X-rays at the ninth and sixteenth days are normally not critical during incubation, but they correspond to periods of usual suppression in the growth of the embryo (*Romanoff, 1929c; Romanoff, 1967, p. 266,* see under growth cycles).

Mechanism of Death. It was observed (*Boland, 1954*), that embryo mortality tends to be concentrated at certain fixed time intervals following X-irradiation. The mortality centered at four characteristic peaks, associated with certain dose levels.

1. Doses in the range from 22,000 to 35,000 R produce death in about *1 hour* after the beginning of irradiation.

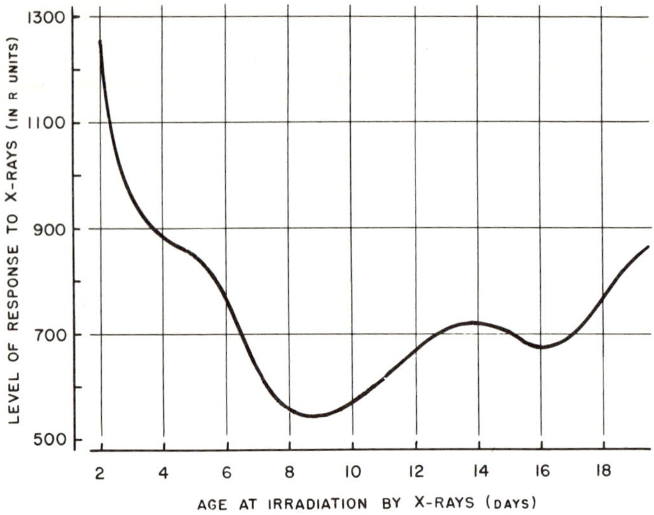

Fig. 65. The acute lethal response of the embryo to different doses of X-rays at dose level necessary to produce 50 per cent embryonic mortality after 20 hours of the postirrradiation period. (After Goff, 1959, with the supplementary data of Clapp, 1964; see also Karnofsky, Patterson, and Ridgway, 1950, and Muller and Moreng, 1966.)

2. Doses in the range from 1200 to 20,000 R produce morbidity in about 3½ hours after the beginning of irradiation.

3. Doses in the range from 950 to 1150 R still leave a few morbid embryos alive in the sixth hour after irradiation.

4. Doses in the range from 800 to 950 R produce maximum mortality on the sixth day; survivors show remarkable recovery from the after-effects of irradiation, although none of them hatch, and all die on the twenty-third day.

The Effect of the Threshold Dose 950 R. The cumulative curve of embryonic mortality after exposure of the 4-day-old embryo to 950 R of X-rays has been traced at frequent intervals, and is shown in the composite Fig. 66*A* and *B*.

The first acute effect of X-rays on embryonic mortality is shown between 2 and 4 hours after irradiation (Fig. 66*A*). The next rise in the incidence of mortality occurs between 8 and 12 days, and the last, delayed lethal effect brings death to the remaining embryos at the time of hatching (Fig. 66*B*).

All acute and delayed lethal effects of X-rays are preceded by various forms of maldevelopment of the embryo and of some of its specific organs.

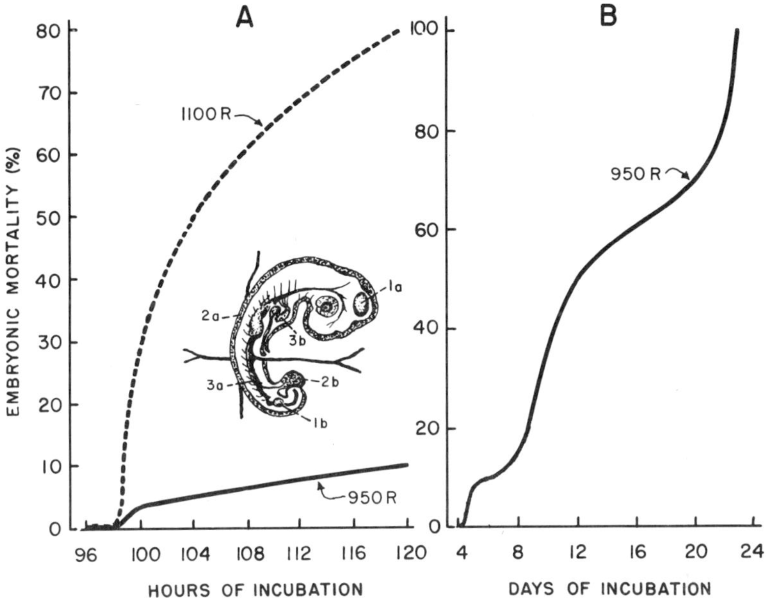

Fig. 66. Effect of a single dose of X-ray irradiation of 4-day incubated eggs on the subsequent cumulative embryonic mortality. (With modifications after Boland, 1954.) *A*, at a threshold dose near 950 R—with *insert*, a diagram of the 5-day-old embryo showing gross lesions appearing during the first 24 hours after irradiation (redrawn from Johansen, 1961); *B*, course of embryonic mortality during the remainder of the postirradiation period of incubation. *1*, hemorrhages in the mesencephalon (*1a*) and in the leg (*1b*); *2*, extravasation of blood into the allantois (*2a*) and into the amnion (*2b*); *3*, accumulation of blood in the aorta dorsalis (*3a*), in the heart, (*3b*), and in the intermyotomic vessels.

The Effect of Shielding. It has been shown (*Schneller, 1951*) that shielding the embryo with lead and exposing the extraembryonic membranes and the yolk to X-rays produces a delayed lethal effect.

X-Ray Dose (R) at 60 Hours	Embryonic Mortality at 6 Days (%)	
	Not Shielded	Shielded
400	0.0	
500	40.0	
600	90.0	
700	100.0	
800		0.0
900		10.0
1000		50.0
1500		90.0

With embryos exposed at 33 hours, the survival value of the shielded embryos was very high. Under exposure to 5000 R, there was only 30 per cent mortality.

Malformations

The pattern of radiation damage by X-ray has been registered by the pathological picture in embryos exposed to different doses and at various developmental stages. Among the manifestations observed are numerous structural injuries and malformations of the body of the embryo and of some of its organs.

The results of the earliest investigations of the teratological effects of ionizing radiation on the chicken embryo are summarized in Table 6.

Recent studies show that the response of the embryo to X-rays varies with the stage of development and dose of irradiation. When doses applied at a given embryonic age are increased in the lethal range, the anomalies increase in frequency and severity.

General Effects. Among the general symptoms observed in the embryo exposed to X-rays are (a) morphological changes and (b) developmental anomalies.

The morphological changes (or diseased conditions) induced by X-rays have been described generally (*Karnofsky, Patterson, and Ridgway, 1950*) as follows: (1) vascular injury and hemorrhage in the allantoic and vitelline vessels, (2) generalized edema, and (3) necrosis of the liver.

The developmental abnormalities which have been noted in irradiated embryos are as follows: (1) beak deformities (underdeveloped or crossed beak, (2) limb abnormalities (thin legs and deformed or absent toes), and (3) abnormal feather patterns. Disproportionately small heads and stunting in embryos which cannot survive have also been observed.

Direct and Indirect Effects on Malformations. It is well recognized that X-irradiation causes damage of the embryo directly and indirectly. The *direct effects* are noticed in inducement of vesicles which do not move freely in the plasma, but migrate within the complicated endoplasmic channel system and within the perinuclear space (*Wendt, 1966*). The *indirect effects* are manifested by the occurrence of toxic humoral substances which are distributed in the body by the bloodstream. Consequently, the toxins can appear in fluids, vessels, or organs which are relatively far removed from the immediate vicinity of the particular cellular structures under consideration (*Tobin and O'Brien, 1952*). Dam-

Table 6 Early Observations of the Teratogenic Action of X-rays on the Development of the Chicken Embryo

Developmental Abnormalities	Gilman and Baetjer 1904	Tur (1904)	Bordier and Galimard (1905)	Ancel and Vintemberger (1925)	Essenberg (1932)	Danchako (1933)
Blastoderm						
Absent				×		
Grossly deformed		×				
Without embryo (anideans)		×		×		
Embryo						
Retarded development			×	×		×
Abnormal curvature of the body	×					
Absence of protovertebrae		×				
Grossly deformed		×				
Destroyed				×		
Head						
Abnormal					×	×
Lesions						
Of the occipital region	×					
Of the brain					×	
Of the eyes					×	
Eyes						
Atrophy	×					
Limbs						
Curved	×				×	
Malformed (ectromelia)					×	
Feathers						
Abnormal distribution	×					

age to the embryo can be mediated through the extraembryonic area and through the yolk sac (*Schneller, 1951*).

In comparison with direct effects, the indirect effects are more delayed and less severe. Depending on the dose, direct effects produce death of the embryo within 24 hours after irradiation. The indirect effects of X-rays produce a delayed mortality with manifestations of various defects and malformations of the embryo and some of its organs.

Dose and Age of the Embryo. The composite data on the effects of X-rays at dosages varying from 40 to 600 R, and at ages ranging from about 1 to 10 days (*Essenberg, 1935*), are shown below.

Little or no development	45%
Death in the shell at full development	24%
Survival up to the 15th day	12%
Paralyzed lower extremities	9%
Deformed lower extremities	6%
Hydrocephalic head deformities	2%
Miscellaneous deformities	2%

The radiosensitivity of the developing embryo to a radiation dose of 600 R at various stages (*Hadjinsky, 1962*) is well demonstrated in the accompanying table.

Age of Embryo (hours)	Type of Malformation
0	Malformations of the cranium, brain, and eyes
22	Micromelia, microphthalmia, and spina bifida
48, 72, 96	Malformations of the limbs (amelia, micromelia, and syndactyly)
120	Lesions of the lower beak (agnathia and micrognathia); malformations of the digits, feet, and plumage
168	Defects of the digits, feet, and plumage
240–320	No abnormalities

Irradiation at various periods of embryonic development results in different and fairly well-defined types of malformations. When irradiation takes place during the early stages of development, beginning with an unincubated blastoderm (*Reyss-Brion, 1956*), it induces malformations of the nervous system and of the sense organs (omphalocephaly), as well as cases of spina bifida and vesiculate and stunted somites (Fig. 67). At 12 and 18 hours of incubation, the anterior region is affected;

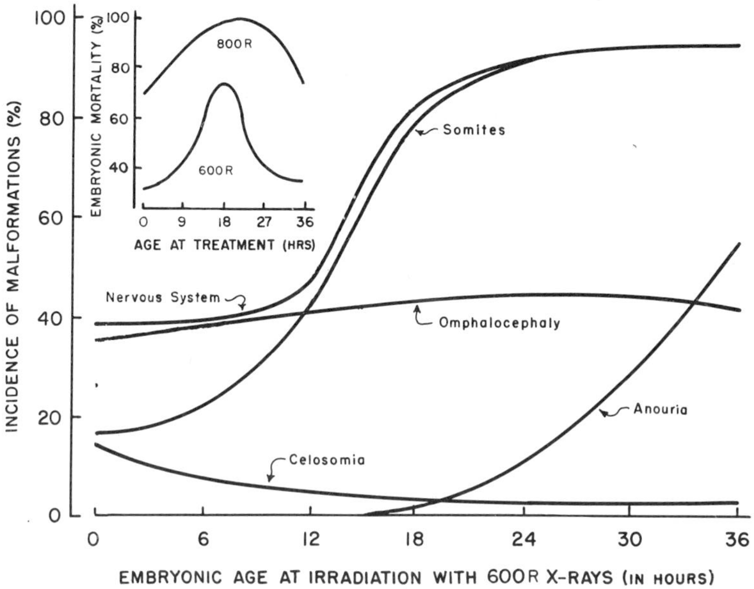

Fig. 67. Teratogenic effect on the embryo of exposure to 600 to 800 R of X-rays at the early stages of development. (Plotted from the data of Reyss-Brion, 1956.) *Insert:* incidence of embryonic mortality with different doses of X-rays.

at 25 hours, the medial region and the trunk; and at 36 hours, the posterior region, resulting in absence of the tail (anouria).

Critical Periods in Development of Leg Bones. Attention has been called (*Puchkov, 1967a*) to the fact that there are critical developmental periods of embryogenesis which correspond to the periods of phylogenetic transformation. Such periods are conspicuous in the development of leg bones under the influence of X-rays (Fig. 68).

At 2 days, irradiation affects the rudiments of the legs as a whole; at 4 days, chiefly the tibia; at 5 days, the tarsometatarsus; and at 6 to 7 days, the development of digits and claws.

The incidence of leg bone anomalies (*Puchkov, 1967a*) at different developmental stages under the influence of X-rays was as follows:

Developmental Age (days)	Incidence of Leg Anomalies (%)
4	81.5
5	86.5
6	91.0
7	67.5
8	2.5

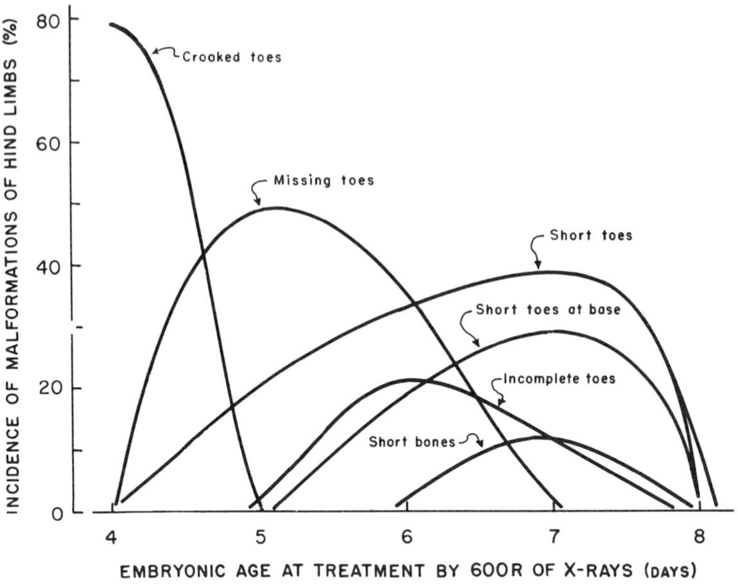

Fig. 68. Incidence of malformations of hind limbs in embryo exposed to 600 R of X-rays at various stages of development; examined on the 16th day of incubation. (Plotted from the data of Puchkov, 1967a.)

Pattern of Development of Limbs. A special study (*Goff, 1962*) based on microscopic observations was made on the effect of X-radiation on the changing developmental pattern of the limbs—legs and wings.

For the *legs,* two different periods of development were delineated by the response pattern (Fig. 69), with the dividing point at 2.5 days. In the early period, the characteristic response was reduction in size or elimination of parts of the appendage. The late period reveals the proximo-distal progression of sensitivity to X-rays with increasing age. There are two types of skeletal abnormalities: (1) reduction in size, which results in the limbs being thin, distorted, short, or completely eliminated; and (2) union of parts at joints or laterally.

In the *wings,* similarly, two periods of development are delineated by the response pattern to X-rays (Fig. 70). In terms of the total number of anomalies, the wing is not as sensitive as the leg at any stage of development. Furthermore, there is a tendency for the response in the wings to be more asymmetrical than in the legs.

Localized Radiation. The technique of localized irradiation was perfected by Ancel and Wolff (*1934*). The fundamental principal of this method consists of irradiating certain embryonic buds selectively, using

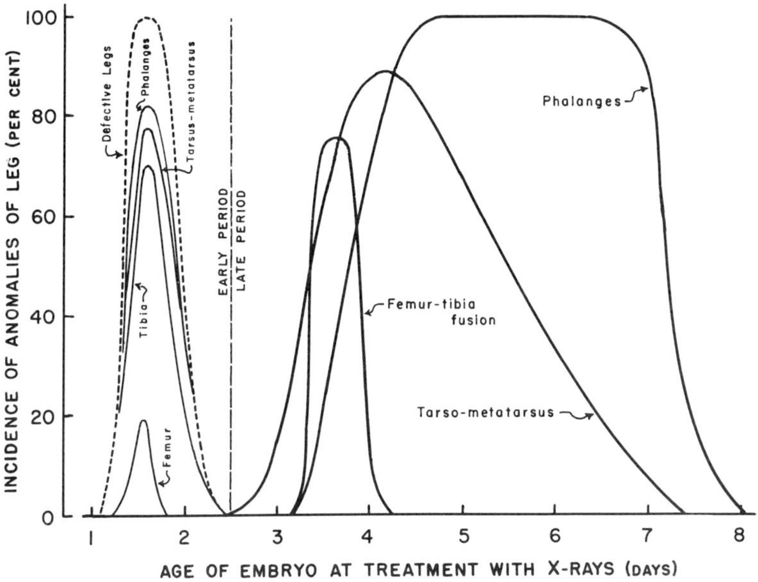

Fig. 69. Effects of X-irradiation (800 R units applied at various stages of development) on the incidence of malformations of the leg in the early and late periods. (Plotted with some modifications from Goff, 1962.)

For the early period, the percentage of defective legs was evaluated at 4 to 12 days of development.

lead screens between the X-ray tube and the target to protect other areas.

Remarkable results were obtained by Wolff (*1936a; 1936b; 1937*) in this type of investigation. The summarized results of the localized radiation are presented in Fig. 71.

For the production of various malformations the selected sections of the blastoderm have to be irradiated at an appropriate stage of development (*Kirrmann and Wolff, 1964*), as follows:

Age (hours)	Selected Site	Malformation
12–15	Primitive streak	Omphalocephaly
20 (in somites)	Cephalic fold	Diplocardia
8–15	Prosencephalon	Cyclocephaly, anophthalmic cyclocephaly
8–15	Mesencephalon	Otocephaly
12–40	Optic vesicle	Anophthalmia
15–23	Lateral part of embryo	Celosomia

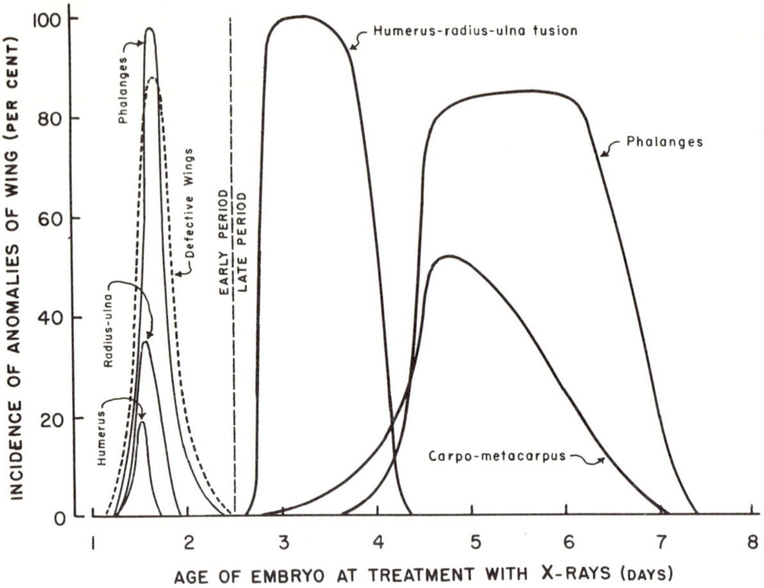

Fig. 70. Effects of X-irradiation (800 R units applied in a single dose at various stages of development) on the incidence of malformations of the wing in the early and the late periods. (Plotted with some modifications from Goff, 1962.) For the early period, the percentage of defective wings was evaluated at 4 to 12 days of development.

Under accurately controlled local irradiation, the monstrosities appear in almost 100 per cent of cases. Examples are shown in Fig. 72 and Fig. 73.

Response of Cells in Vitro. Exposure of choroid and sclerotic tissue from the eyes of 10-day-old chicken embryos to various doses of X-ray *in vitro* (*Lasnitzki, 1943*) shows that there is a breakdown of cells soon after irradiation. The number of degenerate and abnormally dividing cells varies according to the dose employed (Fig. 74). The greatest number of degenerate cells is observed at 3 hours after exposure.

Malformations of Organs

Among the embryonic organs chiefly affected by X-rays are the eyes and the heart. With increased doses during the early stages of development, damage becomes obvious in many other organ systems, as detected both *in vivo* and *in vitro*.

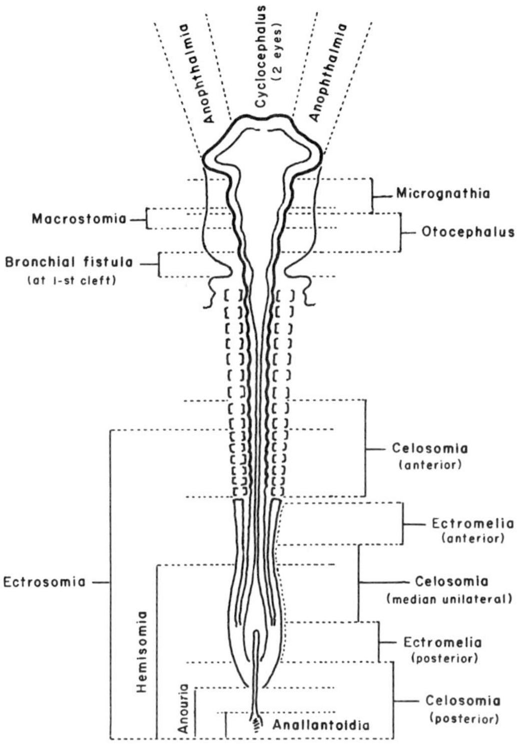

Fig. 71. Effects of localized irradiation with X-rays on development of specific malformations. (With some modifications after Wolff, 1936a, 1936b; Kirrmann and Wolff, 1964.)

Eyes. It has been demonstrated that exposure of the early embryo to X-rays at certain stages interferes with the development of the scleral papillae and ossicles (*Puchkov, 1961; 1965*). On the sixth day of incubation, a dose of 400 R inhibits the development of scleral ossicles about 50 per cent, and a dose of 600 R completely suppresses it. The progressive reduction of the ossicles (from the usual number of 14) at 600 R is shown in the table below.

Age at Exposure to 600 R (days)	Scleral Ossicles (number)
4	12
4½	11
5	8
5½	3
6	0

By local radiation with X-rays (*Wolff, 1933*) 69 cyclops were experimentally produced which had the following distribution of types:

Types of Cyclopia	Distribution (%)
Anophthalmic or microphthalmic cyclocephaly	46
Cyclocephaly with two developed eyes	21
Cyclocephaly with micrognathia and arhinencephalus	16
Asymmetric monophthalmia and hemicyclopia	11
Cyclocephaly with a median eye (2 with triocephaly or opocephaly)	6

The incidence of malformations of the eye induced by local radiation at various doses of X-rays (*Kirrmann, 1962*) is shown below.

	Incidence of Malformations (%)	
Dose of X-rays (R)	Anophthalmic Cyclocephaly	Anophthalmy
2000	0	0
3000	61	71
4000	81	85
5000	95	97
6000	100	100

With the increase in dose of X-rays, the proportional occurrence of various types of eye malformation remains almost the same.

Heart. Irradiation by X-rays at the third day of incubation produced various malformations and anomalies in the hearts of surviving embryos on the thirteenth day (*Le Douarin, 1960*). These morphological anomalies are as follows: (1) Hypertrophy of the heart (left side), and distension of the left atrium; (2) malformations within the heart (77

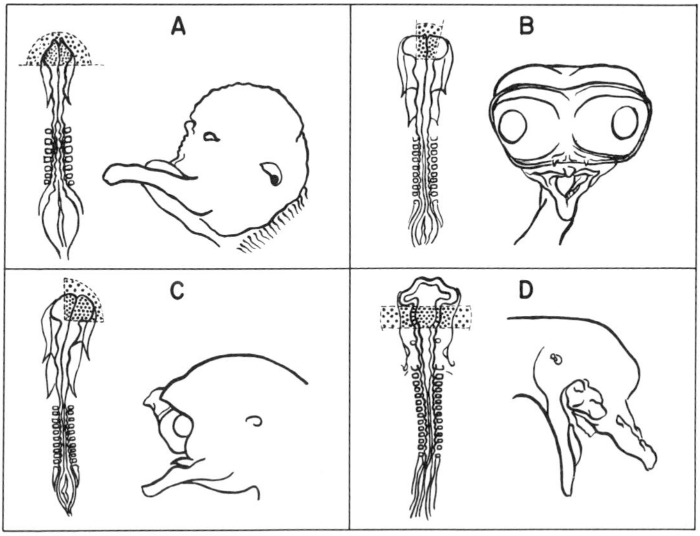

Fig. 72. Effect of controlled localized irradiation with X-rays on the appearance of predetermined monstrosities. (Redrawn with some modification from Wolff, 1936b.) *A*, a scheme of production of *anophthalmic cyclocephaly* (irradiated at the age of 7 somites and examined on the tenth day of incubation); *B*, a scheme of production of *cyclopia* (irradiated at the age of 9 to 10 somites and examined on the tenth day); *C*, a scheme of production of monophthalmic asymmetric cyclopia (irradiated at the age of 1 somite and examined on the fourteenth day); *D*, a scheme of production of atrophy of the upper maxillary buds (irradiated at the age of 13 somites, and examined on the eleventh day of incubation).

per cent)—incomplete partitioning of the bulbus and arterial truncus, persistence of the intraventricular foramen, anomalous rotation and twisting of the arterial bulbus, and incomplete formation of valves; (3) malformations of the aortic arches; and (4) blending of bloodstreams.

It has been observed that such gross malformations are compatible with prolonged survival of the embryo, which can, occasionally, hatch.

The rate of heartbeat may increase markedly after irradiation (*Stearner and Christian, 1968*); rates 300/min and greater were observed shortly after exposure to 1000 R.

Blood Vessels. There is enough experimental evidence to show that total X-irradiation profoundly affects blood vessels, especially those of the external circulation, such as the vitelline (area vasculosa) in the early stages and the allantoic from the time of their appearance.

Four-day embryos, exposed to a dose of 500 R of X-rays (*Boland, 1954*), soon show extraembryonic and then intraembryonic (intermyotomic) hemorrhages. Similar collapse, due to hemorrhages in the

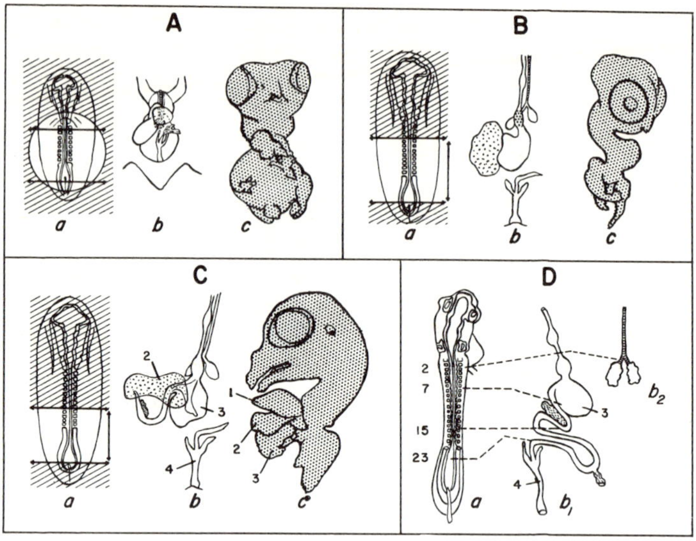

Fig. 73. Effect of localized X-irradiation of the early embryo on the subsequent display of external and internal morphological teratology. (Redrawn with some modifications from Le Douarin, 1964a.) *A,* irradiation of 9-somite (36-hour) embryo: (*a*) scheme of a localized irradiation, (*b*) damage to internal organs, and (*c*) maldevelopment of external parts of the embryo; *B,* irradiation of 13-somite (48-hour) embryo: (*a*) scheme of a localized irradiation, (*b*) damage to internal organs, and (*c*) maldevelopment of external parts of the embryo; *C,* irradiation of 15-somite (about 60-hour) embryo: (*a*), (*b*), and (*c*) as before; *D,* localization of presumptive areas of various organs of gastrointestinal and respiratory systems with localized irradiation of 2-, 7-, 15-, and 23-somite embryos. *1,* heart; *2,* liver; 3, gizzard; *4,* rectum.

extraembryonic vasculature, was observed by Clapp (*1964*), and Stearner and Christian (*1968*).

It has been observed (*Puchkov, 1959*) that the greatest damage to the allantoic circulation is done by a 600-R dose on the sixth day of development—the beginning of its function as a respiratory organ. By this exposure, the vascular area of the allantois is reduced by nearly 50 per cent, with a proportional increase in embryonic mortality.

Nervous System. Observations on the growth and development in tissue culture of spinal ganglia, irradiated with a series of doses ranging from 500 R upward to 128,000 R both *in ovo* and *in vitro,* reveal that sensitivity of the ganglia to X-radiation is greater *in ovo* than *in vitro* (*Goldring, 1956*). This is probably due to an indirect effect through involvement of the circulatory system.

Younger ganglia, from 6-day embryos, become more readily dam-

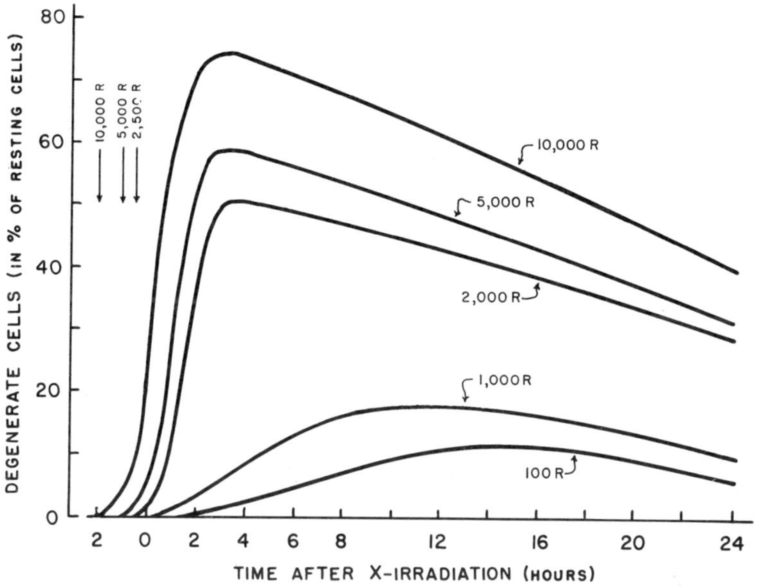

Fig. 74. Effect of various doses of X-rays on the incidence of degenerate cells in tissue from the 10-day-old embryo, cultivated *in vitro* for 24 hours after irradiation. (Based on the data of Lasnitzki, 1943.)

aged when irradiated *in ovo* than older ganglia, from 12-day embryos; 12-day ganglia can also withstand higher doses of X-ray than 6-day ganglia when irradiated *in vitro.*

Gonads. Studies of the effects of irradiation on the gonads of the developing embryo (*Essenberg and Zikmund, 1938*) show that the ovaries have a greater sensitivity to X-rays than the testes.

In eggs irradiated prior to incubation—except in those treated at a very low dosage (80 R or less)—varying degrees of injury could be observed in all ovaries. In the more severely affected ovaries, receiving dosages in the neighborhood of 100 R, there is found a marked reduction of primary follicles. The development of the testes proved to be only slightly affected by X-rays at this stage of development.

In eggs irradiated during the incubation period, between 5 and 12 days, the X-ray injuries of the ovaries vary directly with the age of the embryo and the dosage employed. In the ovaries incurring maximum injury, induced by dosages of 300 to 400 R at the age of 5 to 7 days, the oocytes, which in normal ovaries of comparable age fill the entire cortex, are found to be very scarce. Ovaries of embryos irradiated with dosages less than 200 R are not distinguishable from normal ovaries. The testes, under similar exposures, do not show any effects from the X-rays.

Spermatozoa. Cock spermatozoa are found to be extremely sensitive to X-rays (*Kosin, 1944*). Lethality to the embryo fathered by irradiated spermatozoa is as shown below.

Irradiation of Spermatozoa (X-rays, R)	Incidence of Embryonic Mortality (% of fertile eggs)
0	15
900	45
1800	60
2800	73
3700	82
4600	90
5500	100

Among embryos from irradiated eggs there is an increased frequency of abnormal development.

Extraembryonic Membranes. It has been shown (*Puchkov, 1959*) that irradiation with 600 R of X-rays causes reduction in the area of the allantoic membrane with consequent increase of embryonic mortality, as is shown in the table below.

Time of Irradiation	Area of Allantois (mm^3)	Mortality at 7 Days (%)
0 hours	1110	8
5 hours	930	8
1 day	590	55
2 days	625	26
3 days	800	23
4 days	880	8

The increased embryonic mortality from irradiation at the age of 1 day is caused primarily by reduction of the respiratory function of the allantois.

RADIOACTIVE ELEMENTS

The radioactive elements (e.g., radium) and radioisotopes (e.g., cobalt-60) usually produce gamma rays, which have a shorter wavelength and are somewhat more penetrating than X-rays.

A few observations have been made specifically on the effect of gamma rays on the development of the avian embryo. However, all radioactive elements contain both X-rays and gamma rays in various proportions, differing only in their wavelengths.

For the specific difference in the wavelengths of X-rays and gamma rays, the reader is referred to the section on Relative Biological Effectiveness, on p. 142.

Due to the somewhat higher biological effectiveness of gamma rays, observations of the effects of radioactive substances, especially on embryonic mortality and malformations, deserve consideration.

Radioactive Calcium (^{45}Ca)

Mortality. Embryonic response to 400 μCi of ^{45}Ca is very high. In eggs injected on different days of incubation (*Mraz, Wright, and Ferguson, 1964*), the heaviest mortality occurs at the early stages, nearly 100 per cent from 0 to 6 days of development.

Malformations. A characteristic leg abnormality is observed with doses as low as 100 μCi of ^{45}Ca administered at either 4 or 8 days of incubation (Fig. 75). The abnormal legs and wings are found to be the result of broken or fractured shafts of the long bones. The leg bones (*insert* of Fig. 75) are usually thin in the center of the shafts.

With 400 μCi of ^{45}Ca, severe leg abnormalities reach their peak with injection at 9 or 10 days of incubation. Among surviving embryos (*Mraz, Wright, and Ferguson, 1964*), 74 per cent were found to have abnormal legs and 24 per cent abnormal wings at the time of hatching, while no abnormalities were observed in the controls.

Radioactive Cesium (^{137}Cs)

Mortality. Observations were made (*Hyre, Patrick, and Nestor, 1964*) to determine the effect of irradiating semen with various doses of ^{137}Cs on embryonic mortality of fertilized eggs. Composite values of these results are shown below.

Doses of ^{137}Cs (R units)	Mortality up to Hatching (%)
0	16
2000	29
4000	33
6000	68

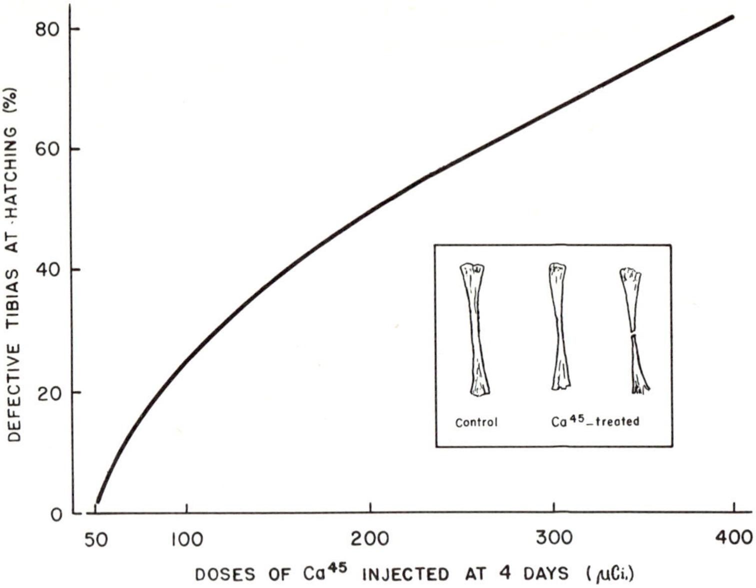

Fig. 75. Effect of various doses of radioactive calcium (^{45}Ca) on the incidence of deformed tibias. (Plotted from the data of Mraz, Wright, and Ferguson, 1964.) *Insert:* appearance of the affected tibias of a hatched chick, injected with 200 μCi of ^{45}Ca on the fourth day of incubation.

The fertilizing capacity of the sperm is lost sooner than the viability of the embryo from irradiated fertile eggs, especially at doses above 800 to 900 R.

Radioactive Cobalt (^{60}Co)—Gamma Rays

In studies of the effects of gamma irradiation by ^{60}Co, applied in continuous flow (*Doina, Onea, and Marcu, 1969*), some cytological modifications were observed in the heart, liver, primitive kidney (mesonephros), and integument of the chicken embryo.

With doses from 12 to 28 R, at a flow of 0.4 R/hour, there is a stimulating effect on 30-hour embryos.

With doses from 50 to 86 R, at a flow of 0.7 R/hour, there is an inhibitory effect on the development of 72- and 124-hour embryos.

With doses from 970 to 1730 R, at a flow of 10 to 18 R/hour, there is a profound cellular and chromosomal alteration in the organs of 97-hour embryos.

This indicates that the various doses of gamma rays have stimulating, inhibitory, and lethal effects (see *Romanoff and Bless, 1942,* on stimulating effects of X-rays).

Growth. It has been shown that irradiation with ^{60}Co prior to incubation (*Knezevic and Karanovic, 1966*) induces a certain depression of the proliferative activity of the primitive hematopoietic cells. This is evident from the table below.

Time after Irradiation with 600-R ^{60}Co (days)	Mitotic Activity of Blood Cells (in % of Normal at Starting)	
	Irradiated	Control
2	23	46
3	22	34
4	21	29
5	20	22
6	7	4
7	1	1

After the fourth day, the values are close to normal, which indicates that a normal level of proliferative activity of blood cells can be reestablished.

Observations indicate (*Stearner, Sanderson, Christian, and Tyler, 1960*) a retardation of development within 3 to 4 days after irradiation of the embryo at the age of 6 days.

Mortality. Extensive tests have been performed (*Sandvik, 1958*) on the effects of various doses of ^{60}Co on embryonic survival (or mortality) and hatchability of exposed eggs (Fig. 76). No embryos survived after irradiation with more than 1650 R or hatched after irradiation with more than 1080 R (see *insert* of Fig. 76). The irradiated eggs hatch more slowly than do the control eggs.

Hatching of surviving embryos is delayed as the ^{60}Co dose increases (Fig. 77). There is an obvious increase in dispersion of the hatch at a higher dose (see also Fig. 10, Chapter II).

Distribution of embryonic mortality during the postirradiation period of 6-day embryos treated with 600 R shows a rapid decline in the death rate with age (see *insert* of Fig. 78). The heaviest mortality occurs at the end of the first day after irradiation, followed by smaller peaks at the third and eighth days.

The relation between dose and exposure time in embryonic mortality has been divided experimentally into three periods (*Stearner, Sanderson,*

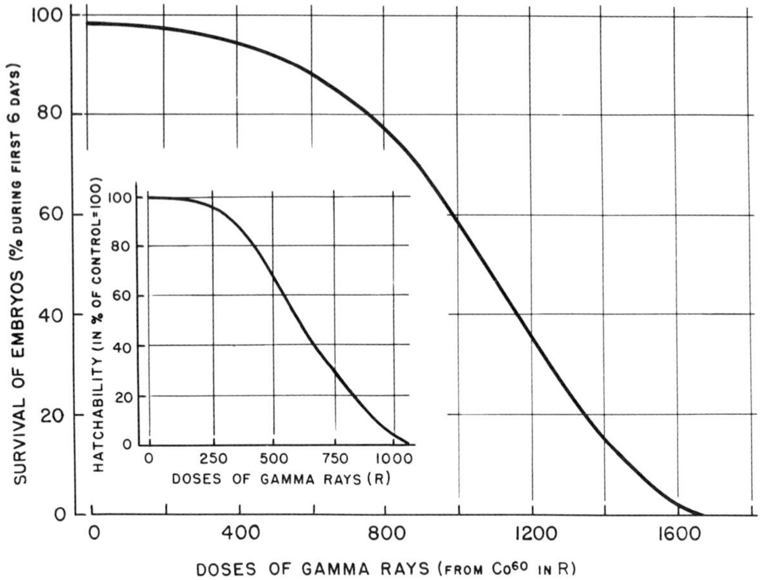

Fig. 76. **Effects of various doses of radioactive cobalt (^{60}Co), on survival and on hatchability (in insert) of chicken embryos.** (Rearranged after the data of Sandvik, 1958.)

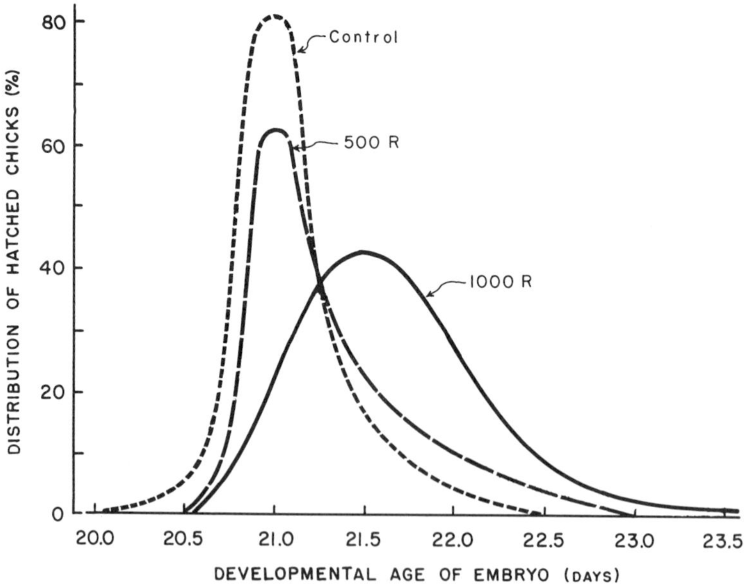

Fig. 77. **Effect of various doses of ^{60}Co on the distribution of hatched chicks, from eggs irradiated at 2 to 3 days of incubation.** (Plotted from the data of Morgan, 1966.)

Christian, and Tyler, 1960). In the first period (0 to 1 day after irradiation of the 6-day embryo), the dose required for an approximate 10 to 90 per cent mortality rate rises from about 1000 to 11,000 R as the exposure time increases up to 28 hours. During the second period (2 to 5 days) the dose required ranges up to 2000 R with a shorter time of exposure, about 8 hours, and does not increase for longer exposure time. During the third period (6 to 12 days) there is no apparent change in dose sensitivity (approximately 1250 R) throughout the entire range of exposure time.

The above observations are summarized in the tabulation below.

Exposure Time (hours)	Total Dose for Mortality up to 90% (R)		
	1st Day	2 to 5 Days	6 to 12 Days
1/4	990	990	—
1	1,020	1020	1250
2	1,280	1280	1300
4	1,800	1750	1320
8	3,900	2180	1280
16	6,900	2060	1250
24	10,800	2200	1260

The effect of the irradiation dose on the lethal response in each of these periods suggests that different injury mechanisms are involved (*Tyler and Stearner, 1960*).

Malformations. Observations during a 12-day postirradiation period of 6-day embryos treated with doses of ^{60}Co from 800 to 13,000 R (*Stearner, Sanderson, Christian, and Tyler, 1960*), disclosed that embryos which die by the eighteenth day of incubation show at least three characteristic syndromes (Fig. 78).

The first syndrome is venous pooling, prevailing during the first few days after irradiation. It is an accumulation of blood within the large veins. In the intermediate period there are localized accumulations of fluid. During the latter half of the observed period, some hepatic lesions are present. The maximum expression of each syndrome corresponds to one of three mortality peaks shown in the *insert* of Fig. 78.

Other indicators of ^{60}Co injury noted include hemorrhages, observed most frequently between 5 and 9 days; hematopoietic failure, seen during the later part of the observed period; and deformities of mandibles and feet, which are marked during the second week of exposure (*Stearner, Sanderson, Christian, and Tyler, 1960*).

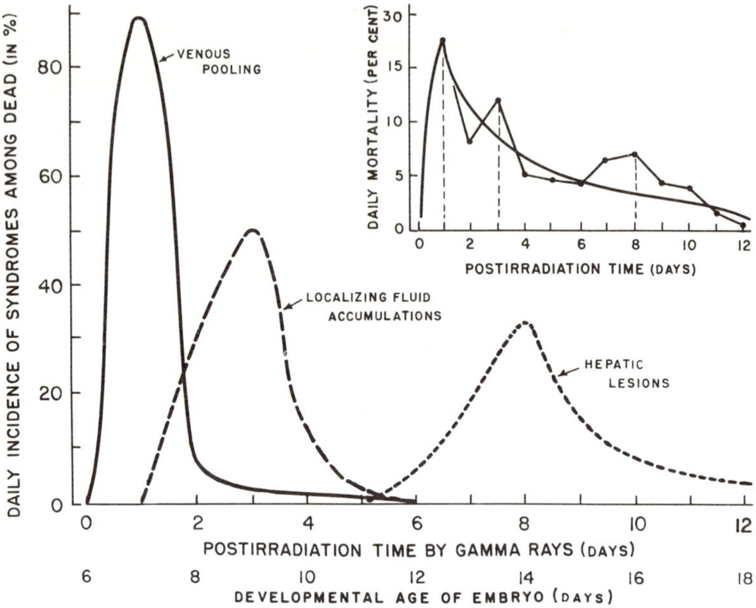

Fig. 78. Daily incidence of the typical radiation syndromes among dead embryos treated with gamma rays (from ^{60}Co isotope) in doses ranging from 800 to 13,000 R. (After Stearner, Sanderson, Christian, and Tyler, 1960.) *Insert:* daily embryonic mortality after the same treatment.

Radioactive Iodine (^{131}I)

Mortality. Survival is reduced in embryos injected with ^{131}I in doses of 25, 50, and 100 μCi or more at 8, 12, and 16 days of development, respectively (*Mraz, 1966*), as is shown in the accompanying tabulation.

Dose of ^{131}I (μCi)	Incidence of Embryonic Mortality (%) Injected at 8 Days	Injected at 12 Days	Injected at 16 Days
5	0	0	0
25	33	0	0
50	50	42	0
100	83	100	35
150	100	—	

Some embryos that received 50, 100, and 150 μCi of ^{131}I are found to be still alive on the twenty-ninth day of incubation.

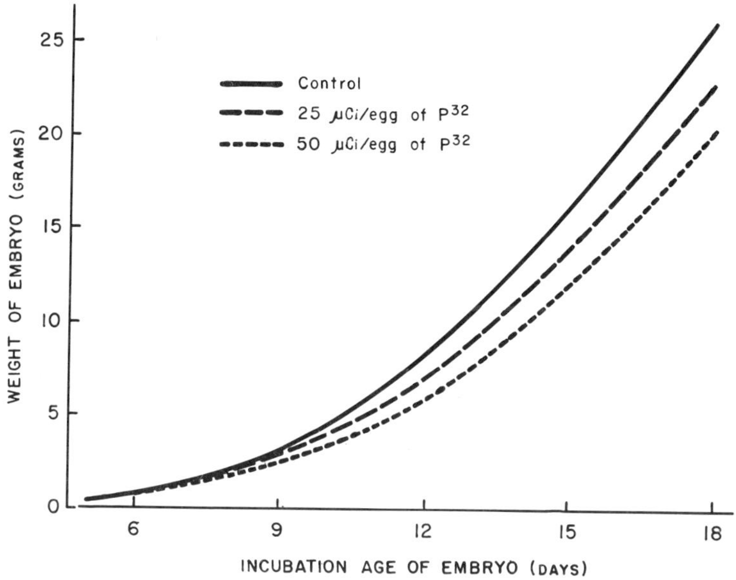

Fig. 79. Weight changes in the chicken embryo after injection into the yolk sac of various doses of ^{32}P on the fifth day of incubation. (Plotted from the data of Chung, 1965.)

Radioactive Phosphorus (^{32}P)

Growth. Internal irradiation by ^{32}P causes overall growth retardation (*Warren and Dixon, 1949a; Chapman, 1954; Chung, 1965*), resulting in a body which is smaller in size, but proportional in dimensions to normal (*Warren and Dixon, 1949b*). Weight changes of the developing embryo after injection with various doses of ^{32}P are shown in Fig. 79. The average inhibition in weight gain of the embryo between 8 and 18 days (*Chung, 1965*) is demonstrated in the table below.

Dose (μCi/egg)	Average Inhibition Rate (%)
5	3.0
10	4.0
25	17.5
50	29.8

This inhibition is due to decreased mitosis of tissues, usually in direct proportion to the dose employed; mitosis returns to normal in a few days after injection of sublethal doses (*Chapman, 1954*), comparable to the return after local radiation with X-rays (see *Le Douarin, 1963;* and Fig. 62). Lethal doses of ^{32}P bring about virtually complete cessation of mitotic activity (*Warren and Dixon, 1949b*). In general, the decrease in mitosis is in direct proportion to the amount of degeneration of tissues.

Lethality and Malformations. Introduction of various doses of ^{32}P into fertile eggs after 2 days of incubation, and examination of embryos 2 days later (*Chapman, 1954*), reveal some specific radiation-induced changes:

1. The embryos exposed to 39 μCi are practically normal.
2. The embryos exposed to 59 μCi show some severe degenerative changes.
3. Most of the embryos exposed to 82 μCi show progressive destruction of tissues.
4. Among the embryos exposed to 117 μCi, three out of four die in 2 days.
5. Most embryos exposed to 165 and 234 μCi are so severely damaged that it is difficult to recognize tissues by their appearance.

Observations of 18-day-old embryos from fertilized eggs injected with various doses of ^{32}P (*Chung, 1965*) reveal distinct histopathological changes in the various organs, such as the gonads, bones, thymus, eyes, thyroid, and spleen. The testis and ovary suffer the most pronounced damage during the early developmental stages, and with the larger doses. The bone tissues show edematous degeneration. The thymus, the eye, the thyroid gland, and the spleen all are sensitive to large doses of ^{32}P, between 50 and 100 μCi.

Radioactive Promethium (^{147}Pm)

Mortality. The highest mortality produced by injection of ^{147}Pm occurs in embryos youngest at injection time and with the larger doses (*Mraz, Wright, and Ferguson, 1964*), as is shown in the table below.

Dose of ^{147}Pm (μCi)	Embryonic Mortality at Hatching (%) Injected at 4 Days	Injected at 8 Days	Injected at 14 Days
100	25	8	0
200	57	57	17
400	75	75	42
800–1600	100	83	50

Malformations. The number of abnormalities is low, and shows no pattern or consistency (*Mraz, Wright, and Ferguson, 1964*). Among the developmental abnormalities observed were deformed toes and beaks, shortened limbs, and weak tendons.

Radium (Ra)

Mortality. Studies by Wilson (*1935*) show that for treatment with radium (by radium needles), the lethal time exposure is different at different stages of embryonic development, as is shown below.

Age of Embryo at Irradiation (days)	Lethal Effect Time (hours)
1	5
2	7
3	10
4	15
5	23
6	25

During the first 5 days there is a continuous decrease in the radiosensitivity of the developing embryo with increase in age. At 5 and 6 days the radiosensitivity becomes nearly constant. The period of survival after irradiation depends upon age—the older the embryo is when irradiated, the longer death is delayed.

Malformations. The first effect of lethal dose of irradiation with radium is upon the blood corpuscles, causing them to bunch together; then follows degeneration of the vessels themselves (*Wilson, 1935*). The smaller vessels degenerate first and the largest last of all.

Abnormalities in mitosis of choroid and scleroid tissues of a 10- to 11-day-old embryo were observed after irradiation with radium (*Simon-Reuss and Spear, 1947*). The relation between dose and incomplete mitosis is shown in the tabulation below.

Dose of Radium (R)	Incidence of Incomplete Mitosis (%)
0	0.0
55	17.2
88	51.5
330	60.7
660	74.2
1320	100.0

Radioactive Strontium (^{90}Sr)

Mortality. It has been demonstrated (*Creger and Colvin, 1967*) that embryonic mortality considerably increases when dose levels of 27 μCi or higher are injected subcutaneously in hens, as is shown in the tabulation below.

Injected Dose of ^{90}Sr (μCi)	Total Mortality at Hatching (%)
0.0	8.3
7.9	11.8
27.0	25.1
48.8	42.1

Malformations. Embryos injected with ^{90}Sr at 8 days may show constricted areas around the center of the long bones, similar to ^{45}Ca damage (*Mraz, Wright, and Ferguson, 1964;* see also *insert* of Fig. 75). The effect is especially noticeable with high doses, as is shown in the table below.

Doses of ^{90}Sr (μCi)	Incidence of Abnormal Bone (%)
100–200	17
400–800	67

The damage to legs is quite severe, particularly to the tissues, since the feet below these constricted areas are swollen, hemorrhagic, and gangrenous.

Some embryos show heavy edema of the head and neck.

Radioactive Sulfur (^{35}S)

Mortality. Observations (*Wright and Mraz, 1965*) show that there is an increase in incidence of mortality among embryos hatched from eggs injected with 100 μCi of ^{35}S at the fourth to eighth day of incubation, as is shown in the table below.

Doses of ^{35}S (μCi)	Embryonic Mortality up to Hatching (%)
50	17
100	33
200	59
400	83
800	100

With successive increases in dose of ^{35}S, the incidence of embryonic mortality increases, finally reaching 100 per cent at a dose of 800 μCi.

Malformations. At a level of 100 μCi of ^{35}S or higher, some incidence of malformations is apparent (*Wright and Mraz, 1965*). With increase in dose, there is an increase in incidence and severity of deformities, as shown below.

Dose of ^{35}S (μCi)	Incidence of Malformations (%)
50	0
100	9
200	17
400	33
800	75
1600	100

Embryos from eggs injected with 800 or 1600 μCi live until the sixteenth to twentieth day of incubation, but are almost all deformed. Many of them are dwarfed and have extreme shortening of the leg and wing bones in relation to body size. The skeletons are complete but disproportioned.

Radioactive Zinc (^{65}Zn)

Mortality. Injection of various doses of radioactive zinc (*Mraz, 1969*) into developing eggs at 4, 8, and 14 days produces the following average incidences of embryonic mortality.

Dose of ^{65}Zn (μCi)	Incidence of Mortality (%)
0	33
50	60
100	70
200	95
400	100

Malformations. No growth abnormalities associated with irradiation were observed, primarily because of the rapid death of the embryos (*Mraz, 1969*).

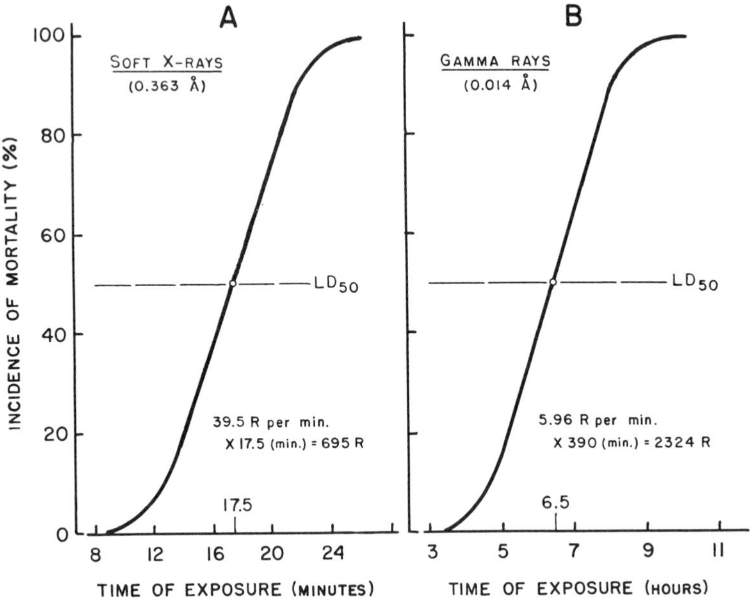

Fig. 80. Relative biological effectiveness (RBE) of ionizing radiation at long and short wavelengths (X-rays and gamma rays, respectively), measured by mortality of embryos treated at the age of 6 days, and recorded 2 to 3 days after radiation. (Plotted from the data of Lea, 1941.)

RELATIVE BIOLOGICAL EFFECTIVENESS

The quality of ionizing radiation has an effect on biological potency. The magnitude of existing variation has only recently been recognized by the experimental biologists.

The inherent variance in this type of investigation is related to control of wavelength and intensity of radiation, among other factors.

Variation in Wavelength. The values for wavelengths of X-rays and gamma rays are given in the accompanying tabulation.

Source of Radiation	Wavelength (λ in cm)
X-ray (softest)	1.4×10^{-7}
Gamma rays (longest)	1.5×10^{-8}
X-rays (hardest)	4.0×10^{-10}
Gamma rays (shortest)	2.0×10^{-10}

As a result of these differences in wavelength, the amount of X-rays needed for effects biologically equivalent to those of 1 roentgen (R) of gamma rays (*Lea, 1941*) was found to be in the ratio shown in the table below.

Radiation	Wavelength (Å)	Equivalents (R)
Gamma rays	0.014	1.00
Medium X-rays	0.15	0.60
Soft X-rays	0.363	0.55

The relative biological effect of radiation by long and short wavelengths on the incidence of embryonic mortality is shown in Fig. 80. The calculated values show a noticeable difference between the effectiveness of soft X-rays and gamma rays.

Table 7 gives a summary of comparative values for doses causing 50 per cent lethality (LD_{50}) in 4-day-old chicken embryos examined 6 to 7 days after irradiation. The variations in LD_{10} from experiment to experiment are generally small. The difference in mortality data between gamma rays and fusion neutrons (ref. 5) is noticeable. Fusion

Table 7 Comparative Data for the LD_{50} in 4-Day-Old Avian Embryos Recorded 6 to 7 Days After Irradiation

Source of Radiation	LD_{50} (R)	LD_{50} (R)	LD_{50} (R)	LD_{50} (Duck) (R)	LD_{50} (R)	LD_{50} (R)
^{60}Co gamma rays					1108	1003
4-MeV X-rays			1178	981		
22-MeVp X-rays						1015
33-MeV gamma betatron		1001				
175-kV X-rays		863				
200-kVp X-rays				884		944
250-keV X-rays	911					
300-keV X-rays			914			
1000-keV X-rays	1005					
Fusion neutrons (CP5)					502	
Reference	1	2	3	4	5	6

[1] Kohn and Kallman, 1956. [2] Oftedal, 1956. [3] Boland and Tranter, 1957. [4] Kallman, Steele, Weissbluth, and Bagshaw, 1958. [5] Vogel and Jordan, 1959. [6] Humphrey and Sinclair, 1962.

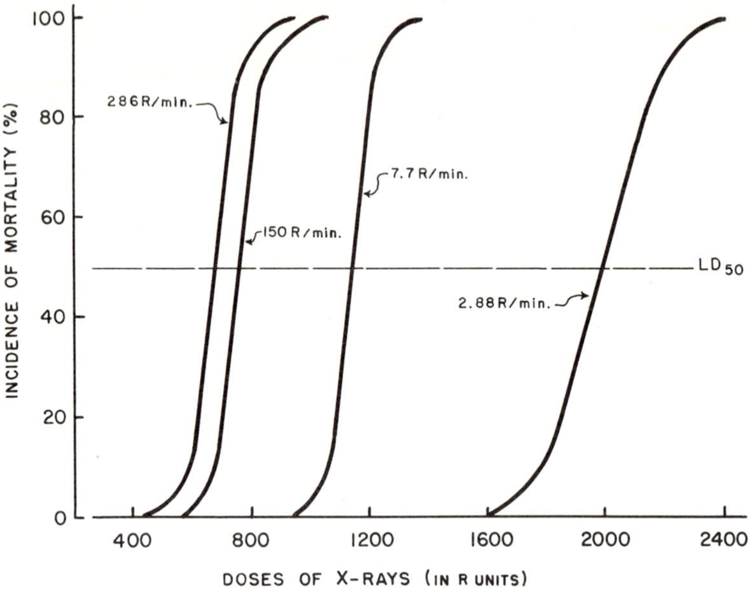

Fig. 81. Relative lethal effect of various intensities of X-rays on chicken embryos treated at the age of 6 days, recorded for mortality 2 to 3 days later. (Reconstructed with modifications from the graphic data of Lea, 1941.)

neutrons are approximately 2.8 times as effective as gamma rays (*Vogel and Jordan, 1959*).

Variation in Intensity. The lethal effects of various intensities of X-rays (with 0.15 Å wavelength) on the developing embryo (*Lea, 1941*) are shown graphically in Fig. 81.

With the change of intensity, there is a noticeable change in lethal dose, as is shown below.

Intensity of Radiation (R/min)	Lethal Dose (R units)
150	800
30	800
8	1150
3	2000

The variation in the effect of radiation with alterations of wavelength and intensity has been studied by a number of workers on various biological materials.

ULTRAVIOLET RAYS

Numerous observations have been made on the effect of unfiltered ultraviolet radiation on birds' eggs. It was found that exposure through the shell causes no noticeable pathological symptoms. On the other hand, it is evident from many observations (*Sheard and Higgins, 1930a; 1930b; Landauer, 1932a; and Romanoff and Ludlam, 1943*) that ultraviolet radiation accelerates embryonic development, resulting in a hatching time 12 to 24 hours earlier than normal.

The effects of various amounts of energy of filtered ultraviolet radiation on the stimulation of the 15-hour chicken blastoderm *in vitro* are shown in Fig. 82. The peak of stimulation at wavelength 3650 Å is nearly twice as high as at wavelength 3100 Å (*Romanoff, 1942; 1943b*). Doses beyond the stimulating range inhibit the embryonic activity (and reduce the potential differences).

It has been shown histologically (*Hinrichs, 1927*) that in some instances inhibition is carried to the point of reducing the embryo to an axiate cell mass. Still greater inhibition results in the production of anaxiate cell masses, particularly following exposures at an early age.

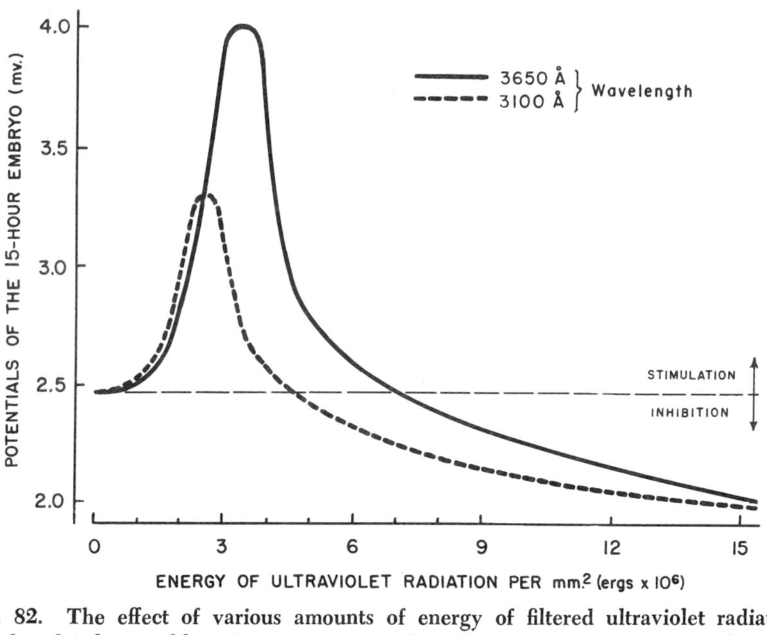

Fig. 82. The effect of various amounts of energy of filtered ultraviolet radiation on the developing blastoderm, measured by bioelectric potentials. (With some modifications after Romanoff, 1943b).

When lethal doses of radiation are used, the exposed regions of the embryo die and disintegrate.

Among the susceptible organs are the nervous system and sensory organs, the heart and circulatory system, and the somatic region (*Hinrichs, 1927*). For each organ the most effective period for the production of abnormalities is the time immediately preceding its morphological differentiation.

The incidence of different abnormalities in various body regions of ultraviolet-treated embryos is shown in the accompanying table (*Hinrichs, 1927*).

Anomaly	Incidence of Anomaly (%)	
Cell mass		10.5
Brain		
Forebrain	9.6	11.6
Midbrain	0.9	
Hindbrain	1.1	
Heart, abnormal		5.3
Circulation, faulty		2.9
Twinning		2.9
Membranes, abnormal		2.4
Heterotaxia		2.2
Hydropsy		1.8

In general, susceptibility is high during the very early stages of the embryo, but decreases during the later developmental period.

MICROWAVE RADIATION

It has been observed (*Ummersen, 1961*) that with radiation at a wavelength of 2450 Mc, regardless of the power or duration of exposure, the temperature at the level of the chicken embryo rapidly rises to approximately 59° C (Fig. 83).

Consequently, at a maximum power of 400 mW/cm^2, irradiation for 4 to 5 min causes abnormalities. An exposure prolonged for another ½ min becomes lethal to the embryo. At lower power levels (280 and 200 mW/cm^2), the difference in time of reaching the lethal point is no more than 2 or 3 min (Fig. 84).

When embryos are irradiated at 2 days and examined at 4 days, there is a general inhibition of growth, especially in the posterior region.

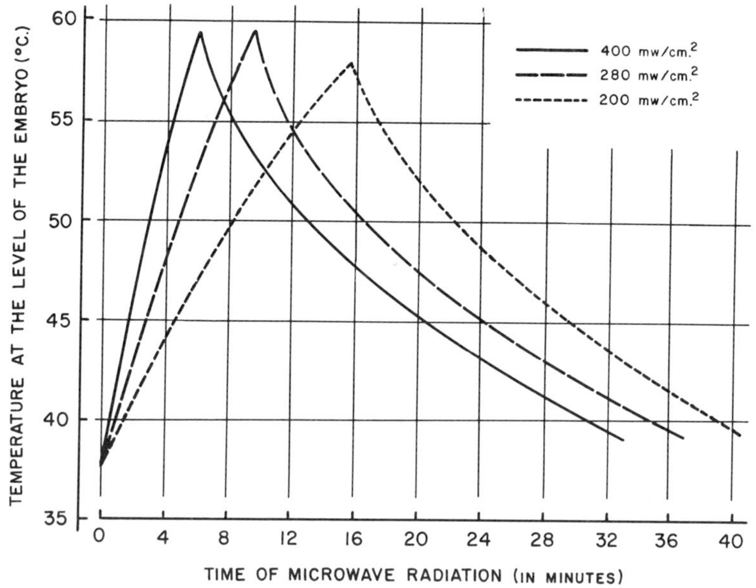

Fig. 83. Temperature measurements at the level of the embryo during irradiation with microwaves of 2450 Mc. (After observations of Ummersen, 1961.)

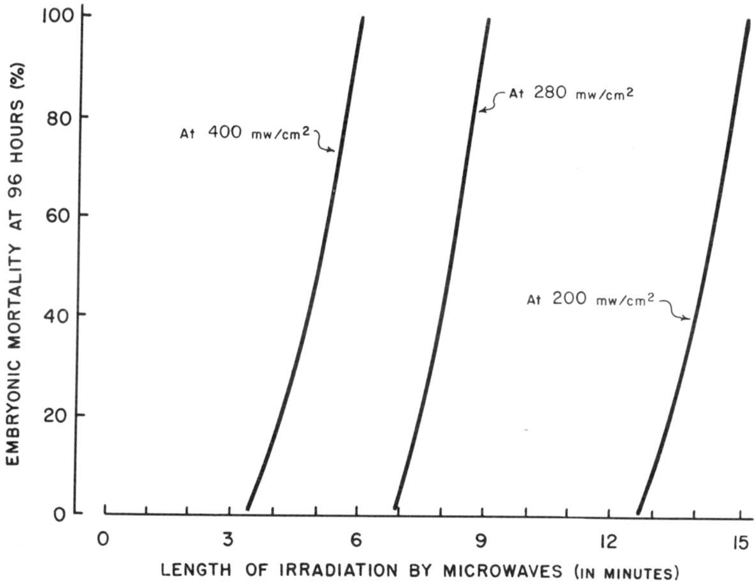

Fig. 84. Effect of microwave radiation frequencies (about 2450 Mc.) on embryonic mortality. (Plotted from the recalculated data of Ummersen, 1961.) *Note:* embryos were irradiated through the intact shell at the age of 48 hours.

In structures which had not yet begun to differentiate before exposure, such as the posterior limb buds, tail, and allantois, further development is totally suppressed.

The question is whether the abnormalities are induced chiefly by a thermal effect, or by some other factors.

ULTRASONIC WAVES

In early observations using large doses of ultrasonic-wave radiation (*Brettschneider, 1952*), it has been noted that the neuroblastic tissues of the chicken embryo are the most susceptible.

Two and one half days after irradiation of fertilized eggs for 10 min at a frequency of 1000 cycles (*Lutz, Bonhomme, and Lutz-Ostertag, 1955*), malformations were observed in the cephalic area, with further appearance of symmelia (fusion of wings and feet) and disorientation of the embryonic axis. Exposure of chicken and duck eggs to ultrasonic waves at frequencies from 500 to 1500 cycles (*Lutz and Lutz-Ostertag, 1957*) resulted in various types of monstrosities. The chief defects were in the nervous system. Local radiation of the anterior part of the blastoderm results in acleiencephaly, and radiation of the posterior part results in rhombencephaly and in pseudosymmelia and pseudosymmetry.

(For the general biological effect of various sources of radiating energy on the development of the avian embryo see Summary Table 8.)

Summary Table 8 General Biological Effect of Radiation on the Development of the Avian Embryo

Source of Energy	Frequency, γ (cm/sec)	Biological Effect of Radiation	Reference
Ionized air	3×10^{3}	Has prophylactic value.	Tchijevsky (1933)—translated by Romanoff (1935)
Radio waves	15×10^{6}	Maintain normal development.	Romanoff and Cottrell (1939)
	25×10^{6}	Stimulate the developing blastoderm.	Ludwig and Ries (1944), Romanoff (1946)
	50×10^{6}		Romanoff and Frank (1941)
	20×10^{7}	Are very effective in early stages; death ensures before the sixth day.	Everdingen and Noyons (1940)
Microwaves	24.5×10^{8}	Have general inhibitory effect on cellular differentiation.	Ummersen (1961)
Radar waves	24×10^{9}	Cause damage to heart.	Paff, Boucek, Nieman, and Diechmann (1963)
Ultraviolet Rays	Unfiltered	Cause early developmental abnormalities.	Hinrichs (1927)
	3.35×10^{16}	First accelerate, then retard the development of blastoderm.	Romanoff (1943b)
X-rays	3×10^{17}	Produce instantaneous stimulation (at 250 R units) followed by inhibition in growth.	Romanoff and Bless (1942), Bless and Romanoff (1943)
Gamma rays	3×10^{19}	Cause general retardation with subsequent monstrous development.	Bursian (1960)
Cosmic rays	3×10^{22}	Are extremely penetrating.	(general knowledge)

VI

Nutritional Deficiency

Nutrition is a cornerstone:
Deficiency or large excess
Makes embryo diseases prone—
Creates defects . . . and lethal stress.

Nutritional requirements of the developing avian embryo are met through maternal feeding and the deposition of nearly all essential chemical elements into the egg during its formation. Actual deficiencies, and some excesses, within the egg can be brought about by controlling the diet of the hen. Experimentally, any nutritive substance can be added by injecting it into the egg before or during incubation.

Deficiencies of organic and inorganic compounds in the egg may seriously disturb embryonic development and lead to various structural malformations and premature death.

Among the groups of substances specifically related to development processes are vitamins and minerals.

VITAMINS

Vitamin content of the egg can be altered by controlled feeding of the hen laying the egg and by addition of vitamin or provitamin substances through injection into the yolk and albumen. Consequently, the effects of vitamins on the incidence of toxicity and morphological modifications in the development of the avian embryo can be studied through nutritional deficiencies and excesses.

Biotin

Experimentally it has been determined (*Couch, Cravens, Elvehjem, and Halpin, 1948a*) that there is a failure of embryonic development when the chicken egg yolk contains less than 50 mg/gram of biotin. Eggs that contain 150 mg/gram can support embryonic development.

Mortality. The effect on embryonic mortality of various doses of biotin injected into biotin-deficient eggs (*Couch, Cravens, Elvehjem, and Halpin, 1949*) is shown below.

Doses of Biotin (μg/egg)	Embryonic Mortality at Hatching (%)
0	100
2	98
10	24
25	12

The effects of various levels of supplemental biotin in the biotin-deficient diet on the distribution of embryonic mortality are graphically presented in Fig. 85. A deficiency of biotin in the incubating egg is associated with a prominent increase in embryonic mortality during the first week of incubation, with a peak on about the fourth day. There is no indication of a second peak on about the thirteenth day (see Fig. 89).

Effect on embryonic mortality of time of injection of 25 μg/egg of biotin into biotin-deficient eggs (*Couch, Cravens, Elvehjem, and Halpin, 1949*) is shown in the accompanying table.

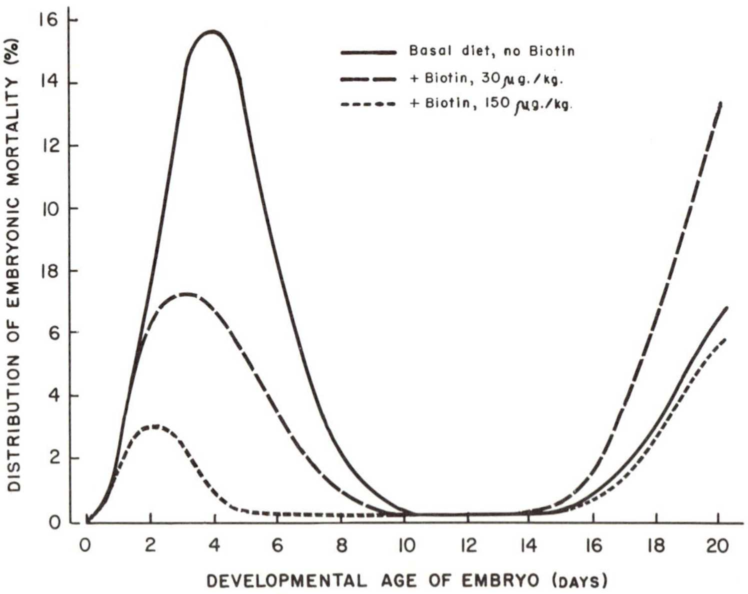

Fig. 85. Effect of biotin deficiency in the diet of hens on the distribution of embryonic mortality during incubation. (Generalized curves are based on the experimental data of Cravens, McGibbon, and Sebesta, 1944.)

Time of Injection (days)	Embryonic Mortality at Hatching (%)
0	11
2	40
3	74
4	91
5	100

Malformations. A biotin deficiency in the diet of the hen is associated with the occurrence of congenital perosis (Fig. 86), which is observed either in the dead embryos after the seventeenth day of incubation or in the day-old chick.

Among skeletal malformations observed are deformed beak and limbs and other anomalies.

(See also Summary Table 9, under Biotin.)

Folic Acid

Growth. It has been observed that experimentally induced folic acid deficiency (*Karnofsky, Patterson, and Ridgway, 1949*) causes profound

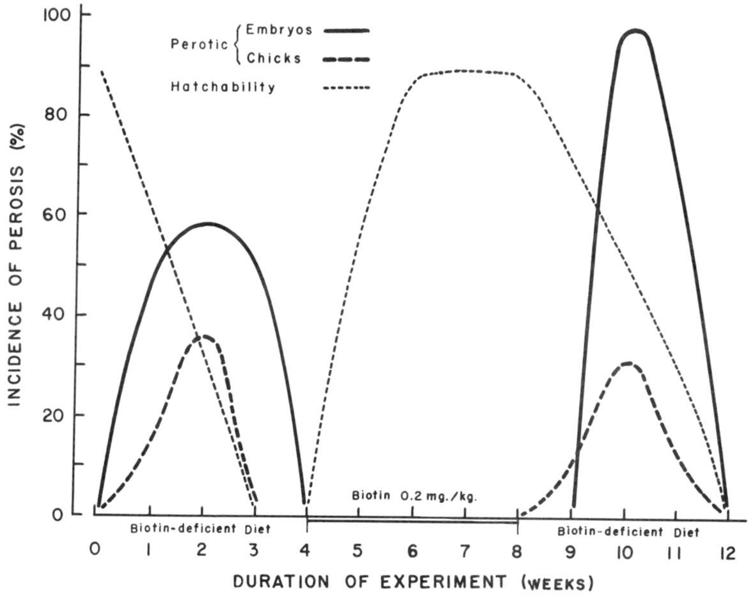

Fig. 86. The incidence of congenital perosis resulting from a maternal biotin deficiency. (With some modifications after Couch, Cravens, Elvehjem, and Halpin, 1948b.)

inhibition of growth in the early embryo. After injection of about 4 μg/egg of a folic acid antagonist (4-amino-pteroylglutamic acid) at 4 days, on the thirteenth day the embryo weighs only 1 gram, while the control embryo of the same age weighs about 6 grams; that is, the weight of the treated embryo is only about 17 per cent of normal.

Mortality. Injection of various doses of folic acid into folic acid-deficient eggs results in a reduction of mortality (Fig. 87*A*), from over 90 per cent to about 30 per cent. A deficiency of folic acid in the hen's diet results in a high incidence of embryonic death on about the twentieth day of development.

The injection of 25 μg/egg of folic acid into folic acid-deficient eggs at different stages of development produces changes in the total embryonic mortality at hatching, as shown in Fig. 87*B*.

Malformations. Some embryos from hens severely depleted of folic acid exhibited at hatching a marked bending of the tibiotarsus (*Sunde, Cravens, Elvehjem, and Halpin, 1950*). Other, less frequently observed symptoms are syndactyly, deformed mandibles, and parrot beak.

(For further details see also Summary Table 9, under Folic Acid.)

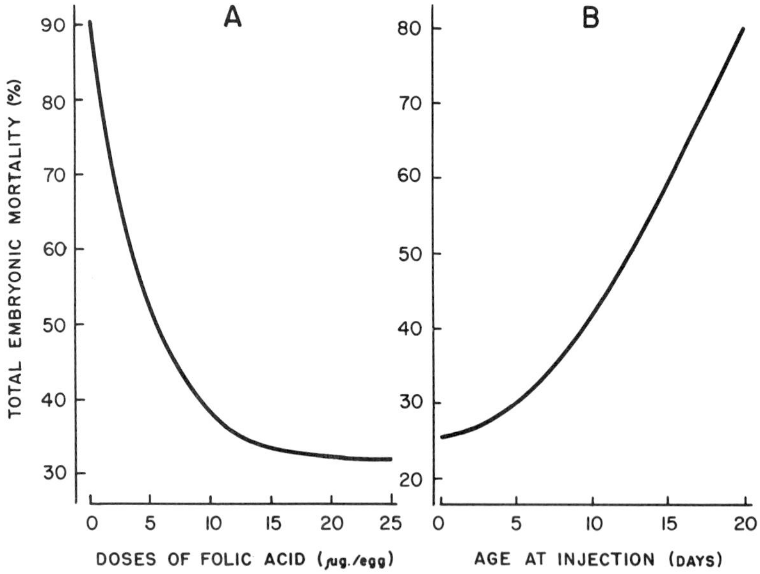

Fig. 87. The effect on embryonic mortality of injecting folic acid into folic acid-deficient eggs. (Plotted with modifications from the data of Sunde, Cravens, Elvehjem, and Halpin, 1950.) *A*, effect of injection of different amounts of folic acid prior to incubation; *B*, effect of injection of 25 μg/egg at various stages of incubation.

Inositol

Mortality. Hens on an inositol-depleted diet are capable of producing fertile eggs with 100 per cent embryonic mortality. When the diet is supplemented by 1 μg/gram of inositol for a few weeks (*Cravens, Sebesta, Halpin, and Hart, 1942*), the results in respect to embryonic mortality are as follows.

Period with Inositol (weeks)	Embryonic Mortality (%)
1	100
3	80
5	30
7	10

In about 6 weeks, embryonic mortality is restored to normal levels.

Nicotinic Acid

(See Summary Table 9, under Nicotinic acid; also see effects of an excess of Vitamin B_3 in Summary Table 12, under Nicotinamide, at the end of Chapter VII.)

Pantothenic Acid

The pantothenic acid content of the diet is reflected in its presence in hens' blood and in the eggs (*Gillis, Heuser, and Norris, 1948*), as is shown below.

	Pantothenic Acid Content	
Pantothenic Acid in Diet (μg/gram)	In Hen's Blood (μg/gram)	In Eggs (μg/gram)
2	0.25	0.30
4	0.26	0.50
6	0.37	0.71
8	0.43	0.95
10	0.52	1.15
12	0.56	1.40
16	0.57	1.80

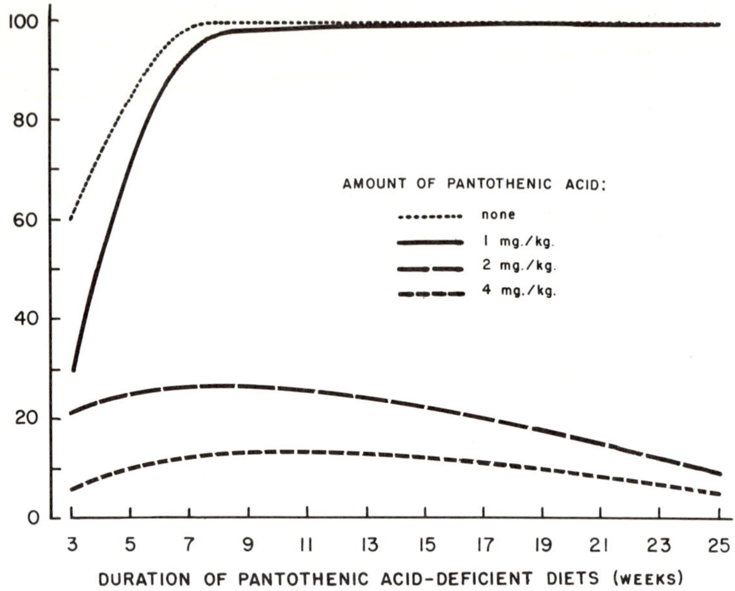

Fig. 88. Effect of different amounts of pantothenic acid, added to the diet of a hen, on the subsequent embryonic mortality. (Simplified curves based on the data of Beer, Scott, and Nesheim, 1963.)

The amount of pantothenic acid stored in the eggs increases with the increase of its dietary level, whereas in the blood it is much lower and tends to reach a certain constant value.

Growth of Organs. It has been observed (*Taylor, Thacker, and Pennington, 1941*) that with increasing pantothenic acid in the egg, there is a relative increase in the size of the brain and heart.

Mortality. There is a direct negative relationship between the level of pantothenic acid in the diet of the hen and the total embryonic mortality at hatching (*Beer, Scott, and Nesheim, 1963*). This is demonstrated in the table below.

Pantothenic Acid in Diet (mg/kg)	Total Embryonic Mortality (%)
0	96.7
1	90.7
2	19.9
4	8.9

The cumulative curves of embryonic mortality in eggs from hens with different levels of pantothenic acid in their diets are illustrated in Fig. 88. With the basal diet (pantothenic acid-deficient), and with only 1 mg/kg of the supplement, mortality increases rapidly to 100 per cent. A diet with 2 or 4 mg/kg of pantothenic acid maintains low mortality.

The daily distributions of mortality with the pantothenic acid-deficient diet and with a supplement of 1 mg/kg are shown for comparison in Fig. 89. Although the total mortality under each condition is almost identical (see text table above) the distribution of dead embryos is different.

In eggs from hens on the deficient diet, mortality occurs during the early period of development, forming two peaks on the fourth and thirteenth days. With the supplement, mortality is shifted to the later period of incubation (cf. normal distribution of mortality, Fig. 13, Chapter II).

Malformations. (For a detailed list see Summary Table 9, under Pantothenic acid, at the end of this chapter.)

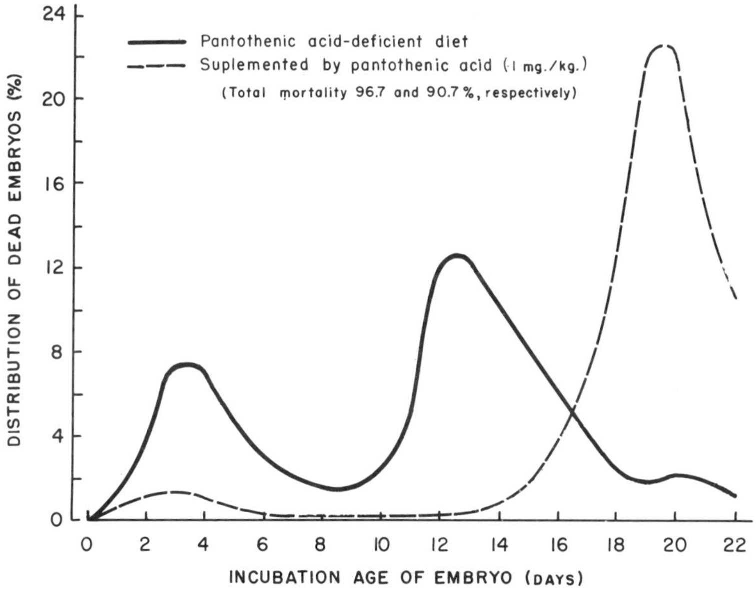

Fig. 89. Effect of pantothenic acid deficiency on the distribution of dead embryos throughout the incubation period. (Plotted from the data of Beer, Scott, and Nesheim, 1963.)

Pyridoxine

Mortality. The effect of excluding pyridoxine from the diet of the hen on the subsequent changes in mortality of the embryo (*Cravens and Snell, 1949*) is shown in the accompanying table.

Duration of Pyridoxine Deficiency (weeks)	Total Embryonic Mortality (%)
0	10
2	20
4	25
5	67
6	72
7	100

Riboflavin (Vitamin B_2)

Growth. Observations on the growth of riboflavin-deficient embryos (*Romanoff and Bauernfeind, 1942*) show that the weight of deficient embryos is consistently below that of normal embryos. This difference in weight is progressively greater with successive stages of incubation. On the ninth day, the morphological age of deficient embryos is less than 1 day behind the incubation age, while on the fourteenth day the difference from normal embryos is nearly 2 days, and on the sixteenth day it is about 2½ days. This is illustrated by the actual weights of the morbid and dwarf embryos in the accompanying table.

Incubation Age of Embryos (days)	Weight of Embryos	
	Control (grams)	Riboflavin-Deficient Morbid and Dwarf Embryos (grams)
10	2.3	2.0
11	3.4	2.7
12	5.2	3.6
13	7.3	5.7
14	11.0	7.0
15	13.1	7.9
16	16.6	9.2

Mortality. Eggs from hens kept continuously on a riboflavin-deficient diet show an increase in incidence of embryonic mortality. This is well demonstrated by the experimental data (*Davis, Norris, and Heuser, 1938*), shown in the accompanying table.

Period of Depletion (weeks)	Embryonic Mortality (%)
0	19.5
3	39.3
6	71.9
9	93.1

The level of riboflavin in the diet of the hen determines the content of riboflavin in the eggs (*Norris and Bauernfeind, 1940*). The total embryonic mortality (*McClymont and Hart, 1947*) is shown below.

Amount of Riboflavin in Diet (μg/gram)	Embryonic Mortality (%)
0.0	100
1.5	96
1.6	85
1.9	68
2.3	29

Of 1268 fertile eggs incubated, from a flock of hens fed a ration known to be deficient in riboflavin (*Romanoff and Bauernfeind, 1942*), 1109 eggs—87.5 per cent—failed to hatch. By daily observations of each embryo, it was established that deaths occur chiefly at three definite periods (Fig. 90): the fourth, fourteenth, and twentieth days of incubation.

The first and last periods occur in normal circumstances as well (see Fig. 13 in Chapter II). The second period, which reaches its peak usually on the thirteenth day, is attributed chiefly to nutritional deficiencies; see under Proteins (*Byerly, Titus, and Ellis, 1933*).

With increased depletion of the riboflavin reserves of individual hens (see Chapter III on genetic factors, p. 48), the major peak of embryonic mortality is shifted progressively from the later critical period to the earlier (Fig. 91).

Embryonic mortality from riboflavin deficiency is not limited to the second period (Fig. 90). With a group of hens in which 99.5 per cent of the eggs failed to hatch, the peak of embryonic mortality oc-

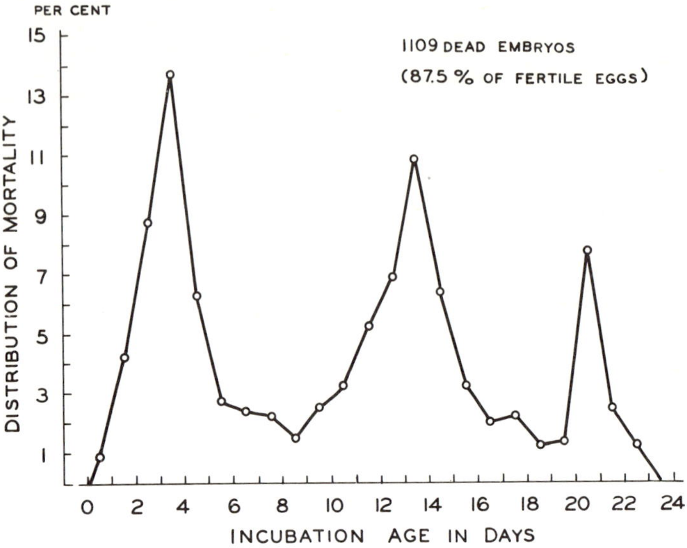

Fig. 90. Distribution of embryonic mortality from riboflavin-deficient eggs. (After Romanoff and Bauernfeind, 1942, based on 1109 dead embryos. See also Davis, Norris, and Heuser, 1938; Lepkovsky, Taylor, Jukes, and Almquist, 1938.)

curred on the fourth day. In a group of hens whose embryonic mortality was only 62.7 per cent, the major peak occurred on the twentieth day. The middle group, with a 91.3 per cent mortality, reached the major peak on the fourteenth day.

Weight of Dead Embryos. There is wide variation in the weight of dead embryos (Fig. 92), indicating a highly individualistic total response to the riboflavin deficiencies in the diet of hens and thus in the riboflavin content of their eggs (*Romanoff and Bauernfeind,* unpublished).

In general, the weight of all dead embryos, expressed in per cent of weight of living embryos, is invariably smaller—on the average, about 50 per cent less. As hatching time approaches, the dead embryos fall into two distinct groups according to weight: the "dwarfs," with weights less than 40 per cent of normal for the corresponding age, and the "giants," weighing at death about 90 per cent of normal.

The lowest weight level of some individual embryos remains at 20 per cent of normal weight. This indicates that these embryos stop growing several days before death (see Fig. 9, in Chapter II, on sporadic occurrence of dwarfs, as detected by the daily measurement of oxygen consumption of intact developing eggs, Fig. 9).

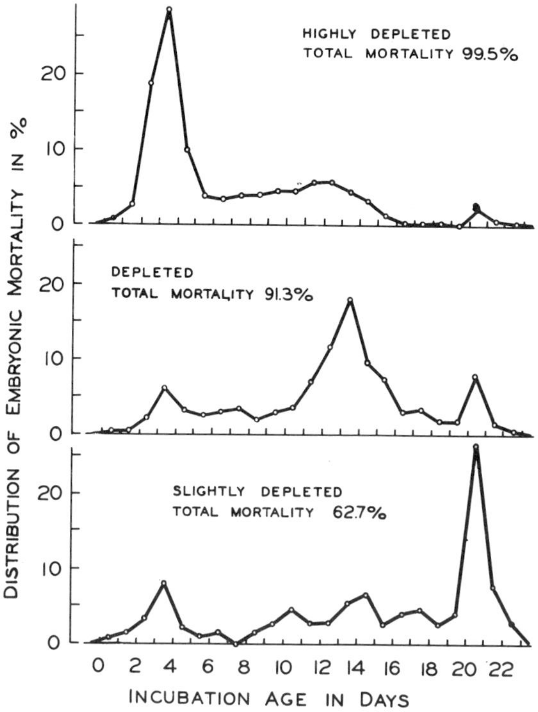

Fig. 91. Influence of the degree of depletion of riboflavin reserves of the hen on the time and distribution of embryonic mortality. (After Romanoff and Bauernfeind, 1942.)

Malformations. Prior to their death on the fourth day, the riboflavin-deficient embryos are markedly reduced in size (*Romanoff and Bauernfeind, 1942*). The blastoderm reveals the beginning of a disorganization in the circulatory system, with the incipient formation of a lethal ring.

The embryos dying on the fourteenth day manifest various forms of pathological changes in the liver, especially in the left lobe. There are noticeable changes in color, from natural light yellow to brown or dark green.

(For further detailed information see Summary Table 9, under Riboflavin.)

Vitamin A (Retinol)

Mortality. It has been observed (*Bearse and Miller, 1937*) that embryonic mortality increases with the decrease of Vitamin A in the hen's diet. The results are presented in the following table.

Vitamin A in Ration (mg/kg)	Embryonic Mortality at Hatching (%)
450	10.0
120	15.0
60	31.7
30	100.0

Free retinoic acid (Vitamin A acid) was found to be extremely toxic to the early chicken embryo (*Thompson, Howell, Pitt, and Houghton, 1965*). As little as 1 μg injected into fertile eggs before incubation prevents development. (This injection appears to cause disintegration of the entire early blastoderm.)

Malformations. Experiments *in vivo* (*Fell, 1960*) show that the growing embryonic skeleton is profoundly disturbed by Vitamin A deficiency. In the skull and spinal column, the normal pattern is drastically altered. Severe deformities result, which in turn cause degenerative changes in the brain, spinal cord, and nerves. These deformative processes extend

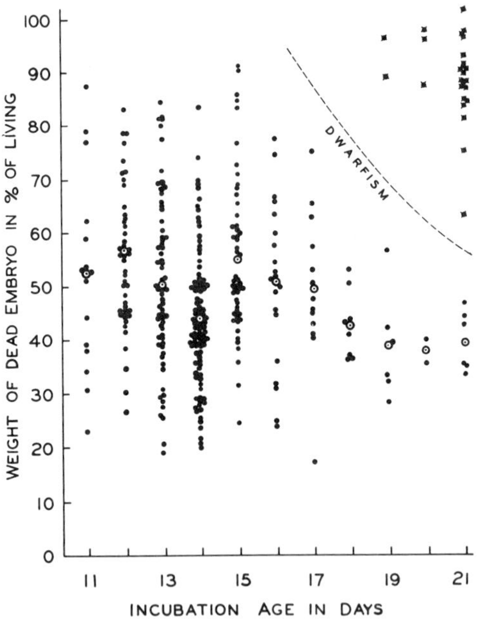

Fig. 92. Weights of 1109 dead embryos from riboflavin-deficient eggs. (After the data of Romanoff and Bauernfeind, unpublished.)

also to the limb skeleton. If the effects of the deficiency have not progressed too far, they can be reversed by restoring Vitamin A to the diet.

The administration of a large excess of Vitamin A is equally injurious to growing bones (*Fell and Mellanby, 1952*). Hypervitaminosis A produces drastic changes in the cartilage matrix of explants from the 5- to 6-day embryo *in vitro*.

(For further details see Summary Table 9, under Vitamin A.)

Vitamin B_{12}

The Vitamin B_{12} received by laying hens is transported directly through the egg (*Halick, Reid, Brown, and Couch, 1953*) to the developing embryo.

Growth. Observations (*Ferguson, Rigdon, and Couch, 1955*) show that embryos from hens fed a Vitamin B_{12}-deficient diet are smaller than either normal embryos or those receiving a supplement of Vitamin B_{12}.

Size of Thyroid. The rate of functional development of the thyroid in the Vitamin B_{12}-deficient embryo is from 1 to 3 days behind that of the normal chicken embryo, in spite of increasing size (*Ferguson, Rigdon, and Couch, 1957*). At the time of hatching, the thyroid gland in Vitamin B_{12}-deficient embryos is twice the size of that in normal embryos of the same age, as is shown in the accompanying table.

Age of Embryo (days)	Dimensions of Thyroid (mm)	
	Normal	B_{12}-deficient
12	1.0	1.0
14	1.1	1.2
16	1.5	2.3
18	1.6	2.9
20	1.6	3.2

The thyroid glands in the Vitamin B_{12}-deficient embryos usually appear more vascular than those of the controls and appear to be hypofunctional.

Mortality. In hens depleted of Vitamin B_{12}, intravenous administration of various doses of this vitamin weekly for 6 weeks has the following effects on total embryonic mortality (*Petersen, Wiese, Lampman, and Dahlstrom, 1950*).

Vitamin B_{12} Injected (μg/hen)	Embryonic Mortality at Hatching (%)
0.0	82.9
0.5	34.9
1.0	8.7
2.0	1.9
4.0	0.0

The level of Vitamin B_{12} administered during eight weekly injections influences the rate of increase of embryonic mortality (*Petersen, Wiese, Dahlstrom, and Lampman, 1952*). Figure 93 shows that doses of either 0.5 μg or 1.0 μg of B_{12} per hen per week result in a very high rate of mortality of the embryo. With successive increase in dose up to 5 μg/hen, the rate of mortality is gradually reduced.

The distribution of dead embryos during the entire period of incuba-

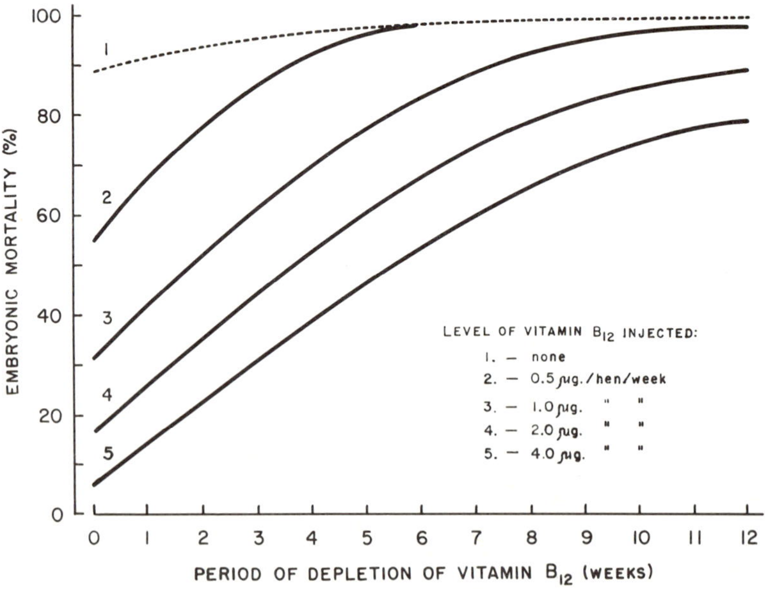

Fig. 93. Rate of increase in embryonic mortality after an 8-week period of depletion of the diet of hens, followed by supplementation with various doses of Vitamin B_{12}. (Reconstructed from the data of Petersen, Wiese, Dahlstrom, and Lampman, 1952.)

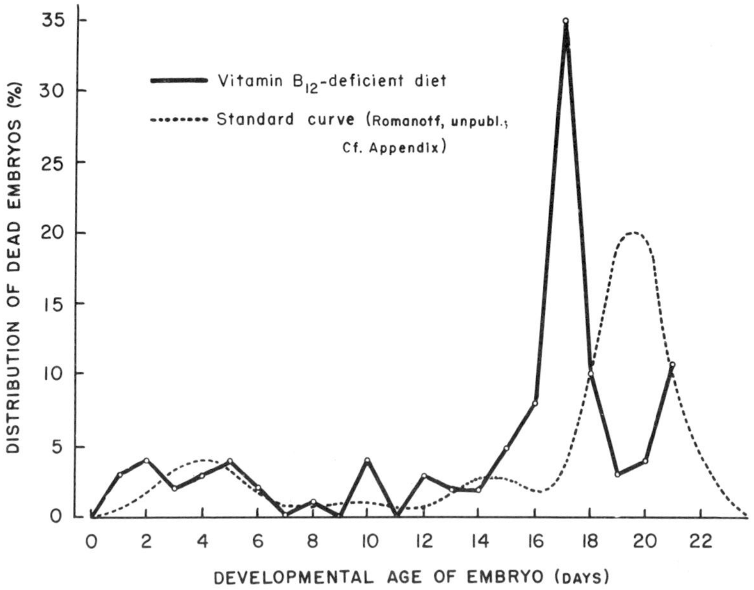

Fig. 94. Effect of Vitamin B_{12}-deficient diet on the distribution of dead embryos during incubation. (Based on the average value of two experiments by Olcese, Couch, Quisenberry, and Pearson, 1950.)

tion of eggs from hens on Vitamin B_{12}-deficient diets is shown in Fig. 94. The maximum number of embryonic deaths (about 34.5 per cent) occurs at the seventeenth day of development.

In comparison with the normal, spontaneous occurrence of embryonic mortality (see Fig. 13 in Chapter II), there is a typical peak at the third critical period, which appears early, on the seventeenth instead of the nineteenth day.

Malformations. The embryos from hens fed a Vitamin B_{12}-deficient diet show a variety of gross pathological changes. Among the prominent ones, besides decrease in size of the embryo (*Ferguson, Rigdon, and Couch, 1955*), are edema and hemorrhages; local areas of necrosis in the liver, brain, and spinal cord; and a marked increase in fat in the parenchymatous tissues.

It has been observed (*Olcese and Couch, 1950*) that embryos from hens on a Vitamin B_{12}-deficient diet frequently show such anomalies as myoatrophy of the legs (39%) and malposition of the embryo with the head between the thighs (26%).

(For further detailed analysis see Summary Table 9, under Vitamin B_{12}.)

Vitamin D

Mortality. In eggs from hens on a Vitamin D-deficient diet, there is a continued decrease in embryonic mortality with increased doses of Vitamin D (*Bethke, Record, Kick, and Kennard, 1936*), as is shown below.

Vitamin D Supplement (mg/100 grams)	Embryonic Mortality at Hatching (%)
0	49
160	28
1,670	16
16,150 (toxic dose)	37

Malformations. For specific pathological responses of the embryo to deficiency and excess of Vitamin D, see Summary Table 9, under Vitamin D.

Vitamin E

Growth. From studies of eggs of turkeys kept on a Vitamin E-deficient diet (*Ferguson, Atkinson, and Couch, 1954*) it was found that about 85 per cent of the experimental embryos are smaller than normal embryos of the same age.

Mortality. In experimental turkeys there is high embryonic mortality between the twenty-fourth and twenty-eighth days of the incubation period.

Malformations. Among malformations observed in turkey embryos (*Ferguson, Atkinson, and Couch, 1954*) due to Vitamin E deficiency in the mother's diet are the following:

Malformation	Incidence of Malformation Among Dead (%)
Cloudy lens	80
Edematous areas on the neck and feet	62
Hemorrhagic vitreous humor	44

(See also Summary Table 9, under Vitamin E.)

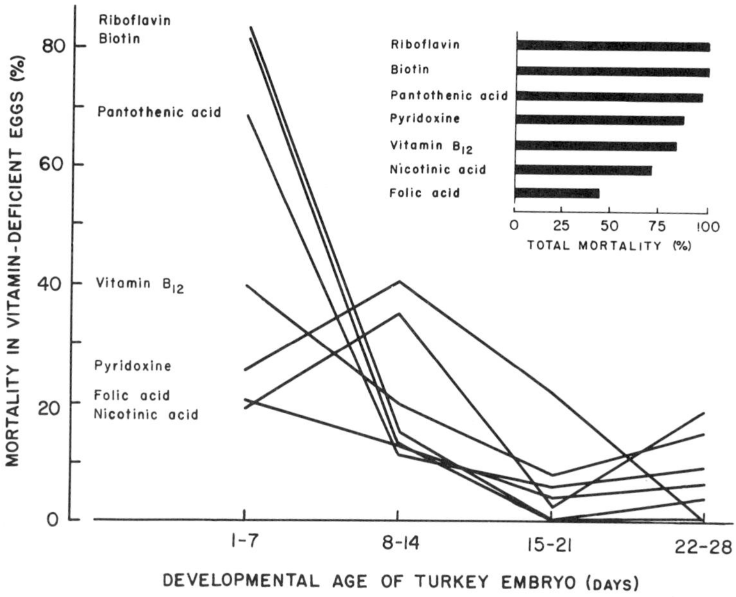

Fig. 95. Effect of vitamin-deficient diets on the distribution of deaths and the total mortality of turkey embryos. (Reconstructed from the data of Ferguson, Whiteside, Creger, Jones, Atkinson, and Couch, 1961.)

Comparative Effects of Vitamins

Mortality. The effect of nutritional deficiency of individual vitamins in the diet of the turkey hen on the distribution of embryonic mortality is shown in Fig. 95.

High embryonic mortality during the first week of incubation occurs in groups deficient in riboflavin, biotin, pantothenic acid, Vitamin B_{12}, and folic acid. High mortality in the second week occurs with deficiencies of pyridoxine and nicotinic acid.

The highest total embryonic mortality (see *insert* of Fig. 95) is shown in groups deficient in riboflavin, biotin, and pantothenic acid, and the lowest with folic acid deficiency.

MINERALS

Calcium

Mortality. The absence of calcium from the diet of the hen is accompanied by a marked decrease in the weight of the eggshell as well as by an increase in embryonic mortality (*Buckner, Martin, and Peter, 1929*). This is illustrated by the experimental data below.

Sources of Calcium in Hen's Diet	At 18 Days of Incubation: Weight of Shell (grams)	At 18 Days of Incubation: Embryonic Mortality (%)
Outdoors, with oyster shell	6.3	47
Confined, with oyster shell	5.6	53
Confined, without oyster shell	3.9	63

Normally, approximately 75 per cent of the calcium in a well-formed embryo comes from the shell during the process of incubation (*Romanoff, 1967, p. 255*).

Malformations. An excess of calcium in the hen's diet produces a high incidence of embryonic abnormalities at hatching (*Kienholz, Turk, Sunde, and Hoekstra, 1961*), as is shown below.

Level of Calcium (% of diet)	Incidence of Abnormal Embryos (in % of fertile eggs)
2 (normal)	8.4
4 (excessive)	16.5

Iodine

The iodine content of the hens' diet determines the amount of this element in their eggs (*Wilder, Bethke, and Record, 1933*), as is shown below.

Iodine (KI) in Hen's Diet (mg)	Iodine in Eggs (μg/egg)
none	0.25
2	8.80
5	16.87

About 95 per cent of the iodine in the egg is found in the egg yolk; the remainder is in the albumen.

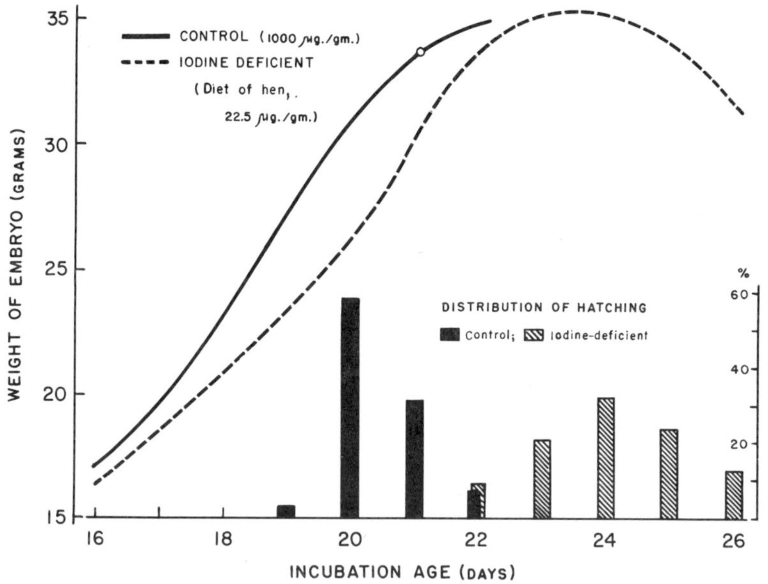

Fig. 96. Effect of dietary iodine deficiency (22.5 μg/gram) in the hen on the body weight of the embryo and the distribution of hatching. (Composed after the data of Rogler, Parker, Andrews, and Carrick, 1959; and Romanoff, unpublished.) See Fig. 10 in Chapter II, on the time and distribution of hatching in various species of birds.

Growth of Embryo. Iodine deficiency in the hen's diet adversely affects the growth of the embryo. The cumulative weight of the embryo at both the deficient and control levels is shown in Fig. 96. With the retardation in growth due to iodine deficiency, there is considerable delay in the development of the embryo, extending to about 6 days over the normal 21 days in chickens.

The distribution of time of hatching is shown in the *insert* of Fig. 96. The embryos delayed in hatching are smaller than those hatched earlier.

Weight of Thyroid. With an iodine-deficient diet, there is a continued increase in the size of the embryonic thyroid during development (Fig. 97A).

When excessive amounts of inorganic iodine are added to the egg on the sixteenth day of incubation, there is a rapid increase in the size of the thyroid—over 300 per cent (*Wheeler and Hoffman, 1949*)—as evidenced by the accompanying table.

Inorganic Iodine (KI) Injected into Egg (μg/egg)	Weight of Embryonic Thyroid at Hatching (mg)
39 (control)	4.3
156 (excessive)	5.8
625 (excessive)	7.3
2,500 (excessive)	8.9
10,000 (excessive)	9.6
40,000 (excessive)	10.0

Concurrently with the increase in thyroid weight, there is a direct relationship between the amount of iodine in the diet of the hen and the iodine content of the thyroid of the embryo at hatching (*Rogler, Parker, Andrews, and Carrick, 1961*). The data are presented in Fig. 97*B* and in the tabulation below.

Dietary Iodine (μg/gram)	Iodine in Embryonic Thyroid (mg)
14.5 (deficient)	0.6
50 (control)	8.4
100 (excessive)	30.0
300 (excessive)	38.0
1000 (excessive)	40.0

Mortality. The effect on embryonic mortality of various levels of iodine in the diet of the hen is shown in Fig. 98.

The incidence of total mortality under both deficiency and excess of iodine in the hen's diet is shown in the accompanying table.

Dietary Iodine (μg/gram)	Embryonic Mortality (%)	Reference
Deficiency (0.95)	60–85	Rogler et al. (1959)
Normal (44)	15–22	
Excess (up to 5000)	75–100	Arrington et al. (1967)

Embryonic mortality is apparently greater at the higher levels of dietary iodine than at the lower.

Malformations. Dosages of inorganic iodine from 0.6 to 39 μg, injected on the sixteenth day into iodine-deficient eggs, fail to produce goiter at hatching (*Wheeler and Hoffman, 1949*). Excessive dosages of inorganic

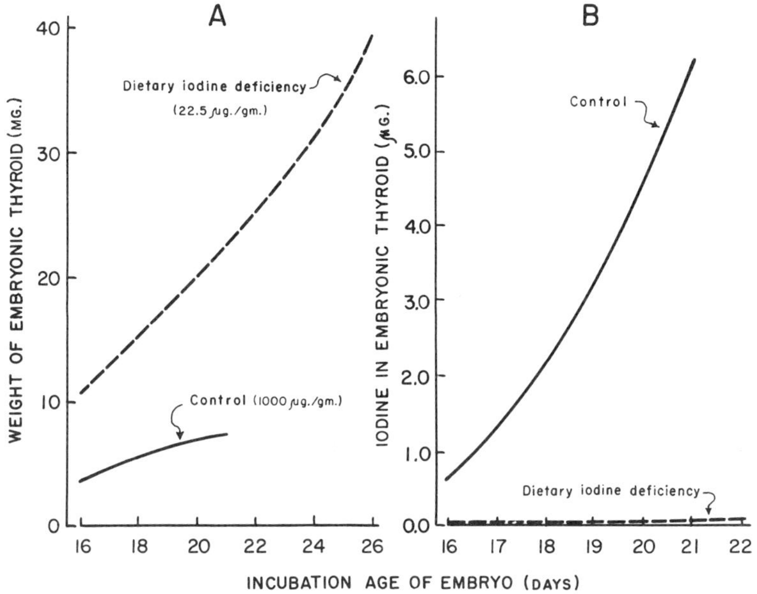

Fig. 97. Effect of dietary deficiency of iodine on the development of the embryonic thyroid. (Rearranged from the data of Rogler, Parker, Andrews, and Carrick, 1959; and Romanoff, 1967, p. 105.) *A*, effect on the weight of the thyroid; *B*, effect on the iodine content of the thyroid.

iodine, from 156 to 40,000 μg per egg, result in progressively more significant thyroid enlargement.

For other malformations see Summary Table 9, under Iodine. See also Summary Table 10, for effect of excesses, under Iodine (potassium iodide).

Magnesium

Mortality. The effect of magnesium deficiency in the diet of a hen is evidenced by a rapid increase in embryonic mortality (*Sell, Hajj, Cox, and Guenter, 1967*), as is shown in Fig. 99.

Malformations. Embryos from eggs laid by hens on magnesium-deficient diets show nervous tremor, gasping, and convulsions at hatching (*Gardiner, Rogler, and Parker, 1960*).

Manganese

(For malformations, see Summary Table 9, under Manganese.)

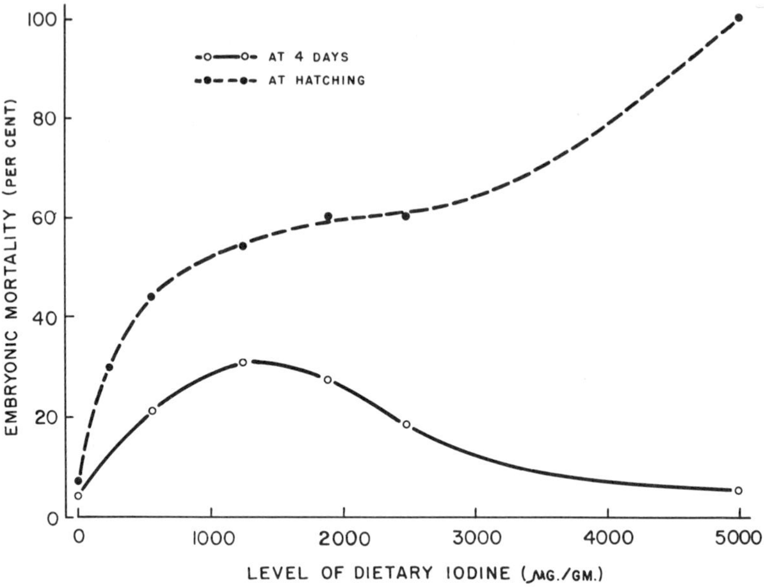

Fig. 98. Effect of the level of iodine in the diet of the hen on incidence of embryonic mortality at 4 days of development and at the time of hatching. (Based on the data of Perdomo, Harms, and Arrington, 1966; Arrington, Cruz, Harms, and Wilson, 1967.)

Phosphorus

Mortality. The effect of phosphorus deficiency, as well as excess, in the diet on embryonic mortality at hatching (*O'Rourke, Bird, Phillips, and Cravens, 1954*), is shown below.

Phosphorus in Diet (%)	Embryonic Mortality (%)
Deficiency (0.19)	90
Normal (0.30)	20
Excess (0.60)	20

Selenium

Mortality. Embryonic mortality increases when grains containing selenium (15 μg/gram) are fed to egg-laying hens (*Poley, Moxon, and Franke, 1937*).

Duration of Feeding Toxic Grains (days)	Embryonic Mortality at Hatching (%)
9	9.8
5	21.4
12	81.9
19	100.0

After restoring a diet containing no selenium, the apparent effects of selenium on embryonic mortality disappear.

Malformations. Malformed embryos are found in eggs laid by hens which have consumed selenium-containing foods (*Poley, Moxon, and Franke, 1937*). Similar abnormalities are produced by the injection of a selenium compound into eggs (*Franke, Moxon, Poley, and Tully, 1936*).

Figure 100 gives the incidence of various types of abnormalities which appear as a result of feeding toxic grains containing about 15 μg/gram of selenium. The curves show that edematous condition of

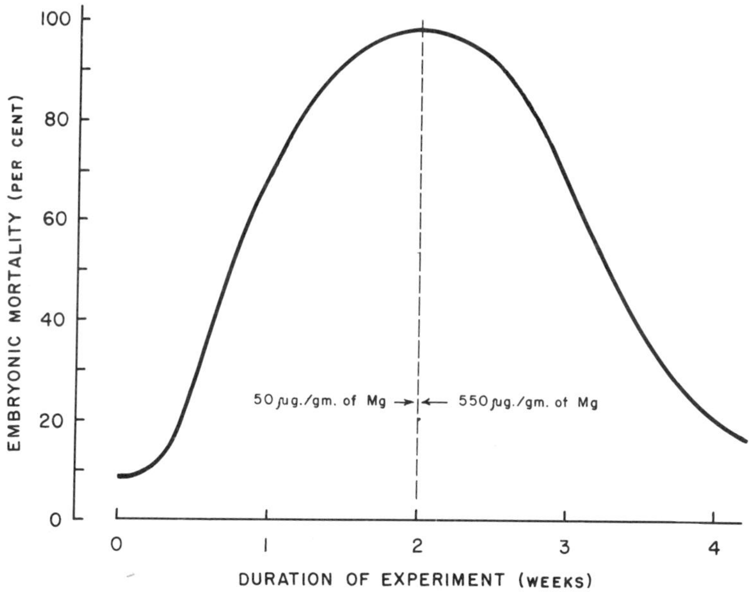

Fig. 99. Effect of dietary magnesium (Mg) deficiency on embryonic mortality. (Reconstructed from the data of Sell, Hajj, Cox, and Guenter, 1967).

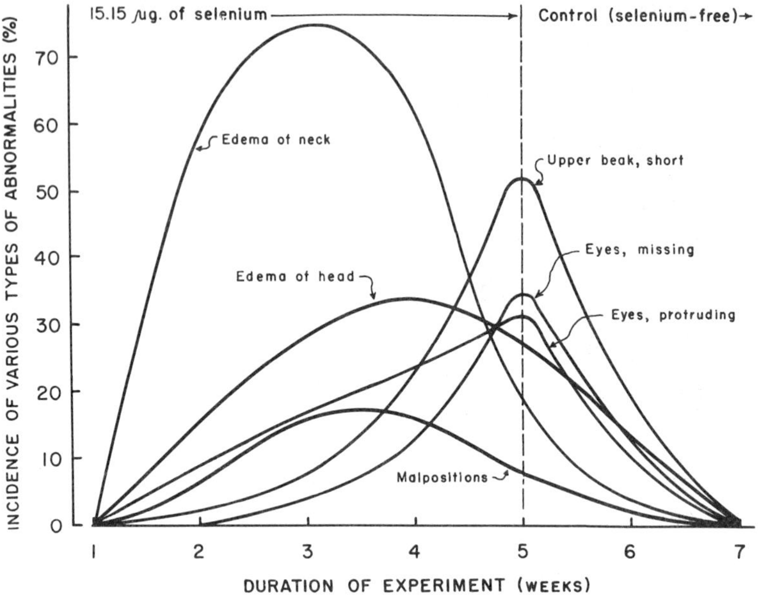

Fig. 100. The incidence of various types of abnormalities of the embryo as a result of feeding the hen toxic grains containing selenium, 15.15 μg/gram of diet. (Plotted with modifications from the data of Poley, Moxon, and Franke, 1937.)

the neck and head and malposition of the embryo begin to decrease before the removal of selenium from the diet of the hen. It has been presumed (*Poley, Moxon, and Franke, 1937*) that this decrease in the effect of selenium results from acquisition of resistance by the hen's body to the selenium grain.

(See also other malformations in Summary Table 11, under Selenium.)

Zinc

Mortality. The rate of embryonic mortality is influenced by the level of zinc in the hen's diet. When the hen is placed on a moderately zinc-deficient diet (*Kienholz, Turk, Sunde, and Hoekstra, 1961*), within 6 weeks the hatchability (inversely related to mortality) drops to a low level, about 12 per cent (Fig. 101). When supplementary zinc is added to the diet, hatchability returns to the control level within two weeks.

Under almost identical experimental conditions (*Blamberg, Blackwood, Supplee, and Combs, 1960*), embryonic mortality parallels that

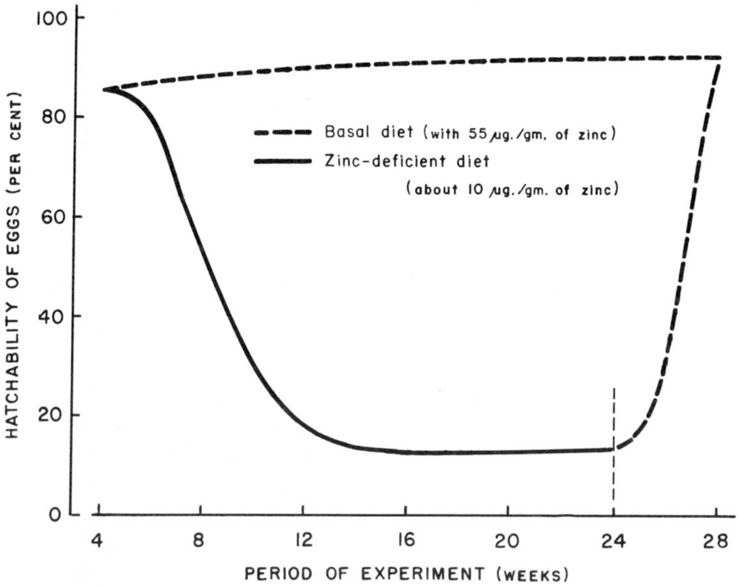

Fig. 101. Effect of a dietary deficiency of zinc on the hatchability of eggs artificially inseminated at weekly intervals. (Plotted from the data of Kienholz, Turk, Sunde, and Hoekstra, 1961.)

shown in Fig. 101 when a change is made from a basal diet to a zinc-deficient diet (Fig. 102).

Malformations. The effect of feeding hens a zinc-deficient diet on the incidence of malformations, with characteristic trunk and limb abnormalities, is shown in Fig. 102 with *insert.*

Zinc-deficient chicks are weak and will not stand, eat, or drink. They have an accelerated respiratory rate and difficult breathing (*Kienholz, Turk, Sunde, and Hoekstra, 1961*).

For a more detailed list of anomalies, see also Summary Table 9, under Zinc.

PROTEINS

Mortality. It has been well established (*Petersen, Wiese, and Lampman, 1948*) that hens fed diets containing proteins only from vegetable sources lay eggs with a high incidence of embryonic mortality. The results of

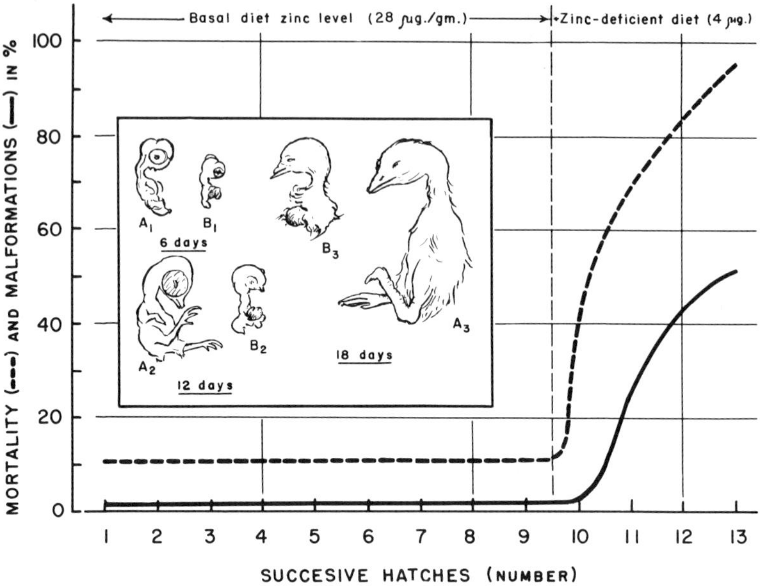

Fig. 102. Effect of a zinc-deficient diet (4 μg/gram) on mortality and malformations of the chicken embryo. (Reconstructed from the data and sketched from the photographs of Blamberg, Blackwood, Supplee, and Combs, 1960.) *Insert: A*'s are controls and *B*'s are malformed embryos at 6, 12, and 18 days of incubation. (Among deformities of the embryos dead at 6 days were observed: ectrosomia, ectromelia, symmelia, lordosis, and microphthalmia; among the embryos dead at 12 days: celosomia, anouria, microphthalmia, ectromelia, ectrodactyly, micromelia, and lordosis; among the embryos dead at 18 days: celosomia, microphthalmia, ectromelia, ectrodactyly, micromelia, deformities of beak, and edema of skin.)

feeding proteins from different sources during a 16-week period are shown below.

Duration of Feeding (weeks)	Embryonic Mortality (in %)	
	With Meat Meal	With Soybean Meal
0	5	5
4	10	60
8	14	80
12	17	90
16	19	100

The distribution of mortality during the period of incubation indicates that with diets containing proteins from vegetable sources, hens produce embryos many of which die during the second week (about the eleventh day), which is often a symptom of nutritional deficiency (*Byerly, Titus, and Ellis, 1933*).

Malformations. Among the anomalies attributed to feeding only proteins of vegetable origin is so-called chondrodystrophy, a collective term for such symptoms as shortened and recurved legs, parrotlike beak, and sparse down (described by *Dunn, 1927a,* and *Landauer, 1927,* and occurring as a genetic factor).

The incidence of chondrodystrophy due to diets containing proteins from different sources (*Byerly, Titus, and Ellis, 1933*) is shown below.

Source of Protein	Chondrodystrophic Embryos (% of fertile eggs)
Animal origin	
Meat meal	0.56
Dried buttermilk	0.34
Vegetable origin	
Soybean meal	5.03
Cottonseed meal	5.25

A deficiency of proteins in the diet of a hen is due chiefly to the absence of specific amino acids; this lack alone may induce various abnormalities in the embryo.

Amino Acids

Growth. Extensive investigations using heart fibroblasts from 8-day-old chick embryos (*Kieler, 1954*) have shown that various amino acids have a definite stimulating effect on cell division *in vitro.*

The mere presence of the necessary amino acids is not sufficient to secure normal mitosis; their relative concentrations are equally important for normal development.

Observations *in vitro* (*Biggers, Webb, Parker, and George, 1957*) indicate that the absence of arginine, threonine, and tryptophan causes retardation in growth of the tibia and femur in 6- to 7-day old embryos.

Malformations. The teratogenic activity of some amino acids, injected in doses of 1 μg/egg into the yolk sac of the chicken embryo before incubation (*Aleksandrov, Bogdanova, and Chernukh, 1966*), is shown quantitatively in the accompanying table.

Amino Acid	Incidence of Malformations (% of all embryos)
Cystine	10
Phenylalanine	12
Methionine	12
Cysteine	15
Tryptophan	17
Glycine	20
Glutamic acid	25
(Control)	3.6

The largest numbers of defective embryos were caused by injections of tryptophan, glycine, and glutamic acid.

Gross anatomical malformations were as follows: in the brain, disorganized protrusions in various regions; in the beak, reduction in size and maldevelopment of upper or lower parts; often, unclosed or protruding viscera (celosomia); twisted and shortened limbs; twisted spine (lordosis) and shortened embryonic body.

Except for their quantitative variation, these malformations are common for all amino acids, and appear at identical stages of embryonic development.

It is possible to cause degeneration of the eye by tyrosine deficiency in the egg (*Grau, Austic, and Matteson, 1965*). Subjecting a 4-day-old chicken embryo to a yolk sac medium lacking tyrosine resulted in arrest of retinal pigmentation and degeneration of the neural retina, as is shown below.

Tyrosine in the Egg (mg/liter)	Pigment Score of the Eye (arbitrary numbers)
0	0.5
40	1.6
160	2.7

(See also the effect of excess of amino acids in Summary Table 12, under L-Arginine, Ethionine, and Leucine.)

CARBOHYDRATES

Hexoses

The linear growth of tibia from the 8-day-old embryo, cultivated for 5 days in a medium with 2 per cent consisting of various hexoses

(*Kieny, 1958*) is shown in the tabulation below.

Hexose	Relative Growth of Tibia (in % of control)
Glucose	46
Mannose	46
Fructose	5
Galactose	4
Sorbose	3

The rate of growth of tibia is influenced by the concentration of glucose in the medium (*Kieny, 1958*), as is shown below.

Dose of Glucose (% in medium)	Rate of Growth of Tibia (%)
0.0	3
0.2	32
0.5	47
2.0	49

FATS

Linoleic Acid

Growth. Some embryos from eggs of hens on a linoleic acid-deficient diet have a considerable delay in development (*Calvert, 1967*).

Mortality. The effect of a dietary deficiency of linoleic acid on embryonic mortality (*Menge, Calvert, and Denton, 1965b*) is shown below.

Linoleic Acid in Diet (mg/day)	Embryonic Mortality (%)	
	Early	Total
none (low fat)	39	100
20	31	95
80	15	80
1600	14	43

The major peaks of embryonic mortality occur between 0 and 4 days and 20 and 22 days (Fig. 103). A small peak of mortality is also noted at the fourteenth day of incubation.

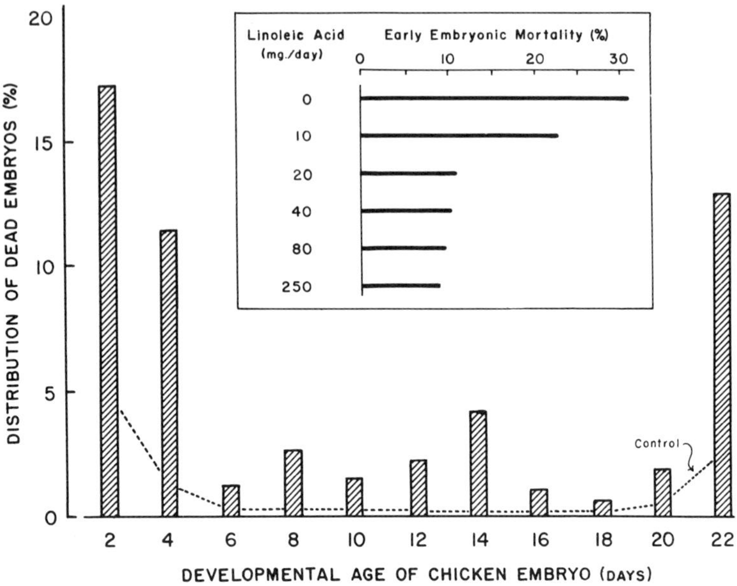

Fig. 103. Effect of dietary deficiency of linoleic acid on the distribution of embryonic mortality. (Plotted from the data of Calvert, 1967.) *Insert:* effect of different levels of dietary linoleic acid on early embryonic mortality. (After Menge, Calvert, and Denton, 1965a.)

The *insert* of Fig. 103 shows the effect on early embryonic mortality of various doses of linoleic acid in the diet of the hen (*Menge, Calvert, and Denton, 1965a*).

Malformations. At the time of hatching, about 75 per cent of embryos from linoleic acid-deficient hens are found in an abnormal position (head over the right wing instead of under the wing), as compared with only 8 per cent of embryos from control hens. (See Fig. 15 on malpositions, in Chapter II.)

Summary Table 9 Pathological Effect of Vitamin and Mineral Deficiencies on the Developing Avian Embryo

Age (days) Treatment	Age (days) Examination	Malformations (%)	Mortality (%)	Incidence of Typical Syndromes	Reference
				VITAMINS	
Biotin					
Hen's diet			25–95	(Doses: from 0 to 150 μg/kg hen's diet) Hemorrhage—body and allantois (18%); retardation in growth (14%); chondrodystrophy (12%); crooked tibia (12%); short twisted tarsus (12%); syndactyly (8%); down short and clubbed (5%); edema (3%). (See Fig. 85 on distribution of mortality.)	Cravens, McGibbon, and Sebesta (1944)
Hen's diet			82	(Doses: from 0 to 0.2 mg/kg hen's diet) Parrot beak (17%), syndactyly (16%), crooked tibia (13%), short twisted tarsometatarsus (13%).	Couch, Cravens, Elvehjem, and Halpin (1948a)
Hen's diet				(Doses: 0.2 mg/kg hen's diet) Congenital perosis, ataxia, characteristic skeletal deformities. (See Fig. 86 on perosis.)	Couch, Cravens, Elvehjem, and Halpin (1948b)
2–5	21		12–100	(Doses: from 2 to 25 μg/egg from deficient diet) Parrot beak, shortened and twisted tibiotarsus, very short tarsometatarsus.	Couch, Cravens, Elvehjem, and Halpin (1949)
Folic Acid					
4	12–14		0–100	(Doses: from 0.003 to 20 mg/egg, in yolk sac) Profound inhibition of growth (0.8 to 1.2/5 to 7 grams at 12 days), flattened head, small eyes, absent or undeveloped lower beak, long and twisted neck, exteriorized viscera, abnormal or absent toes, generalized edema.	Karnofsky, Patterson, and Ridgway (1949)
0	21		32–90	(Doses: from 0 to 22 μg/egg, in albumen) Syndactyly, deformed mandibles, and parrot beak; marked bending of tibiotarsus.	Sunde, Cravens, Elvehjem, and Halpin (1950)

Summary Table 9 (*Continued*)

Age (days) Treat-ment	Age (days) Exami-nation	Malfor-mations (%)	Mor-tality (%)	Incidence of Typical Syndromes	Reference
Nicotinic acid				(Doses: 20 mg/egg, in albumen)	Hansborough (1947b)
0	2–4	49–55	34–41	Abnormalities in the nervous and vascular systems, including the heart.	
Pantothenic acid				(Pantothenic acid-deficient turkey eggs)	Dawson, Ferguson, Deyoe, and Couch (1962)
	17, 21			Edema, hemorrhage, fatty liver, a pale dilated heart, a short lower beak, hydrocephalus, twisted legs, opacities of the eye.	
				(Pantothenic acid-deficient eggs)	Beer, Scott, and Nesheim (1963)
			60–100	Subcutaneous hemorrhage (19 to 27%), edema (15%).	
Riboflavin				(Riboflavin-deficient eggs)	Lepkovsky, Taylor, Jukes, and Almquist (1938)
Hen's diet			67–86	Dwarf-type embryos, increased incidence of edema, degeneration of embryonic kidney (mesonephros), defective down.	
				(Riboflavin-deficient eggs)	Engel, Phillips, and Halpin (1940)
Hen's diet	15–21	97		Gross anomalies: abnormally short legs, long curved beak. Histological examination: degenerative changes in the myelin sheath of the sciatic nerve (60%), kidney lesions, light color of liver.	
				(Doses: various degrees of deficiency in hen's diet)	Romanoff and Bauernfeind (1942)
Hen's diet			63–100	Retardation in growth (40 to 60%; at 16 days 9/17 grams), micromelia, shortened legs, prognathism, de-	

				colorized liver, edemic condition of the neck, undeveloped down.	
			55	(Riboflavin-deficient chicken eggs) Clubbed down (12.4%), edema (4.9%), parrot beak (4.4%), chondrodystrophy (1.8%) of fertile eggs.	McClymont and Hart (1947)
Thiamine					
0	2–5			(Doses: from 0.8 to 1.0 mg/egg) Accelerated differentiation and growth.	Salazar (1950)
Vitamin A—Retinoic acid					
0				(Doses: 1 μg/egg) Lack of development; disintegration of the entire early blastoderm.	Thompson, Howell, Pitt, and Houghton (1965)
Vitamin A—Retinol					
1	2–3			(Doses: from 22.5 to 450 μg/egg, in yolk sac) Delayed development; structural abnormalities: pronounced defect of the nervous and vascular systems.	Hansborough (1947a)
5–6	13–17			(Doses: from 541 to 1,680 mg/100 ml plasma, *in vitro*) Hypervitaminosis: femur, tibia, humerus, ulna, and radius are smaller, shorter, and thinner than control (with 541 μg); constriction at both ends of all bones.	Fell and Mellanby (1952)
0				(Doses: 0.5 mg/egg) Lack of development; disintegration of the entire early blastoderm.	Thompson, Howell, Pitt, and Houghton (1965)
Vitamin B_{12}					
		36	75	(Doses: vitamin-deficient diet) Myoatrophy of the leg (35%), malposition of the embryo (head between thighs, 38%), hemorrhages in body (28%). (See Fig. 94 on distribution of mortality.)	Olcese, Couch, Quisenberry, and Pearson (1950)
	17			(Doses: vitamin-deficient diet) Enlargement of left thyroid (49%), abnormal heart (19%), fatty liver (14%), and other defects observed previously.	Ferguson and Couch (1954)

Summary Table 9 (*Continued*)

Age (days)		Malfor-mations (%)	Mor-tality (%)	Incidence of Typical Syndromes	Reference
Treat-ment	Exami-nation				
	17			(Doses: vitamin-deficient diet) Decrease in size of embryo; edema (82%); fatty liver (61%); hemorrhages: in body (75%), in heart (33%), in kidney (18%), in brain (12%); local areas of necrosis in the liver, brain, and spinal cord; enlarged pericardial sac (7%).	Ferguson, Rigdon, and Couch (1955)
Vitamin D					
	21		37	(Doses: 16,150 mg/100 gram of ration—excessive dose) Microscopic examination: small areas of calcification in the kidneys.	Bethke, Record, Kick, and Kennard (1936)
1	2–3			(Doses: from 1.5 to 30 mg/embryo in culture) Delayed development, maldevelopment of heart, reduction in brain size, incomplete closure of the neural tube.	Hansborough (1947a)
	13–15			(Doses: Vitamin D-deficient diet) Malformed embryos with anemia, bent or broken legs, general edema; symptomatic occurrence of rickets in embryos.	Karb (1969)
Vitamin E					
	1–18	rare	100 at 4 days	(Vitamin E-deficient eggs) Retardation in development, failure of vascular system, tendency to hemorrhagic condition.	Adamstone (1931)
	7–24		58	(Vitamin E-deficient turkey eggs) Cloudiness in the lens (80%), edematous areas on	Ferguson, Atkinson, and Couch (1954)

			the neck and feet (62%), hemorrhaged vitreous humor (44%).	
		52	(Vitamin E-deficient turkey eggs) Smaller size, apparent blindness: a cloudy lens or a cloudy spot under the cornea.	Atkinson, Ferguson, Quisenberry, and Couch (1955)
Vitamin E	15–29		(Vitamin E-deficient turkey eggs) Keratoconus-like change; bilateral cataracts at hatching.	Ferguson, Rigdon, and Couch (1956)
		46–48	(Vitamin E-deficient turkey eggs) Cataracts, extensive liquefaction of lens protein and focal areas of degeneration in the cornea; opacities (39%); keratoconus (17%).	Ferguson, Swanson, Couch, Feldman, and Rigdon (1960)
			MINERALS	
Iodine (KI)	12–21	15	(Doses: from deficient diet (14.5) to 1000 μg/gram) Retardation of growth (Fig. 96), delay in hatching (5 days), increase in size of thyroid, which inhibits hypertrophy of epithelial cells with a block of colloid follicles.	Rogler, Parker, Andrews, and Carrick (1959)
Hen's diet	21	56	(Doses: in diet, 14.5 (deficient) to 500 μg/gram) At the lowest dose: delay in development from 2 to 2.5 days, increase in thyroid size.	Rogler, Parker, Andrews, and Carrick (1961)
Manganese		95	(Manganese-deficient diet) Chondrodystrophy: parrot beak (96%) and short legs (52%); edema (76%).	Lyons and Insko (1937a; (1937b)
			(Manganese-deficient diet) Micromelia	Caskey and Norris (1939)
			(Manganese-deficient diet) Ataxia (of newly hatched chick).	Caskey, Norris, and Heuser (1944)

Summary Table 9 (***Continued***)

Age (days) Treat-ment	Age (days) Exami-nation	Malfor-mations (%)	Mor-tality (%)	Incidence of Typical Syndromes	Reference
Zinc					
		64	88	(Zinc-deficient diet, about 4 μg) Defective trunk and limb development. Trunk: anouria, ectrosomy, and lordosis. Limbs: ectromelia, ectrodactyly, and symmelia. Abnormal brains and visceral arches and small eyes (microphthalmia); exposed viscera (celosomia); abnormalities of beak and external head structures; sometimes extreme edema.	Blamberg, Blackwood, Supplee, and Combs (1960)
				(Zinc-deficient diet) Presence of phocomelia, micromelia, lordosis, shortened abnormal thorax and pelvic skeleton, edematous hatching muscle, absence of any skeleton posterior to the cervical vertebra; undeveloped eye.	Kienholz, Turk, Sunde, and Hoekstra (1961)

VII

Chemical Teratogenecity

Most chemicals and drugs possess
Enormous power to destroy,
Put living tissues in distress . . .
And let morbidity deploy.

Each chemical compound, in certain doses, could be viewed as an essential food substance, or as a drug useful in treatment, or as a teratogen causing diseases, malformations, and prenatal death. There is wide variation in the duration of sensitivity to a particular substance and in the peaks of teratological and lethal response.

The anomalies induced in the chicken embryo by a chemical compound are often similar to those caused by other, unrelated, teratogens. Yet the general characteristics of the sensitive period vary greatly with each compound used. Some compounds cause a variety of developmental effects which are related to the stage of development at the time of treatment.

INORGANIC COMPOUNDS

Boric Acid

Growth. Increase in body size is considerably retarded by injection of 0.05 cm^3 of 5 per cent boric acid at 4 days of development (*Landauer, 1952*). The mean weight of treated embryos at 18 days of incubation is about 15 grams, whereas the weight of controls is about 21 grams. The treated embryos weigh only about 70 per cent as much as the controls.

The length of the tarsometatarsus at the age of 13 days varies with the dosage of boric acid (*Lane, Darlington, and Coursin, 1954*), as is shown below.

Dosage of Boric Acid (ml/egg)	Length of Tarsometatarsus (mm)
0.05	17.5
0.10	13.9

Mortality. The toxic effect of boric acid and the increase in embryonic mortality on the thirteenth day with increased dosages (*Lane, Darlington, and Coursin, 1954*) are shown in the tabulation below.

Dosage of Boric Acid (ml/egg at 4 days)	Incidence of Mortality (%)
0.0	16.6
0.05	35.8
0.075	64.6
0.1	89.5

Malformations. It has been demonstrated (*Landauer, 1952*) that the maximum incidence of the various malformations results from injection of boric acid at 4 days of incubation. The effects also depend upon the dose level employed, as is shown below.

	Incidence at Various Doses (%)	
Malformation	At 0.05 cm^3	At 0.1 cm^3
Beak defects (all types)	69.9	85.3
Abnormal toes	14.7	35.2
Tarsometatarsus		
Bent	11.9	26.5
Shortened	5.6	13.2

The relationship between dose of boric acid and total incidence of malformations of the facial and appendicular skeleton is shown in the *insert* of Fig. 104.

The incidence of various malformations among embryos treated with 2.5 mg of boric acid between 0 and 5 days of incubation is demonstrated by the data (*Landauer, 1954*) plotted in Fig. 104. The peak incidence of anouria (rumplessness) occurs after injection at 1 day; of beak defects and long-bone defects, after injection at 4 days.

(See, for a further detailed list of malformations, Summary Table 11, under Boric acid.)

Lithium Salts

Mortality. Injections of identical doses of lithium chloride, lithium sulfate, and lithium bromide (2 cm^3 of $\frac{1}{10}$ *M*) into eggs at 6 days of incubation (*Landauer, 1929*) resulted in different rates of embryonic mortality (Fig. 105*A*). Lithium chloride caused the highest initial mor-

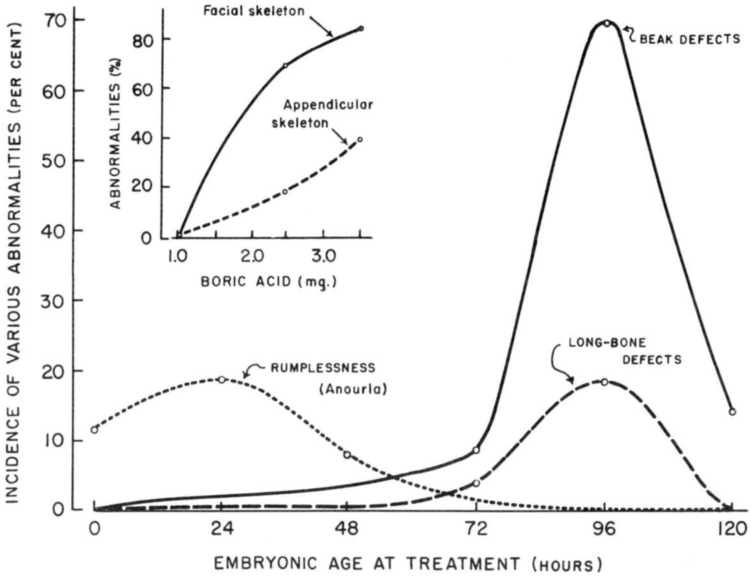

Fig. 104. Incidence of various abnormalities among chicken embryos treated with 2.5 mg of boric acid between 0 and 120 hours of incubation. (After Landauer, 1954.) *Insert:* effect of various dosages of boric acid on incidence of facial and appendicular skeleton abnormalities among survivors on the thirteenth day of incubation. All embryos were treated at the age of 96 hours.

tality, and lithium bromide the lowest mortality. However, with all lithium salts, the survival of embryos at hatching is very low.

Injections of different doses of lithium chloride produce considerably different rates of embryonic mortality, as is shown in Fig. 105*B*.

Malformations. The effect of lithium chloride on malformations of the chicken embryo is shown in Summary Table 11, under Lithium chloride. The malformations that occur, such as those of the neural tube, inhibition in development, defects of the eye, and enlargement of the aorta (*Noto, 1967a*), have been explained as resulting from mechanical inhibition of morphogenetic movements, caused specifically by lithium chloride.

Sodium Molybdate

Mortality. When molybdenum, as a solution of the sodium molybdate salt, is injected into the yolk sac in different amounts and at various stages of embryonic development (*Lepore, Miller, and Siegel, 1962*), its effect is less lethal as embryonic age advances beyond 8 days (Fig. 106).

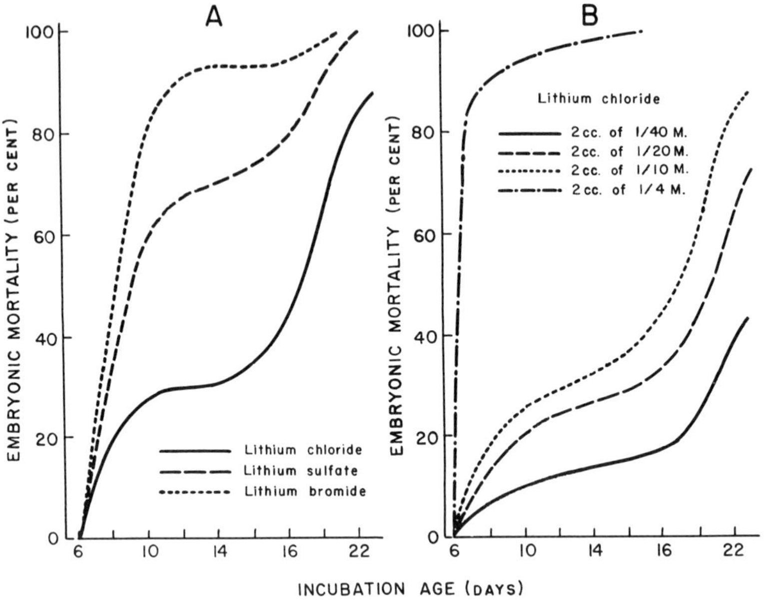

Fig. 105. Effect of lithium salts on mortality of chicken embryos as a function of incubation time. (Recalculated from the data of Landauer, 1929.) *A*, relative effect of lithium chloride, lithium sulfate, and lithium bromide (with 2 cm³ of 1/10 *M*); *B*, effect of various doses of lithium chloride.

Thallium Salts

Growth. Delayed development has been observed after injection of thallium acetate (*Shipkowitz and Clarke, 1955*), thallium nitrate (*Cervenansky, Kalman, Micek, Maar, Makovsky, and Skrovina, 1964*), and thallium sulfate (*Karnofsky and Lacon, 1964*).

Mortality. The incidence of embryonic mortality increases with increase in the amount of thallium sulfate injected into 4-day-old embryos (*Karnofsky, Ridgway, and Patterson, 1950*), as is shown below.

Dose of Thallium Sulfate (mg/egg)	Incidence of Mortality (%)
0.2	20
0.5	27
1.0	46
2.0	88

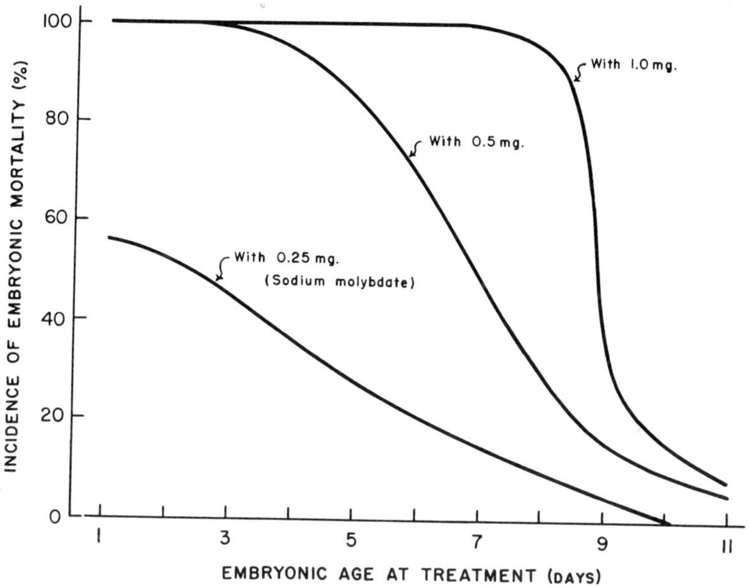

Fig. 106. Incidence of embryonic mortality as influenced by dose levels and by age at injection of molybdenum (sodium molybdate). (Summarized from the data of Lepore, Miller, and Siegel, 1962.)

Malformations. The typical symptoms following treatment with thallium salts are as follows.

	Incidence (%)		
Malformation	Thallium Acetate	Thallium Nitrate	Thallium Sulfate
Beak	88	83	—
Legs	80	90	100
Tail	65	—	

For a detailed analysis of teratogenic effects of thallium salts, see Summary Table 11, under Thallium acetate, Thallium nitrate, and Thallium sulfate.

It has been demonstrated (*Karnofsky, Ridgway, and Patterson, 1950*) that thallium sulfate can produce, in a highly consistent manner, a severe achondroplasia (Fig. 107).

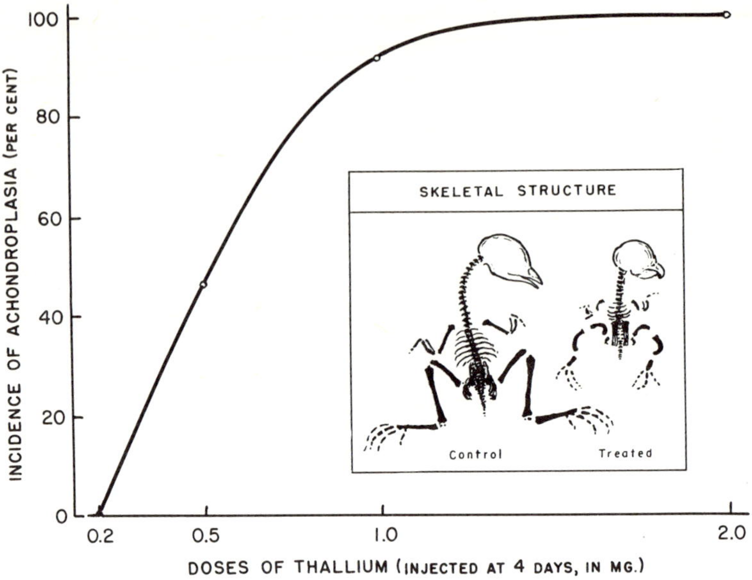

Fig. 107. Effect of injection of various doses of thallium sulfate at the fourth day on the incidence of achondroplasia in the chicken embryo. (Plotted from the data of Karnofsky, Ridgway, and Patterson, 1950.) *Insert:* skeletal structure of control and thallium-treated (with 1 mg/egg) 21-day-old embryos. (Redrawn from the photograph of the above authors.)

ORGANIC COMPOUNDS (CLASSIFIED)

Antibiotics

It has been demonstrated that the toxic effect of different antibiotics varies greatly (*Gentry, 1958; Chernukh and Aleksandrov, 1965*). The amounts of individual soluble antibiotics required to cause 50 per cent mortality (ELD_{50} = end point of lethal dose at 50%), after inoculation by different routes on the tenth day (*Gentry, 1958*), are shown in the accompanying table.

On the other hand, insoluble antibiotics, namely, furazolidone and micronized furazolidone, give almost identical results except that the air cell is the least effective route and the yolk sac the most effective.

(See also Summary Table 12, under Oxytetracycline, Penicillin, and Tetracycline.)

Antibiotic	Amount Required for ELD_{50} (mg) Air Cell	Allantoic Sac	Yolk Sac
Tetracycline	0.96	3.2	5.2
Chlortetracycline	1.5	12.0	6.5
Oxytetracycline	2.5	10.9	8.8
Neomycin	8.7	20.7	45.5
Bacitracin	9.9	30.0	50.0
Penicillin	15.6	88.4	85.8
Dihydrostreptomycin	46.8	80.6	85.8

Dactinomycin. Injection of various doses of dactinomycin into the yolk sac of eggs incubated 48 hours resulted in the following incidence of *mortality* in the 5-day-old embryos (*Pierro, 1961*).

Dose of Dactinomycin (μg)	Incidence of Mortality (%)
0.0625	23
0.125	72
0.25	80

The incidence of embryonic mortality decreases with increase in the age of the embryo at the time of injection.

Dactinomycin has a pronounced effect on the incidence of some *malformations* of the chicken embryo. The effect of various dosages of the drug, injected at the age of 48 hours, is shown below.

	Incidence of Malformations (%)	
Dose of Dactinomycin (μg/egg)	Trunkless (at 5 Days)	Tailless (Anourous) (at 17 Days)
0.0625	4	34
0.125	15	53
0.25	18	34

The trunkless embryos did not survive the eighth day of incubation; the tailless were still living on the seventeenth day.

The effect of time of injection on the incidence of malformations is shown in the tabulation below.

Age at Injection with 0.0625 μg (days)	Incidence of Malformations (%)	
	Trunkless (at 5 Days)	Tailless (Anourous) (at 17 Days)
1	6	3.9
2	6	40.7
3	0	0.0

It is evident that dactinomycin, at the doses given, is weakly teratogenic to chicken embryos injected at the age of 1 day and apparently nonteratogenic at 3 days.

(See also Summary Table 12, under Dactinomycin.)

Mitomycin C. The effect of injecting various doses of mitomycin C into 1-day incubated eggs on the incidence of malformations among survivors 3 days after treatment (*Cury and Montenegro, 1969*) is shown below.

Dose of Mitomycin C (μg/egg)	Incidence of Malformations (% of survivors)
0.000	5.1
0.060	37.0
0.083	41.2

(For specific malformations, see Summary Table 12, under Mitomycin C).

Tetracycline. The effect of tetracycline on *growth* is manifested by an overall decrease in size, particularly marked in the limbs (*Rolle and Bevelander, 1966*).

Cumulative *mortality* curves for embryos under the influence of various doses of tetracycline are shown in Fig. 108.

The toxicity of this antibiotic is noticeably influenced by the route of injection. It is five or six times as toxic when inoculated at 10 days into the air cell as when injected into the yolk sac (*Gentry, 1958*), as is shown below.

Site of Injection	End point Lethal Dose (ELD_{50}) (mg)
Air cell	0.96
Allantoic cavity	3.2
Yolk sac	5.2

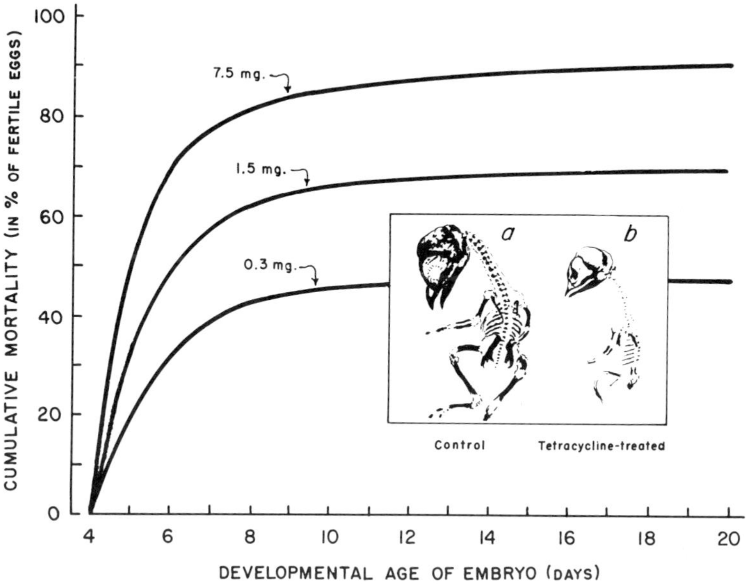

Fig. 108. Effect of different doses of tetracycline on cumulative mortality of the chicken embryo. (After Genazzani, Pelagalli, Pagnini, and DiCarlo, 1965.) *Insert:* effect of tetracycline on the development (ossification) of the skeleton—*a*, the 14-day embryo, control; *b*, the 14-day-old embryo, tetracycline-treated on the eighth day. (After Rolle and Bevelander, 1966.)

The tetracyclines—tetracycline, oxytetracycline, and methacycline—have a similar effect on the *malformation* of the skeleton (*Genazzani, Pelagalli, Pagnini, and Di Carlo, 1965*).

Among modifications of the skeleton are inhibition of mineralization (see *insert* of Fig. 108) and erosion of the long-bone cartilage.

(See also Summary Table 12, under Tetracycline.)

Antivitamins (Antimetabolites of Vitamins)

Aminopterin (Antagonist of Folic Acid). Inoculation of aminopterin, in a dose of 100 μg, over the blastoderm (*Blattner, Williamson, Simonsen, and Robertson, 1960*) results in complete inhibition of growth. The inhibitor becomes relatively less toxic as embryonic development progresses.

The effect of various doses of aminopterin prior to incubation (*Snell and Cravens, 1950*) on the distribution of embryonic *mortality* is shown in the accompanying table.

Amount Aminopterin Injected (μg)	Distribution of Mortality (%) 0–1 Day	2–5 Days	6–20 Days	Total
0	11.7	4.7	2.3	18.7
2	26.7	13.3	6.7	46.7
5	36.7	23.3	16.7	76.7
10	53.4	33.3	13.3	100.0
20	73.9	22.6	3.5	100.0
50	96.7	3.0	—	100.0
100	100.0	—	—	100.0

The data show that doses of 10 μg or more of this compound produce 100 per cent mortality.

The injection of aminopterin produces profound changes in development, with numerous malformations of the embryo. (For a detailed list of malformations see Summary Table 12, under Aminopterin.)

Desoxypyridoxine (Metabolic Analogue of Pyridoxine). The effect of injection of various doses of desoxypyridoxine prior to incubation on the incidence of embryonic *mortality* at hatching (*Cravens and Snell, 1949*) is shown in the tabulation below.

Dose of Desoxypyridoxine (μg/egg)	Incidence of Mortality (%)
200	71
500	89
1000	100

The effect of 500 μg of desoxypyridoxine, injected into egg albumen at different stages of incubation, on the incidence of embryonic *mortality* at the time of hatching (*Cravens and Snell, 1949*) is presented below.

Time of Injection (days)	Incidence of Mortality (%)
0	81
3	32
6	10
15	5

Methyl Pyridyl Ketone (Antagonist of Nicotinic Acid). The typical effects of injections of methyl pyridyl ketone at 4 days of incubation on the incidence of embryonic mortality and malformations (*Landauer and Clark, 1962b*) are shown in the accompanying table.

Dose of Compound (μg/egg)	Mortality (%)	Malformations (%)
375	50.9	21.6
500	74.7	50.7

(For types of anomalous development, see Summary Table 12, under Methyl pyridyl ketone.)

Pantethine (Antagonist of Pantothenic Acid). Injection of unincubated turkey eggs with 43 mg of pantethine (*Goetinck, Abbott, and Kratzer, 1957*) sharply increases embryonic *mortality.* The distribution of mortality during the incubation period is shown in the table below.

Time of Incubation (days)	Incidence of Mortality (%)	
	Injected	Control
1–6	54.9	16.7
7–14	0.0	2.8
15–24	1.6	1.4
25–28	29.0	16.6
Total	85.5	37.5

The first peak of mortality (1 to 6 days) is more than triple, the second peak (25 to 28 days) almost double that in the control group (see also critical periods, in Chapter II, Spontaneous Malformations).

(For the occurrence of malformations, see Summary Table 12, under Pantethine.)

Pyrimethamine (Antagonist to Thiamine). Pyrimethamine is relatively toxic to the developing embryo even at later stages of development (*Naber, Cravens, Baumann, and Bird, 1954*), as is shown below.

Dose of Pyrimethamine (mg/egg)	Incidence of Mortality (%)	
	Injected at 10 Days	Injected at 15 Days
0.25	30	10
0.50	40	90
1.00	90	90
2.00	100	100

The incidence of *malformations* depends upon the dose and time of injection of the drug (*Stanzhevskaya, 1966a*), as shown in the tables below.

Dose of Pyrimethamine (mg/egg at 4 days)	Incidence of Malformations (% at 10 days)
0.0	24
0.5	73
1.0	77

Time of Injection (days)	Incidence of Malformations (% at 10 days)
0	58
2	56
4	35

(For a detailed list of malformations see Summary Table 12, under Pyrimethamine.)

Biologics

Blood. Intravenous injection of adult or embryonic (12 days or younger) whole blood into 11- to 12-day embryos (*Brandly, Thorp, and Prickett, 1949*) affects the incidence of mortality and occurrence of lesions, as is shown below.

Inoculum	Mortality (%)	Lesions (%)
Adult chicken blood	64	46
Embryonic blood	43	0

The effects of the route of injection of adult chicken blood on the incidence of embryonic mortality and disease are as follows:

Route of Injection	Mortality (%)	Lesions (%)
Intravenous	64	46
Chorioallantoic	71	32
Intra-amniotic chamber	44	11
Allantoic chamber	47	2
Yolk sac contents	31	0

The intravenous route induces the highest percentage of gross lesions, and the yolk sac route appears not to be effective in inducing embryonic disease.

Blood Serum. Among 3-day-old embryos given intravenous injections of 1 μl of blood serum from 9-day-old embryos, 17 per cent died within 24 hours, and 28 per cent were abnormal (*Kaplan and Johnson, 1970*).

Normal human blood serum, injected intravenously in amounts from 0.25 to 5 ml into 5- to 12-day-old embryos (*Lustig, 1954*), inhibits development and in certain cases stops it completely.

Tissue Extracts. It has been shown (*Lenicque, 1959*) that when extracts of organs such as brain are injected into the embryo after the primitive streak has started to form (8 to 12 hours), or at the moment of its regression, characteristic anomalies appear. The changes caused by injection of brain extract (from the 24-hour-old embryo) at various stages (between 12 and 32 hours) are shown in Fig. 109.

When the injection was made after regression of the primitive streak (48 hours), no definite effect could be noted.

Experiments have also shown that brain extract, when in contact

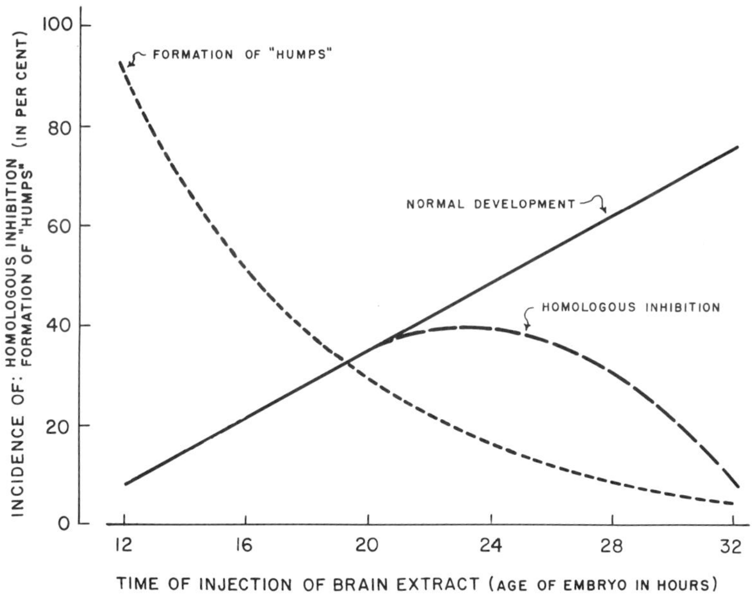

Fig. 109. Effect of injection of 0.5 mg of brain extract into eggs between ages 12 and 32 hours, on embryonic development. (Rearranged after Lenicque, 1959.)

with the chicken blastoderm, inhibits the normal formation of the brain and its parts.

Inhibitory action of the brain extract from donors of increasing age is shown by the data (*Lenicque, 1959*) presented below.

Age of Donor (days of incubation)	Brain Inhibition (%)	Undeveloped Embryos (%)
4	12	12
5	12	20
6	28	20
12	44	16

The inhibitory activity of the brain distinctly increases between ages 6 and 12 days in the donor.

Endotoxin. The susceptibility of the 10-day-old chicken embryo to endotoxin (of *Escherichia coli*), injected intravenously (*Gruninger and Spink, 1967*), is demonstrated in the tabulation below.

Endotoxin Injected (μg)	Embryonic Mortality (%)
0.01	11
0.05	81
0.10	96

Bee Venom. When doses of 0.4 cm^3 of bee venom are injected into egg albumen at 0 to 3 days of incubation (*Ruch, Kremer, Schneegans, and Rohmer, 1961*), the incidence of *mortality* and *malformations* decreases with increasing age of the embryo at injection, as is evidenced by the data below.

Time of Injection (days)	Mortality (%)	Malformations (%)
0	74	17
2	46	7
3	15	2

Among specific *malformations* observed are those of the nervous system (exencephalia, anencephalia, spina bifida, microphthalmia, and anophthalmia), the thoraco-abdominal region (complete celosomia), the upper beak, the caudal region (anouria), the posterior limbs (micromelia), and the joints.

The incidence of cardiac malformations (*Ruch, Kremer, Schneegans, and Rohmer, 1962a*) may reach 59 per cent of all malformed embryos. The appearance of these anomalies in the 11-day-old embryo is shown graphically in Fig. 110.

For further description of malformations see also Table 10, under Bee venom.

Snake Venoms. The venoms of various species of snakes differ in their content of toxic principles: hemorrhagins, hemolysins, and neurotoxins.

Injections of 0.5 μg of cobra (genus *Naja*) venom into 5- and 7-day-old embryos produce on the tenth day some noticeable cardiac malformations (*Ruch, Zahnd, Kremer, Rumpler, and Gerlinger, 1965*). (For a detailed list of malformations, see Table 10, under Cobra venom.)

The hemorrhagic effect of 1.0 cm^3 of 1% solution of moccasin (*Agkistrodon piscivorus*) venom (*Witebsky, Peck, and Neter, 1935*) is illustrated in Fig. 111.

The effects of serial dilutions of 1 cm^3 of moccasin venom on the

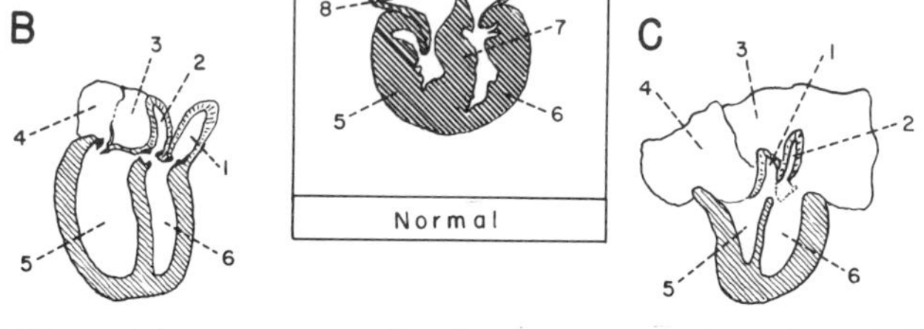

Fig. 110. Effects of bee venom on the development of the embryonic heart. (Redrawn from schematic illustration of Ruch, Kremer, Schneegans, and Rohmer, 1962a; normal, from Romanoff, 1960, p. 700.) The most common forms of cardiac malformations: A_1 and A_2, complete or partial interference in interventricular communication; *B*, malposition of bulbus aortae; *C*, interference in auricular communication. *1*, aortic arch; *2*, pulmonary arch; *3*, right auricle; *4*, left auricle; *5*, left ventricle; *6*, right ventricle; *7*, interventricular septum; *8*, interatrial septum.

Table 10 Pathological Effect of Various Venoms on the Development of the Avian Embryo

Age (days) Treatment	Age (days) Examination	Malformations (%)	Mortality (%)	Incidence of Typical Syndromes	Reference
Bee venom				(Doses: from 0.2 to 0.4 cm^3/egg, in albumen)	Ruch, Kremer, Schneegans, and Rohmer (1961)
0	11	12	74	Atrophy of posterior organs; developmental disturbances of nervous system (exencephalia, anencephalia, spina bifida, microphthalmia, anophthalmia); in thoraco-abdominal wall (celosomia); cephalic extremities: eye, neck (shortened), beak (parrot); posterior extremities: limbs, feet (or micromelia).	
2	11	4	46		
3	11	2	15		
0–3	11	59		(Doses: from 0.2 to 0.4 cm^3/egg, in albumen) Disturbances in morphogenesis of the heart: induction of necrosis and other anomalies: 59% with cardiac defects (in bulbobronchial region—35%, in arterial communication—24%, malposition of bulbus—24%, and diverse malformations—17%)	Ruch, Kremer, Schneegans, and Rohmer (1962a)
0–3	11–15	85		(Doses: from 0.2 to 0.4 cm^3/egg, in albumen) Retarded ossification in areas of the spinal cord (85%), and muscular atrophy (amyotrophy).	Ruch, Kremer, Schneegans, and Rohmer (1962b)
0–3	11–15			(Doses: from 0.2 to 0.4 cm^3/egg, in albumen) Multiplicity of lesions, injuries to neural tube.	Ruch and Kremer (1963b)
3	7–11	75		(Doses: 0.2 ml/egg, in albumen) Bulbus bearing to the right (40%), affected thoracic aorta (27%), deranged communication between arteries (27%), and interventricular communication (6%).	Ruch, Kremer, Rumpler, and Gerlinger (1965)

				(Doses: 0.2 ml/egg, in albumen)	Ruch, Zahnd, Kremer, Rumpler, and Gerlinger (1965)
0–3	11	58		(Similar to the above)	
Cobra venom				(Doses: from 2 to 20 $\mu g/0.5\ cm^3$ Tyrode solution/egg, in albumen)	Ruch and Kremer (1962)
0	7–11	10–22	53–90	Celosomia, spina bifida, exencephalia (4%), anophthalmia (3%), brachymelia, microphthalmia, various malformations of heart: exteriorization of heart and blood vessels.	
				(Doses: 20 μg/egg)	Ruch and Kremer (1963a)
5–7	11	85		Cardiac ectopia, misshapen heart, lack of closure of septum.	
				(Doses: 20 μg/egg, in albumen)	Ruch, Kremer, Rumpler, and Gerlinger (1965)
3	7–11	80		Displacement of bulbus (38%), damage of bulbothoracic area (38%), interventricular communication (12%), and communications between small arteries or arterial branches (12%).	Ruch, Zahnd, Kremer, Rumpler, and Gerlinger (1965)
Moccasin venom				(Doses: 1 cm^3/1% saline solution/egg)	Witebski, Peck, and Neter (1935)
3	3+			Petechial hemorrhages at various sites in the vascular network (Fig. 111).	
Rattlesnake venom				(Doses: from 12.5 to 25 μg/egg, in allantois of pigeon)	Biocca (1942)
12–18	15–21		Early	Hemolysis, permanent halt of respiratory movements.	

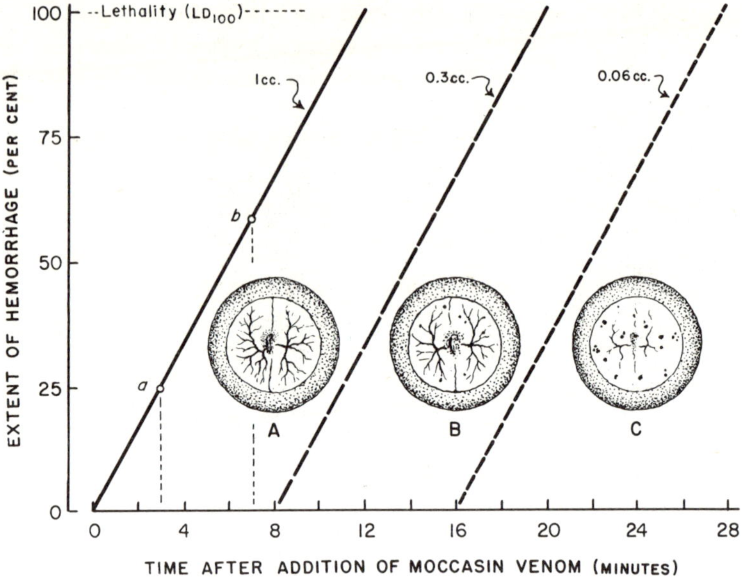

Fig. 111. The petechial hemorrhagic effect of moccasin venom on the vascular network of the 3-day-old chicken embryo. (Reconstructed from the data of Witebsky, Peck, and Neter, 1935.) *Inserts:* graphic illustrations of the effect of 1 cm^3 of the hemorrhagin—*A*, a 3-day-old embryo before treatment; *B*, at 3 min after treatment; and *C*, at 7 min after treatment.

delay in occurrence of hemorrhagic conditions and death of the chicken embryo after injection are shown in the table below.

Dilution of Venom (%)	Hemorrhages (min)	Mortality (min)
1	8	12
0.5	12	14
0.25	13	16
0.125	21	24

The latent period increases with the decrease of venom concentration. (For further details see Table 10, under Moccasin venom.)

Dyes

Janus Green. Following injection of Janus green (monoazo dye) into the amnion of 6-day-old embryos in doses ranging from 2 to 50 μg

(*Menkes and Deleanu, 1964*) there is a retardation of *growth* and development in various degrees in nearly 80 per cent of surviving embryos. Some embryos attain only one half or less the size of controls by the eleventh day of incubation.

Incidence of *embryonic mortality* increases noticeably with doses of Janus green exceeding 10 μg, and no survivals are seen with doses of 20 μg (*Menkes and Deleanu, 1964*), as is shown below and in Fig. 112.

Doses of Janus Green (μg/egg)	Mortality at 10 Days (%)
2	13
5	15
7	19
12	41
15	60
20	90
25	100

The relationship between the time of injection of 2 μg of Janus green and the embryonic *mortality* is demonstrated by a curve in the *insert* of Fig. 112.

The effect of doses of Janus green on the incidence of syndactyly (associated in various degrees with hypophalangism in all survivors) is also graphically illustrated in Fig. 112.

Trypan Blue. Trypan blue is a very potent teratogen among the group of diazo dyes. When injected into the yolk sac of the 36-hour embryo, it gives the highest incidence of embryonic mortality of any of these dyes (*Beaudoin, 1961*), as is shown below.

Diazo Dye	Incidence of Embryonic Mortality (%)
Trypan blue	56.5
Niagara blue	29.9
Evan's blue	14.4
Azo blue	10.7

The relative potencies of the diazo dyes remain the same as above in the incidence of some of the most common types of defects. Comparative effects of subgerminal injection of the various dyes on the incidence

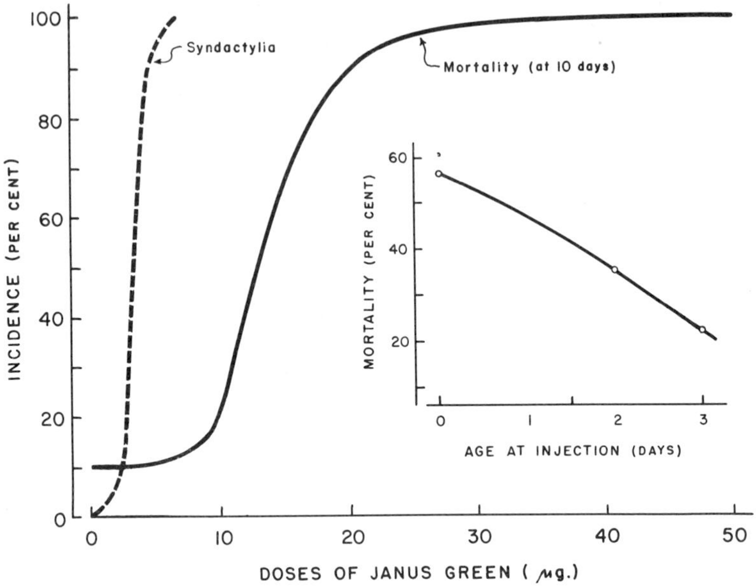

Fig. 112. Effect of various doses of Janus green on incidence of syndactylia and mortality. (Plotted from the data of Menkes and Deleanu, 1964.) *Insert:* incidence of embryonic mortality as influenced by the time of injection of a 2 μg dose. (After the data of Elias, 1963.)

of anouria among surviving 10-day-old embryos (*Beaudoin, 1961*) are presented below.

Dye	Incidence of Anouria (%)
Trypan blue	69.1
Evan's blue	51.2
Niagara blue	20.6
Azo blue	14.9

Injections of 0.1 ml of Trypan blue into the yolk sac at various stages of development (*Beaudoin, 1961*) indicate that this teratogen is most effective on the early embryo, until about the fourth day. Figure 113 shows that the curves for mortality and for malformed survivors on the tenth day tend to parallel one another.

A small *insert* of Fig. 113 also lists the occurrence of various types of defects. (For further details on the occurrence of various malformations, see Summary Table 12, under Trypan blue).

It was observed that fewer deaths and abnormalities occur in em-

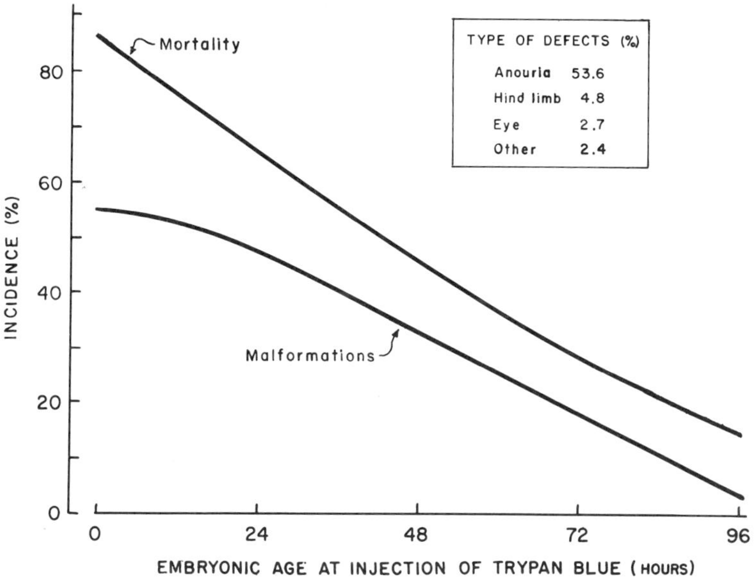

Fig. 113. Effect of Trypan blue on the incidence of mortality and of malformations in the embryo after injections at various stages of development. (Reconstructed from curves of Beaudoin, 1961.) Eggs were injected with 1 mg of Trypan blue into the yolk sac; survivors were examined on the tenth day.

bryos from eggs injected in the yolk sac than from eggs injected subgerminally (*Beaudoin, 1961*), as is shown below for Trypan blue.

Incidence of Mortality and Malformations

Time of Injection (days)	Mortality (%)		Malformation (%)	
	Subgerminal	Yolk Sac	Subgerminal	Yolk Sac
0	100	85	—	50
1	75	67	70	50
2	18	38	50	48
3	5	31	2	8
4	—	18	—	3

Pesticides (and Insecticides)

Chlorinated hydrocarbons (aldrin, dieldrin, and endrin) in the diet of female pheasants affect the incidence of total embryonic *mortality* at hatching (*DeWitt, 1956*), as shown below.

Level in Diet (μg/gram)	Embryonic Mortality (%)		
	With Aldrin	With Dieldrin	With Endrin
1	44	40	43
2	61	48	52
10	70	57	59

Embryonic mortality increased in all groups of pheasants receiving 10 μg of aldrin, dieldrin, or endrin.

Injections of the same chlorinated hydrocarbon compounds—aldrin, dieldrin, and endrin—into the albumen of fertile eggs prior to incubation (*Smith, Weber, and Reid, 1970*) cause increases in embryonic mortality of from 55 to 70 per cent.

In the evaluation of pesticide toxicity in chicken and duck embryos at several embryonal ages (*Khera and Lyon, 1968*), it was observed that at midincubation (10 days in chick and 13 days in duck embryos) both species showed a marked mortality response, occurring primarily within 24 hours of injection.

Twenty-five insecticides have been tested at various concentrations for their toxicity to chicken embryos (*Dunachie and Fletcher, 1969*). Of the two major groups, the organophosphorus compounds (such as ethion, malathion, and parathion) are much more toxic than the organochlorine.

In testing the teratogenic effect of insecticides in chicken and duck embryos (*Khera, 1966*), insecticides representative of different chemical groupings were injected in doses of 10 to 1000 μg/egg at 0, 7, 10, and 13 days of incubation.

When parathion, diazinon, carbophenothion, and ruelene are injected at an early stage of embryonic development, the following *malformations* have been observed: cartilaginous and osseous skeletal dwarfism (Fig. 114), micromelia, ectrosyndactyly, and irregular beak growth.

After injection of phosdrin, diazinon, and DDT (1,1,1,-trichloro-2,2-bis(*p*-chlorophenyl)ethane) near the middle of development in both species, there are congential foot deformities, some of which are permanent.

Solvents and Diluents

Chemical solvents and diluents are commonly used in experiments with various chemical teratogens. None of these substances is entirely inert

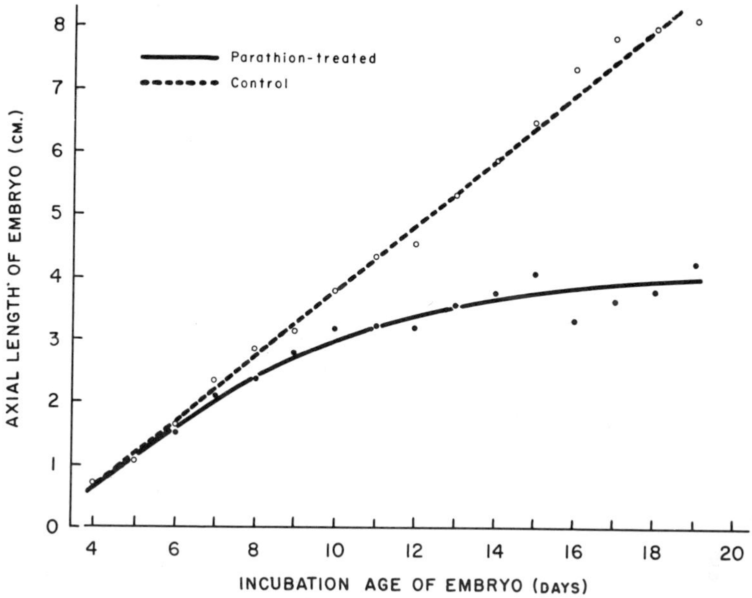

Fig. 114. Effect of 1.0 mg of parathion, injected into the yolk sac after 4 days of incubation, on the progressive inhibition of the vertebral column. (With some modifications after Khera and Bedok, 1967.)

and, in conjunction with other chemicals, each may introduce new deterrent factors in teratological research, especially when the avian embryo is used as an experimental material.

Among inorganic solvents, the most commonly used are distilled water and sodium chloride (a saline solution), and other physiological preparations, such as Ringer's and Tyrode solutions.

Organic substances which have been used as solvents and tested for toxicity and teratogenic effects are mineral, sesame, and corn oils, among others.

Corn Oil. It has been observed (*Smith, Weber, and Reid, 1970*) that corn oil, injected in doses over 0.05 ml into the albumen of fertile eggs prior to incubation, produces excessive embryonic mortality, as is shown in the accompanying table.

Corn Oil Injected (ml/egg)	Total Mortality at Hatching (%)
0.00	4.5
0.05	13.8
0.10	72.2
1.20	97.2

Mineral Oil. The toxicity of mineral oil (*Newcomer, Wright, and Tamblyn, 1952*) injected daily into the allantoic cavity of 10-day-old embryos is shown in the tabulation below.

Dose of Mineral Oil (ml/egg)	Mortality on the 17th Day (%)
0.05	10
0.10	10

Ringer's Solution. The injection of 2 cm^3 of Ringer's fluid into the air cell on the sixth day of incubation (*Landauer, 1929*) results in a higher embryonic mortality following the injection, with a correspondingly lower death rate at the end of the incubation period.

Saline Solution. Physiological salt solution (8.5 per cent of sodium chloride) was found not to be detrimental to the developing chicken embryo *in vivo* (*Conti and Milio, 1964b*). After injection of doses up to 1 cm^3, observations during the following 3 days do not reveal any toxic or teratogenic effects.

In another study (*Clegg, 1964*), injection of various volumes of 0.9 per cent saline into the yolk sac caused some increase in mortality as the volume of saline increased, as is shown in the tabulation below.

Volume of Saline (ml/egg)	Mortality on the 18th Day (%)
0.05	0
0.10	5
0.15	20
0.20	25

Injection of 0.2 ml of saline into the yolk sac of 4-day-old embryos (*Quan, 1963*) resulted in cumulative mortality as follows:

Time after Injection (days)	Total Embryonic Mortality (%)
3	14
11	25
13	34
14	55

Injections of 0.05 ml of saline immediately beneath the blastoderm of embryos about 1 day old (*Cruz, Campillo-Sainz, Munoz-Armas, Pascual-Chavez, and Zamudio-Bandera, 1963*) produced various structural malformations in 28.7 per cent of the embryos.

(For further information about the effects of saline see Summary Table 11, of inorganic compounds, under Sodium chloride.)

Sesame Oil. Experiments on sesame oil (*Lakshmi, 1967a*) indicate that this substance augments embryonic mortality. Injections of 0.4 mg/0.05 ml of sesame oil into the vitelline membrane on the third day of development produced the following incidence of mortality.

Incubation Age (days)	Incidence of Mortality (%)
4	3
8	18
12	24

During the above experiments no malformations were detected.

Tyrode Solution. Injection of Tyrode salt solution in doses up to 1 cm^3, into egg albumen of 0- to 5-day embryos (*Conti and Milio, 1964c*), has some effects on embryonic mortality, as is shown in the accompanying table.

Dose of Tyrode Solution (cm^3/egg)	Mortality of Embryos (%)	
	Injected at 0 Day	Injected at 5 Days
0.05	16	25
0.1	10	49
1.0	34	92

Injections at early stages of development are less harmful than those at later stages, especially as the amount of substance injected increases.

When Tyrode solution is injected into the bloodstream of young chicken embryos (*Martinovitch, Kanazir, Knezevitch, and Simitch, 1962*), there are noticeable symtoms of *anomalous* development.

Anomaly	Incidence (%)	
	With 0.01 ml	With 0.04 ml
Malformations of the beak	3	0
Stumpy legs	3	43
Missing dorsal cranial bones	0.8	57

In addition, at high doses of Tyrode solution there is retardation in *growth.*

***Water* (*Distilled*).** It has been stated (*Ancel, 1956*) that 1 drop of distilled water dropped directly on the embryo has both lethal and teratogenic effects.

Some quantitative observations on the effects of injecting various doses of distilled water into the egg albumen on the incidence of embryonic mortality during the following three days of incubation (*Conti and Milio, 1964a*), are shown below.

Age at Injection (days)	Incidence of Mortality (%) 0.0 cm^3	0.1 cm^3	1 cm^3
0	4	11	39
2	8	11	41
4	5	2	78

The highest doses of distilled water do not show any significant teratogenic effect on the embryo.

Sulfa Drugs (Sulfanilamides)

Many synthetic organic antibacterial (bacteriostatic) drugs related to sulfanilamide have been tested on the avian embryo. Since all these drugs are closely related chemically, they can produce somewhat similar developmental disorders. (See Summary Table 12, under individual Sulfanilamides and their derivatives, such as Sulfamethazine, Sulfamide, Sulfaphenazole, Sulfapyridine, Sulfathiazole, and others whose effects on the avian embryo have been studied.)

Growth of Embryo. Sulfanilamide injected into the yolk sac produces retardation in growth, and increases the incidence of mortality and of some malformations.

The *weight* at the fourteenth day of the embryo injected with various doses of sulfanilamide (*Zwilling and DeBell, 1950*) is shown in the accompanying table.

Sulfanilamide (mg)	Weight of Embryo at 14 Days (grams)
0.0	9.4
0.5	8.9
1.0	7.9
1.6	6.9
2.0	5.8

Retarded growth in body size is accompanied by shortening of the long bones.

As regards the length of embryos, the effects of treatment with various sulfanilamides (*Borsuk, 1964*), especially during the later part of development, are shown below.

Sulfanilamide	Reduction in Trunk Length at 21 Days (% of control)
Sulfapyridine	16
Sulfathiazole	20
Sulfamethazine	22
Sulfaphenazole	30

Mortality. When sulfanilamide, in amounts of 0.5 to 2.0 mg/egg, is injected into the yolk sac of 5-day embryos, indications are that the dose has no relationship to the rate of embryonic mortality during the first 2 weeks of development (*Zwilling and DeBell, 1950*). There is a uniform mortality of about 32 per cent in treated eggs, as against 6 per cent for control groups.

When an identical dose of sulfanilamide (1.6 mg/egg) was injected into embryos of different ages, mortality dropped considerably with increasing embryonic age (*Zwilling and DeBell, 1950*), as is shown below.

Age of Treatment (days)	Incidence of Mortality (%)
1	57
2	53
5	38
7	8
9	8
Control	9

Malformations. The teratogenic effect of sulfanilamide is independent of its toxic effect. With increasing doses of the drug, there is increased incidence of extreme micromelia, as is shown below.

Dose of Sulfanilamide (mg/embryo)	Extreme Micromelia (%)
0.5	0
1.0	27
1.5	60
2.0	66

The incidence of micromelia after injections of sulfanilamide (1.6 mg/egg) at various stages of development is shown in the accompanying tabulation.

Age at Treatment (days)	Extreme Micromelia (%)
1	65
2	77
5	42
7	0

Sulfanilamide may also produce anouria (4.0 to 7.6%), deformed beak (parrot), and other anomalies. (See Summary Table 12, under Sulfanilamide.)

Volatile Chemicals

The relative toxicity of various volatile solvents to the chicken embryo (*McLaughlin, Marliac, Verrett, Mutchler, and Fitzhugh, 1963*) is shown in the tabulation below.

Volatile Chemical	50 Per Cent Lethal Dose (LD_{50}) (mg/kg egg, in yolk sac)
Allyl alcohol	4
Toluene	86
Benzene	88
n-Butyl alcohol	160
Methyl propyl ketone	180
n-Butyl acetate	180
Isopropyl alcohol	400
Carbon tetrachloride	640
Acetone	780
Methanol (methyl alcohol)	780
Alcohol, anhydrous (ethyl)	780

Butyl alcohol is the only alcohol in this group that was found to have teratogenic effects in the chicken embryo, causing damage to the eyes, kidneys, and nerves.

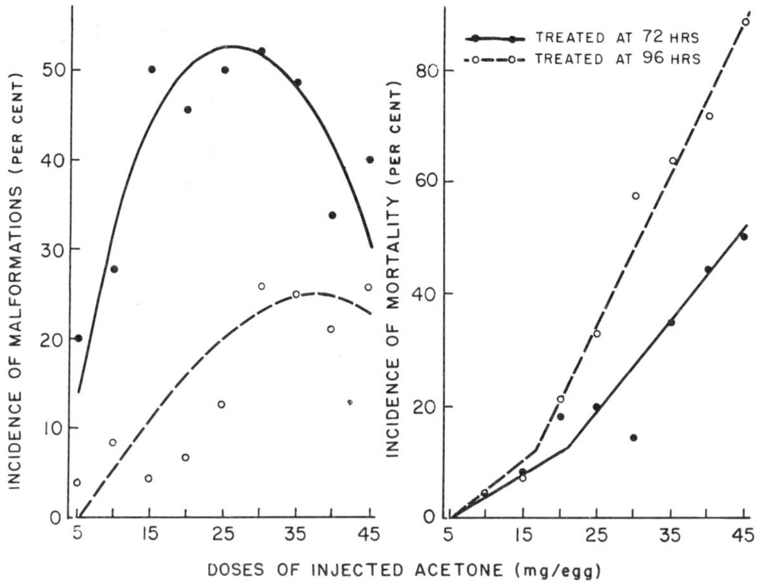

Fig. 115. Incidence of embryonic malformations and mortality resulting from the injection of various doses of acetone; observed after 24 hours of further incubation. (Plotted after the data of Caujolle, Caujolle, Cros, Dao-Huy-Giao, Moulas, Tollon, and Caylus, 1965.) *A,* incidence of embryonic malformations from injections at 72 and 96 hours of age; and *B,* incidence of embryonic mortality from injections at 72 and 96 hours of age.

ORGANIC COMPOUNDS (UNCLASSIFIED)

Acetone. The effects of single injections of various doses of acetone at 72 and 96 hours of incubation on the incidence of malformations and mortality are shown graphically in Fig. 115.

During the first 24 hours after treatment, the incidence of *mortality* increases linearly with the increase in dose of acetone. This increase is more rapid in older than in younger embryos (Fig. 115*B*).

The incidence of *malformations* during the 24 hours after injection is much lower in older embryos than in younger embryos (Fig. 115*A*).

For various forms and relative frequency of observed malformations see Summary Table 12, under Acetone.

Aflatoxin B (Toxic Metabolite of the Fungus Aspergillus flavus). The injection of about 0.05 μg of aflatoxin into the air cell of the fresh egg (*Shibko, Arnold, Morningstar, and Friedman, 1968*) resulted in reduction of the weight of the embryo at hatching to almost one half that of the control. The retardation in *growth* begins on the twelfth day of development.

There is also a decrease in weight of the liver parallel to the decrease in weight of the embryo (see Summary Table 12, under Aflatoxin B).

The effect of various doses of aflatoxin B on the incidence of embryonic mortality is shown below.

Dose of Aflatoxin B (μg/egg)	Total Mortality (%)
0.5	4
1.2	68
3.0	98

Variations related to the time of injection in the incidence of embryonic mortality (*Clegg, 1964*) are demonstrated in the accompanying table.

Age at Injection (days)	Incidence of Mortality (%)
3	100
4	90
7	50
9	25
11	0

Alcohol, Anhydrous. In early experimental work (*Féré, 1894; Stockard, 1914;* among others) it was demonstrated that ethyl alcohol has a deleterious effect on the development of the avian embryo, causing a noticeable increase in the incidence of mortality and malformations.

Recent studies (*Sandor and Elias, 1968*) on injection of alcohol into the air cell at 0 and 23 hours of incubation show the quantitative effect on embryonic *mortality* (Fig. 116*B*, with *insert*). The injections at day 0 produce two periods of high mortality—an initial peak at the first day followed by another peak at the sixth day. The injections at 23 hours of incubation induce only one continuous rise in mortality, initiated soon after injection.

The incidence of *malformations* in embryos from eggs injected with 0.4 to 0.6 ml of ethyl alcohol at 0 and 23 hours (*Sandor and Elias, 1968*) is shown in Fig. 116*A*. In both cases, the peak incidence (about 45 per cent) occurred between 30 and 40 hours of incubation.

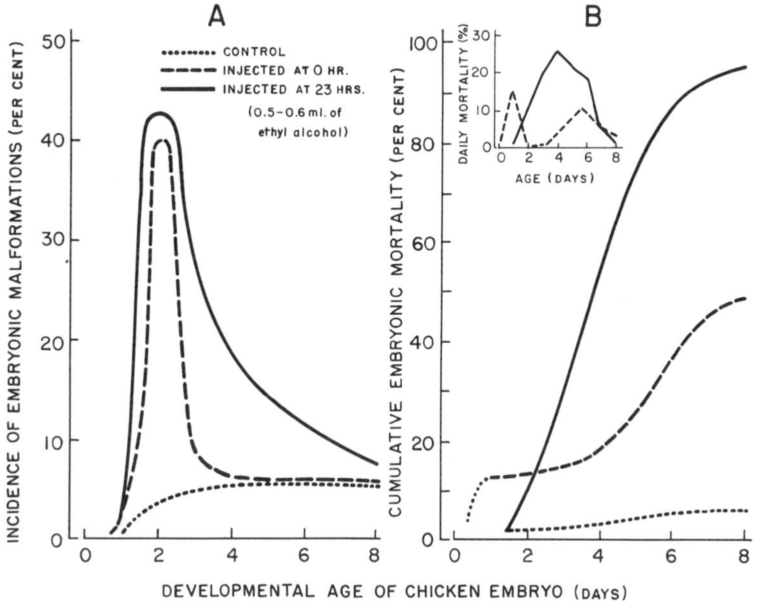

Fig. 116. Effect of ethyl alcohol on the early development and mortality of the chicken embryo. (Redrawn with modifications after Sandor and Elias, 1968.) Ethyl alcohol (64%) was injected into the air chamber in the amount of 0.05 to 0.6 ml at 0 hour and/or 23 hours of incubation; then lots of 15 to 20 embryos were examined at different stages of development. *A*, incidence of embryonic malformations, among which deformities of the brain vesicles, caudal defects, and generalized malformations are most frequent (see Summary Table 12, under Alcohol); *B*, cumulative curve of embryonic mortality—with *insert* showing the distribution of mortality of embryos in both groups of eggs.

Various types of abnormalities were observed. The incidence of each malformation is shown in the tabulation below.

	Incidence (%)	
Type of Malformation	Injected at 0 Hours	Injected at 23 Hours
Generalized malformations	61	89
Deformed brain vesicles	15	24
Caudal malformations	12	24
Malformations of the heart	9	0
Abnormal flexure	3	8
Double or triple monsters	0	2

There is the greatest possibility of alcohol injury in the early stages of embryonic development (*insert* of Fig. 116*B*).

Aminoguanidine. Of all organs, the liver is most susceptible to aminoguanidine. Injections of various doses into the yolk sac of 4-day embryos (*Neuman and McCoy, 1955*) resulted in reduction in liver weight at 14 days, as follows.

Aminoguanidine Injected (mg/egg)	Weight of Liver (mg)
0	175
2	139
10	75
20	45

Effects of embryonic age at the time of injection of various doses of aminoguanidine are graphically illustrated in Fig. 117. The *insert* of this figure also shows the continuous growth of livers under both experimental and control conditions.

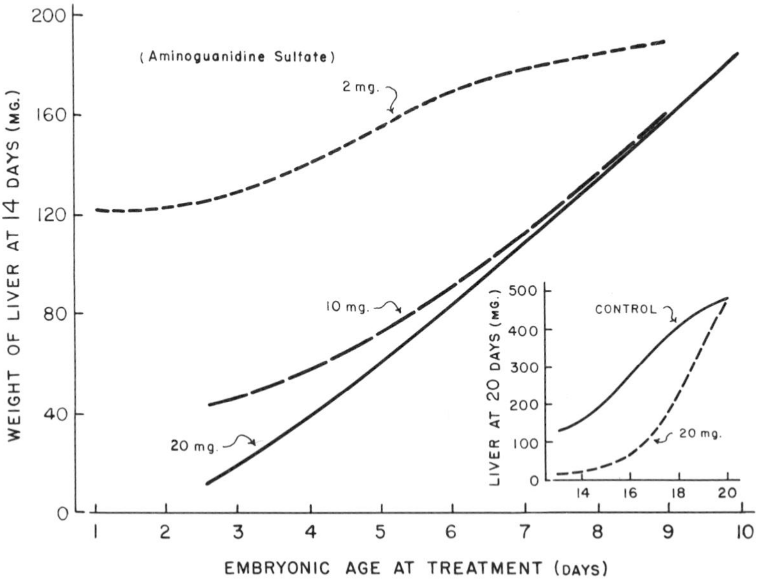

Fig. 117. Effects of various doses of aminoguanidine sulfate on the weight of the liver of embryos treated at early stages of development. (After the data of Neuman and McCoy, 1955.) *Insert:* the same at later stages of embryonic development. (After Nosal and Watterson, 1959.)

Injection of various doses of aminoguanidine into 3-day-old embryos (*Stoll, Bodit, and Maraud, 1968*) produced incidences of malformations at 14 days as follows.

Dose of Aminoguanidine (mg/egg)	Incidence of Malformations (%)
0.0	6.0
0.5	7.5
7.5	10.5
25.0	42.8
50.0	100.0

For further details on malformations, see Summary Table 12, under Aminoguanidine.

6-Aminonicotinamide. Injection of various doses of 6-aminonicotinamide into the yolk sac of the 4-day-old embryo affects the incidence of embryonic *mortality* (*Landauer and Clark, 1962b*) as shown in the accompanying table.

Dose of 6-Aminonicotinamide (μg/egg)	Mortality at Hatching (%)
2.5	36
5.0	82
10.0	100

The incidence of total and individual malformations under the influence of various doses of the compound is shown below.

	Incidence of Malformation (%)			
Malformation	At 2.5 μg	At 4 μg	At 5 μg	At 10 μg
Micromelia	9.1	21.4	59.6	97.5
Short upper beak	0.0	0.0	0.0	0.2
Parrot and/or short lower beak	0.0	0.0	20.5	79.9
Miscellaneous defects	6.3	1.0	0.6	0.0
Total defective embryos	15.5	22.4	59.6	97.5

3-Aminopropionitrile. Toxicity of 3-aminopropionitrile (a lathyrogenic agent), as shown by embryonic mortality in the accompanying table, depends in degree upon the dosage administered (*Naber and Blackwood, 1962*).

3-Aminopropionitrile Injected (mg/egg)	Embryonic Mortality (%)		
	0–5 Days	6–20 Days	Hatching
0.1	10	10	20
0.2	20	40	60
0.4	35	65	100
0.6	50	100	
0.8	95		
1.0	100		

Observations show that embryonic death takes place only after formation of an extensive vascular system. Injections of the substance before incubation or prior to 2 days do not inhibit primary differentiation, and do not produce a lethal effect.

From the above table it is also obvious that normal hatchability is possible at the level of 0.1 mg/egg; at the 0.2 mg level hatchability is only 40 per cent, and there are no survivals at hatching time at 0.4 mg or higher.

The feeding of 3-aminopropionitrile to laying hens, in amounts of about 0.05% of the diet, for 2 to 3 weeks reduces the hatchability of fertile eggs almost to zero (*Barnett, Richey, and Morgan, 1957*). The hatchability immediately improves (Fig. 118) when the hens are placed on a basal diet.

Supplementation with 0.03 to 0.06 per cent aminopropionitrile in the hens' diet causes a high incidence of soft-shelled and malformed eggs—a typical symptom of metabolic disturbance of the reproductive organs. Furthermore, from 40 to 80 per cent of dead embryos from these hens are in a hemorrhagic condition.

Eggs injected with 0.1 to 0.2 μg/egg of 3-aminopropionitrile produced embryos with highly specific lesions, primarily affecting the development of the legs (*Chang, Witschi, and Ponseti, 1955*). Alizarin-stained skeletons of 17-day-old embryos (Fig. 119) reveal thickening of the femur and tibia with dislocation of joints (*Levene and Gross, 1959*) and possibly twisting and shortening of the lower beak (*Naber and Blackwood, 1962*).

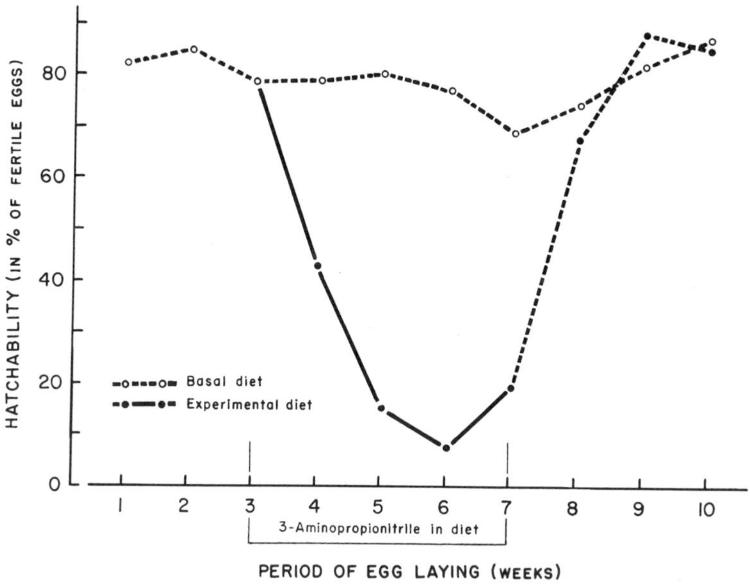

Fig. 118. Effect of feeding 3-aminopropionitrile (0.05%) to laying hens on the hatchability of eggs. (Based on the data of Barnett, Richey, and Morgan, 1957.)

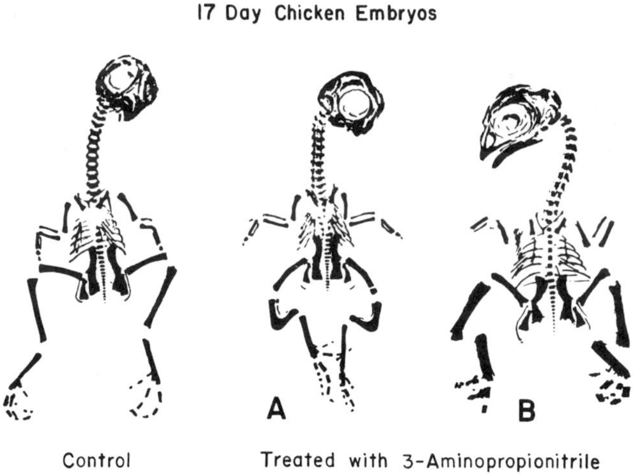

Fig. 119. Appearance of alizarin-stained skeletons of 17-day-old embryos treated with 3-aminopropionitrile. (After Levene and Gross, 1959.) *A,* injection at four days with 0.32 mg results in extreme bowing of tibia and fibula; *B,* injection at 14 days with 20 mg causes dislocation of joints without bowing of long bones.

Atropine. Injection of various doses of atropine into the chicken egg affects the incidence of embryonic mortality (*Grüter, 1916*) as shown below.

Dose of Atropine (mg/egg)	Embryonic Mortality (%)
0.000	29.9
0.005	37.5
0.01	50.0
0.02	66.1
0.03	75.0
0.04	87.5
0.05	100.0

Azaserine. Embryos injected with azaserine in doses of 0.012 mg or more show general *growth* retardation (*Blattner, Williamson, and Simonsen, 1958*). The average body weight of abnormal embryos is lower than that of unaffected embryos receiving the same dose. Among

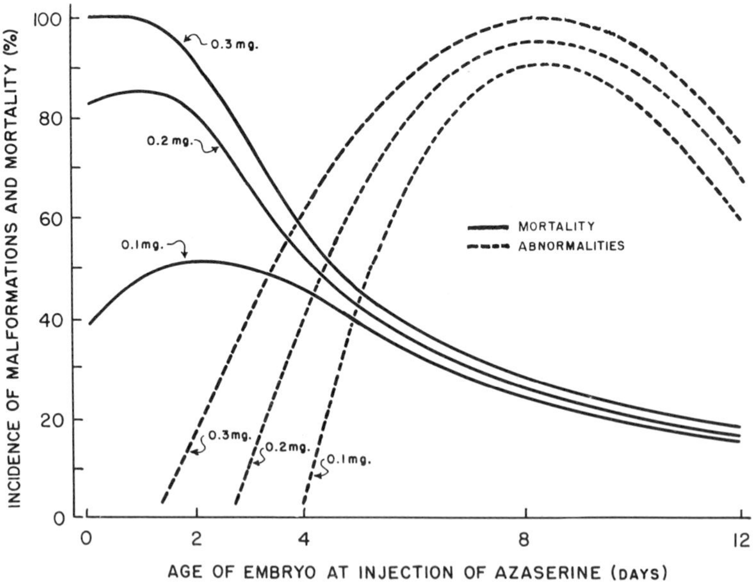

Fig. 120. The relationship between the toxicity (incidence of mortality) and the damage (incidence of abnormalities) from various doses of a teratogen (azaserine). Embryos were injected via yolk sac at various stages of development (at 0, 2, 4, 8, and 12 days) and grossly examined on the sixteenth day. (The curves are constructed from the experimental data of Dagg and Karnofsky, 1955.)

16-day-old embryos which have received 0.2 mg at 4 days, weight of the abnormal embryos, as compared with the unaffected embryos (*Dagg and Karnofsky, 1955*), is as follows:

Abnormal embryos	7.5 grams
Unaffected embryos	11.3 grams
Saline-control embryos	12.4 grams
Normal embryos (*Romanoff, 1967, p. 14*)	18.3 grams

Injection of azaserine into the yolk sac of 4-day embryos in doses of 0.15 to 0.2 mg per egg (within the range of the LD_{50}) causes the majority of injected embryos to die by the twelfth day of development (*Murphy, Dagg, and Karnofsky, 1957*).

The amount of azaserine tolerated by the embryo increases with the age of the embryo at the time of injection. The relationship between the incidence of mortality and the incidence of malformations is graphically shown in Fig. 120.

Azaserine produces a characteristic syndrome of abnormalities with great regularity when injected into the yolk sac. The nature of the defects is dependent upon the time of administration, as is shown below in a summation from experimental data (*Dagg and Karnofsky, 1955*).

Occurrence of Various Abnormalities

Abnormality	Time of Injection of Azaserine		
	At 4 Days	At 8 Days	At 12 Days
Growth inhibition	yes	—	—
Microphthalmia	yes	—	—
Coloboma of upper eyelid	yes	—	—
Defects of face			
Shortened lower beak	yes	some	—
Parrot beak	yes	some	—
Facial coloboma	yes	—	—
Cleft palate	yes	—	—
Defects of appendicular skeleton			
Micromelia	yes	some	—
Hypophalangism	yes	—	—
Ectrodactyly	yes	—	—
Hemimelia	yes	—	—
Edema	yes	yes	yes
Feather inhibition	yes	yes	yes
Down cysts	yes	yes	yes

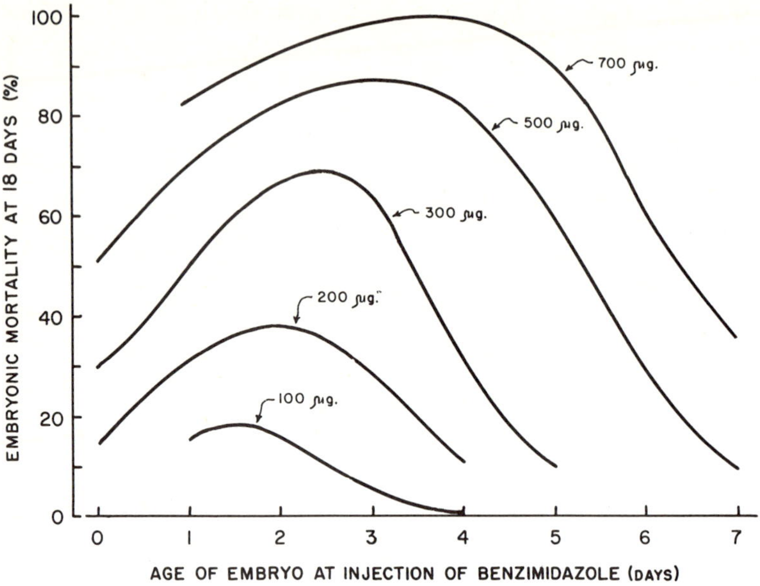

Fig. 121. Effect of embryonic age at treatment on dose effectiveness of benzimidazole and its derivatives. (After the data of Blackwood, 1961, with some modifications.)

(For further details on malformations see Summary Table 12, under Azaserine.)

Benzimidazole. In a study of the effect of benzimidazole on the incidence of embryonic *mortality* (*Blackwood, 1961*) it was learned that age at treatment influenced the dose effect. When the results for different doses per embryo were plotted (Fig. 121), the age of peak effectiveness shifted gradually as the dose was increased.

The effect of various doses of benzimidazole, injected into the albumen prior to incubation, on the incidence of *malformations* at three developmental periods is shown in the table below.

Dose of Benzimidazole (mg/egg)	Incidence of Malformations At 1–3 Days (%)	At 2–7 Days (%)	At 6–14 Days (%)
0.25	27	53	20
0.35	41	46	14
0.50	57	32	11

Further studies (*Blackwood, 1962*) indicate that, as the age of the embryo at injection time increases, malformations decrease in cephalic and increase in caudal regions, as is shown below.

Malformations	Time of Injection (hours) 0	24	36	48
Cephalic region—head (%)	24	23	15	9
Caudal region—trunk (%)	0	8	38	43

(For the incidence of various malformations, see Summary Table 12, under Benzimidazole.)

Carbutamide. There is some evidence (*Lane and Slepecky, 1960*) that carbutamide retards growth in general body size of the embryo before the ninth day of incubation.

Examination of the data also reveals that growth in length of the leg bones (tibiotarsus and tarsometatarsus) is affected relatively more than body size.

The incidence of embryonic *mortality* (from the seventh through the seventeenth day) among embryos treated at 5 days (*Lane and Slepecky, 1960*) with various doses of carbutamide is shown in the table below.

Carbutamide Injected (ml)	Incidence of Mortality (%)
0.05	0
0.1	4
0.2	13
0.3	23
0.4	35

Embryonic age at carbutamide treatment with 0.4 ml/egg affects the incidence of mortality within the same period (*Lane and Slepecky, 1960*) as demonstrated in the accompanying tabulation.

Age at Injection (days)	Incidence of Mortality (%)
1	0.0
2	3.4
3	7.5
4	12.8
5	35.0

With administration of the same dosage (0.4 ml/egg) later, at 7 and 9 days of development, there is a rapid decline in the incidence of embryonic mortality.

A syndrome of which the most common *malformations* are micromelia and beak defects has been observed among embryos surviving 17 days after carbutamide treatment.

The incidence of micromelia is affected by the dosage administered (*Lane and Slepecky, 1960*), as is shown by the data below.

Dose of Carbutamide (ml/egg)	Incidence of Micromelia (%)
0.05	10.3
0.1	31.3
0.2	54.5
0.3	74.1

The effect of time of treatment on the incidence of micromelia is shown below.

Age at Injection (days)	Incidence of Micromelia (%)
1	33
2	40
3	75
4	95

The incidence of beak defects in embryos treated with various doses of carbutamide is given in the accompanying table.

Dose of Carbutamide (ml/egg)	Incidence of Beak Defects (%)
0.1	0.0
0.2	5.5
0.3	10.5
0.4	36.4

It has been noted (*Ancel 1950; Zwilling and DeBell, 1950*) that the abnormalities produced by carbutamide are comparable to those caused by sulfanilamides, because of the presence of SO_2NH_3 in their structural formulas.

(See also Summary Table 12, under Carbutamide.)

Cytosine (Arabinoside). It has been shown (*Karnofsky and Lacon, 1966*) that cytosine arabinoside inhibits the development of the chicken embryo. Degree of inhibition depends upon the dose of the compound; data for injections into the yolk sac on the fourth day of development follow.

Dose of Cytosine (mg/egg)	Weight of Embryo at 18 Days (grams)
0.00	18.3*
0.010	10.8
0.025	11.5
0.050	8.4
0.100	3.4

* Value for the 17th day (*Romanoff, 1967, p. 24*).

The toxic effect of cytosine arabinoside progressively decreases with the age of the embryo at injection (*Karnofsky and Lacon, 1966*), as is shown (Fig. 122) by adjustment of each dose to a level of 50 per cent lethality (LD_{50}).

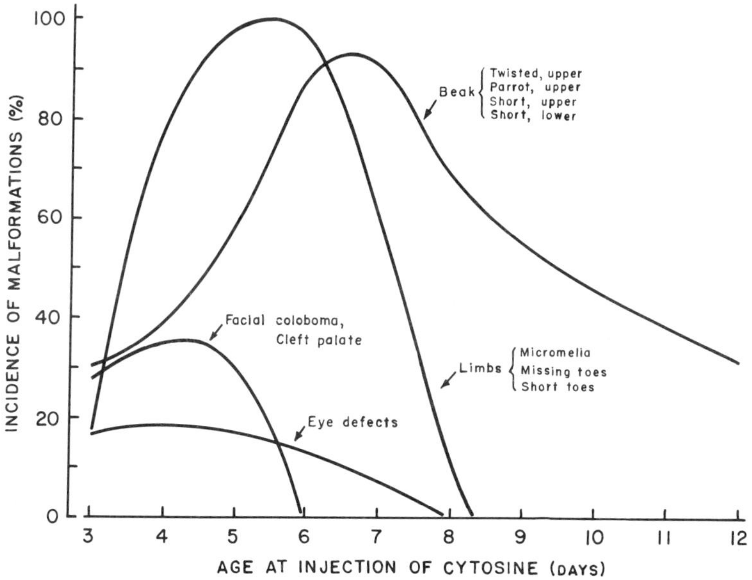

Fig. 122. Effect of cytosine arabinoside, in the range of the LD_{50}, injected into the yolk sac of the chicken embryo at various stages of development: observed at 18 days. (Recalculated from the data of Karnofsky and Lacon, 1966.)

Age at Injection (days)	Approximation LD_{50} (mg/egg)
3	0.03
4	0.04
5	0.05
6	0.06
7	0.09
8	0.15
12	0.75

Gross abnormalities observed under the above conditions are presented in Summary Table 12, under Cytosine arabinoside.

2-Deoxyguanosine. Effect of injection of various doses of 2-deoxyguanosine at different stages on the incidence of embryonic mortality (*Pierro and Spiggle, 1966*) is illustrated by the data below.

Age at Injection (hours)	Mortality (%) from Various Doses		
	With 50 μg	With 100 μg	With 200 μg
48	49		
72	35	56	
96		5	10
120			8

The total incidence of skeletal defects after injection of 2-deoxyguanosine at different stages is shown below.

Age at Injection (hours)	Total Skeletal Defects (%)		
	With 50 μg	With 100 μg	With 200 μg
48	30.7		
72	17.7	49.3	
96		22.4	65.7
120			66.3

The incidence of malformations increases with increase in dose of the substance, and decreases with increase in age of the embryo at injection (Fig. 123).

(For a detailed list of malformations see Summary Table 12, under 2-Deoxyguanosine.)

5-Fluorouracil. Embryos treated with 0.045 to 1.0 mg of 5-fluorouracil on the third day (*Karnofsky and Lacon, 1966*) had marked retardation

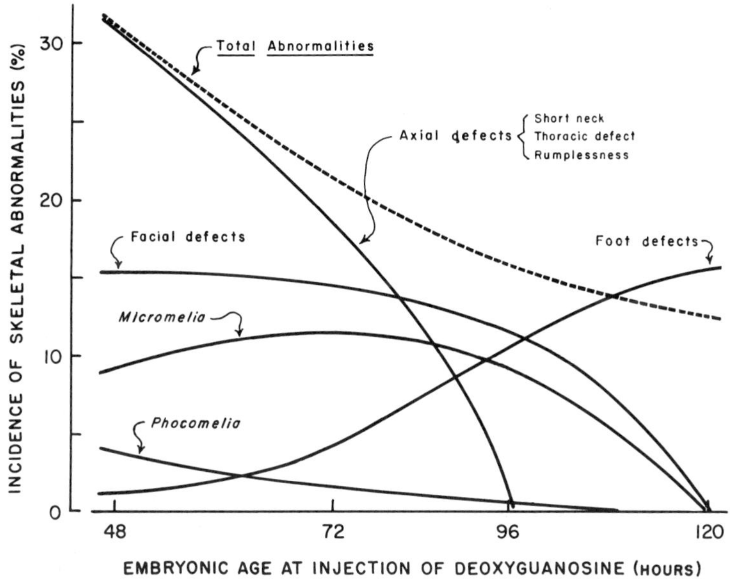

Fig. 123. Effect of deoxyguanosine on skeletal development of the chicken embryo, injected with 50 μg at 2, 3, 4, and 5 days of incubation. (Plotted from the recalculated data of Pierro and Spiggle, 1966.)

of *growth* of body and extremities. Injection of 0.2 mg on the fourth day resulted in delayed closure of the abdominal wall and moderate retardation of growth.

The incidence of mortality is definitely related to the dose (*Conti and Aita, 1966*), as well as to the age of the embryo.

The most frequent anomalies involve the beak, the eyes, the brain, and the limbs. (See Summary Table 12, under 5-Fluorouracil.)

The teratogenicity of fluorouracil is especially manifest in the production of complex monstrosities, similar to those illustrated in Fig. 124.

Glycols. Maximum sensitivity toward glycols (propylene and trimethylene glycols, and glycerol) was observed on the fourth day of development (*Gebhardt, 1968*), when 90 per cent of embryos treated died within 2 hours (Fig. 125).

Among surviving embryos, about 20 per cent have asymmetric malformations of the limbs, and only trimethylene glycol causes a specific type of malformation (micromelia) in a large number of the embryos.

Guanazolo. It has been observed that injections of this drug in the developing embryo (*Youngner, Ward, and Salk, 1950*) inhibit growth and produce high mortality and developmental malformations.

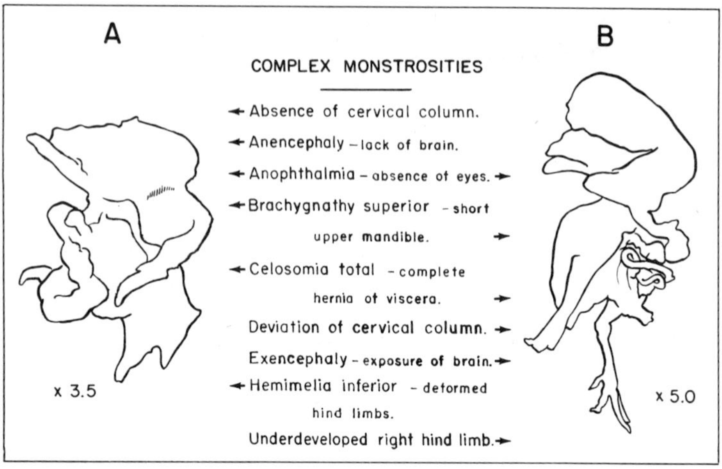

Fig. 124. Complex monstrosities observed after injection of 5-fluorouracil into the egg albumen before incubation. (After Conti and Aita, 1966, with modifications.) *A*, appearance of the 14-day-old dead embryo, from an egg injected with 1000 μg of 5-fluorouracil; *B*, appearance of the 12-day-old dead embryo from an egg injected with 200 μg of 5-fluorouracil.

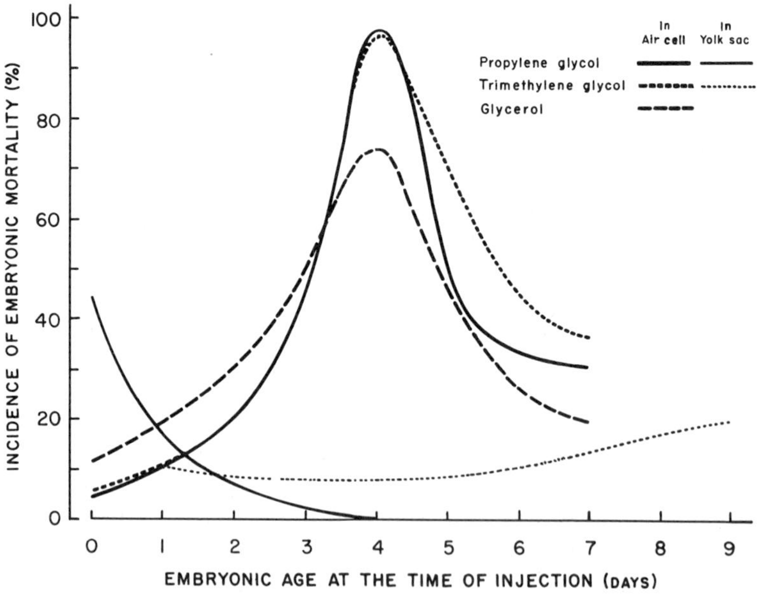

Fig. 125. Effect of glycols and their homologues (propylene glycol, trimethylene glycol, and glycerol) on the incidence of embryonic mortality. (Plotted with modifications from the data of Gebhardt, 1968.)

Surviving embryos at the sixteenth day of development are much smaller than controls of the corresponding age. Embryos treated at the fourth day of age with 0.125 mg of guanazolo have an average *weight* of 9.6 grams whereas embryos in the control group weigh 12.8 grams. Thus there is about a 30 per cent reduction in the weight of experimental embryos.

The incidence of embryonic *mortality* is closely related to the time of treatment and the dosage employed. Embryos younger than 4 days of age do not survive toxic doses. Injection of different amounts of guanazolo into the yolk sac of 4-day-old embryos shows that the mortality increases with further development (Fig. 126).

The embryos which survive up to the sixteenth day of incubation show a high rate of *malformations* (30.3 per cent). The abnormalities observed are abnormal beaks, edema of the skin, exteriorization of viscera, deformed extremities, and hemorrhage. (See Summary Table 12, under Guanazolo.)

Mechlorethamine. *Growth* is inhibited in 25 per cent of embryos injected with mechlorethamine at 2 days (*Blattner, Williamson, Simonsen, and Robertson, 1960*), and the embryos are visibly smaller than controls (Fig. 127).

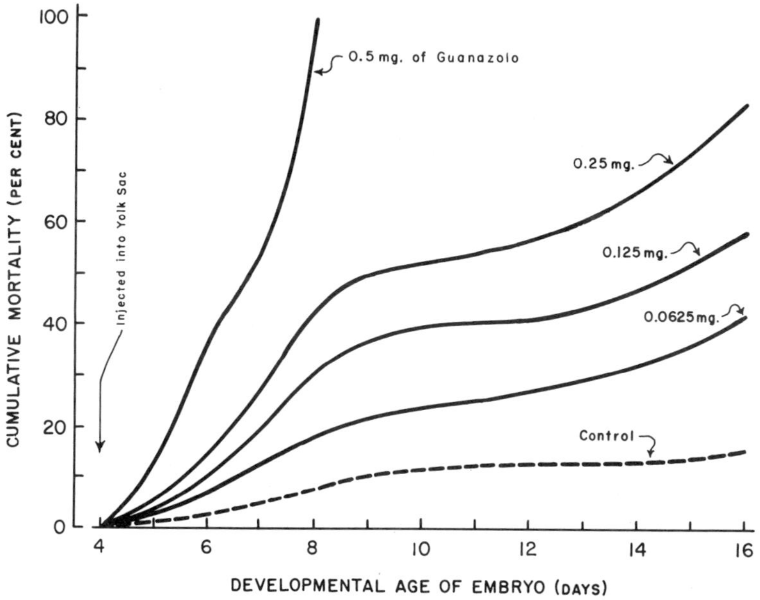

Fig. 126. Effect of various doses of guanazolo on cumulative mortality of the chicken embryo. (Redrawn after Youngner, Ward, and Salk, 1950.)

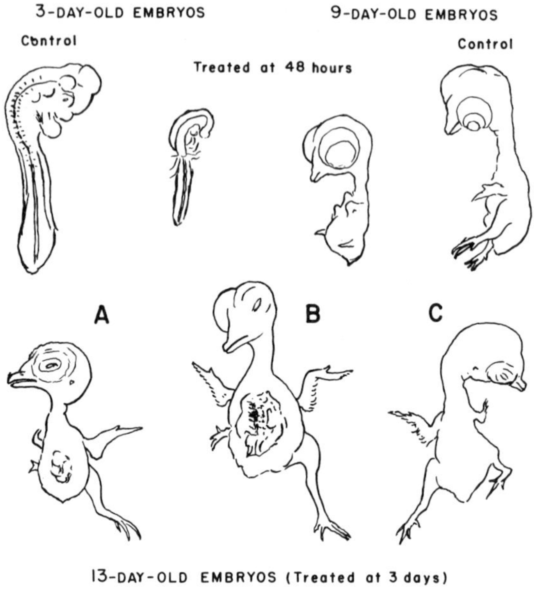

Fig. 127. Effect of mechlorethamine (nitrogen mustard) on early growth and incidence of phocomelia. (Redrawn from Blattner, Williamson, Simonsen, and Robertson, 1960—on growth; and from Salzgeber, 1962, 1966—on phocomelia.) *A,* a phocomelic embryo with considerable reduction in size of the right wing and the right leg; *B,* an embryo with only the right leg affected; *C,* an embryo with somewhat similar defects as in *A.* In addition to this specific anomaly, there is some celosomia (especially in *B*) and microphthalmia. (Embryos in *A, B,* and *C* are about one third natural size.)

The cumulative *mortality* of embryos injected with 6 to 9 μg at the age of 3 to 4 days (*Salzgeber, 1962*) was found to be as follows.

Age of Embryo (days)	Cumulative Mortality (%)
5	19
7–10	40
12–19	41

The most characteristic *malformation* is phocomelia (*Salzgeber, 1966*) which appears in about 49 per cent of embryos (Fig. 127*A, B,* and *C*).

The limbs, legs and wings, are much reduced in size. This is obvious in individual parts of the legs and wings, as is shown by the incidence of malformations.

Part of the Leg	Reduced (%)	Absent (%)
Femur	26	7
Tibia	54	7
Fibula	11	83
Metatarsus	80	15

Part of the Wing	Reduced (%)	Absent (%)
Humerus	47	0
Radius	50	19
Carpus	41	9
Metacarpus	55	16

Femur and humerus persist in most cases and are always less altered than the other segments.

(For a further detailed list of malformations, see Summary Table 12, under Mechlorethamine.)

3-Methylcholanthrene. Application of 3-methylcholanthrene crystals to the allantoic membrane at various stages of incubation affects the incidence of embryonic mortality (*Beohar, Goldhaber, and Hertig, 1964*), as is shown in the table below.

3-Methylcholanthrene Applied (age in days)	Incidence of Mortality at 18 Days (%)
4	94
6	90
8	67
10	30

(For observed malformations see Summary Table 12, under 3-Methylcholanthrene.)

Nicotine. Injection of nicotine sulfate in a dose of 2.5 mg/egg at 2 days of incubation produces marked reduction in size of the embryo and severe skeletal malformations (*Strudel, 1971*). The cervical region is most affected, sometimes completely disappearing. The neural tube is deranged or, in some cases, entirely absent.

Physostigmine. Injection of 0.5 mg of physostigmine into the yolk sac of 4-day-old embryos resulted in 37 per cent mortality on the thirteenth day and 100 per cent at the time of hatching (*Landauer, 1957*).

From this treatment there were 98.5 per cent abnormal embryos.

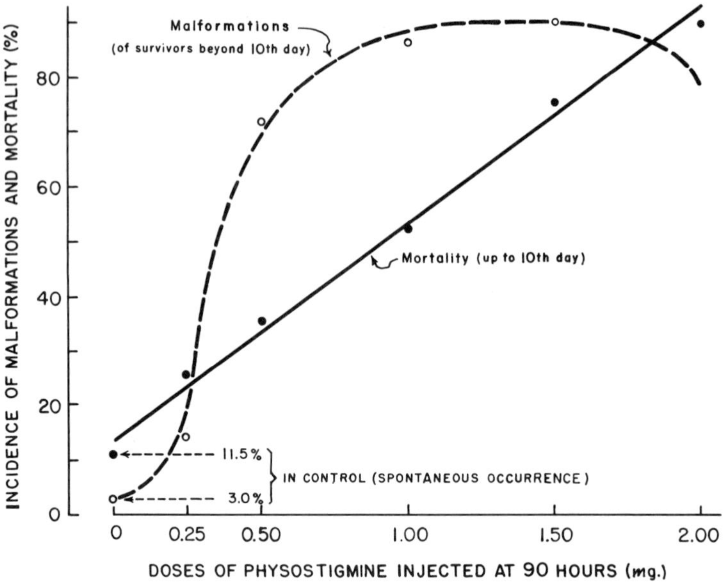

Fig. 128. Effect of physostigmine sulfate (eserine sulfate), injected at 90 hours, on the incidence of malformations and mortality among survivors beyond the tenth day of incubation. (Plotted from the data of Agarwal, Monga, and Agarwal, 1959, with modifications.)

The interrelation between toxic and teratogenic responses in embryos with increasing doses of physostigmine is shown graphically in Fig. 128.

For types of malformations see Summary Table 12, under Physostigmine salicylate and Physostigmine sulfate.

Pilocarpine. The data on the means of body weight and of length of tarsometatarsus of 18-day chicken embryos which had been treated with different amounts of pilocarpine at 96 hours of incubation (*Landauer, 1954*) are shown in the accompanying table.

Dose of Pilocarpine (mg)	Body Weight (grams)	Length of Tarsometatarsus (mm)
0	20.7	18.6
3	20.0	17.5
6	16.6	14.7
12	11.1	9.9

With the highest dose of pilocarpine, the body weight was only 69 per cent and the length of tarsometatarsus 48 per cent of the values for unexposed embryos.

The post-injection embryonic *mortality* with various doses of pilocarpine (*Landauer, 1953b*) is shown below.

Pilocarpine Injected (mg)	Incidence of Mortality at 18 Days (%)
3	21
6	28
12	60

Among chemical compounds which produce developmental *abnormalities* in the chicken embryo, pilocarpine is of particular interest because of its localized effects, especially with low dosages.

The principal location of abnormalities is the skeleton of the face and of the extremities. Typical symptoms in chicken embryos injected with pilocarpine on the fourth or fifth day of incubation (*Landauer, 1954*) are as follows: *Facial skeleton* (depending on dose: short upper or lower beak, facial coloboma, and cleft palate); *long bones of leg* (micromelia—shortening and bending of tarsometatarsus); *syndactylism* (shortening or absence of first toe).

The effect of different doses of pilocarpine on the incidence of various malformations is shown in Fig. 129*B*.

The effect of time of treatment with various doses of pilocarpine on the incidence of syndactylism is shown in Fig. 129*A*.

Figure 130 illustrates the time of appearance of various abnormalities in embryos treated at various stages of development. The peak of anouria (rumplessness) occurs at 24 hours and that of syndactylism at 120 hours, whereas defects of long bones and of beak gradually increase beyond the 144-hour period of experimentation.

Tests with various pilocarpine and pilocarpinelike compounds, at two dosage levels (*Landauer, 1956e*), show that their effects on the occurrence of syndactylism differ somewhat, as is obvious from the tabulation below.

Pilocarpine	Incidence of Syndactylism (%) With 3 mg/egg	Incidence of Syndactylism (%) With 6 mg/egg
Pilocarpine HCl	7.4	58.1
Pilocarpine HNO_3	11.5	40.3
Pilocarpidine HNO_3	13.9	40.6
Isopilocarpine HNO_3	25.4	88.1
Isopilocarpidine HNO_3	39.8	79.5

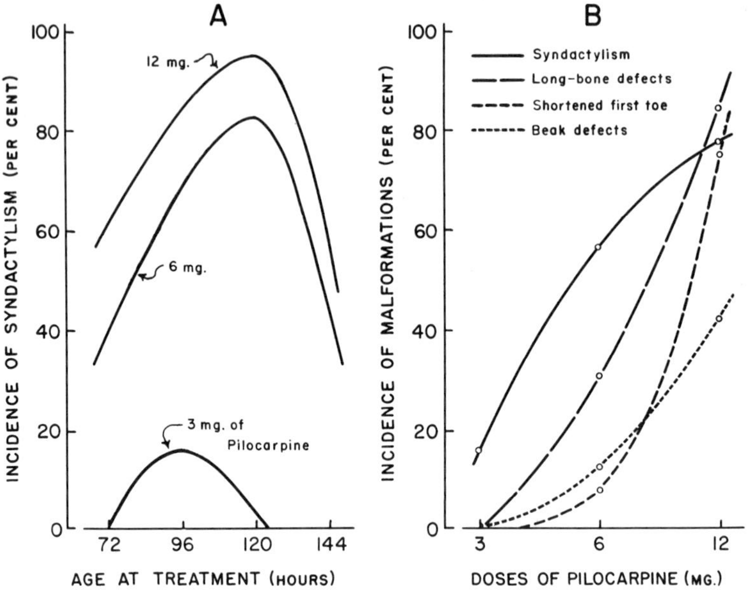

Fig. 129. Effect of pilocarpine on the incidence of various malformations. (Plotted with modifications from the data of Landauer, 1954.) *A*, effect of time of treatment with different doses on the incidence of syndactylism; *B*, effect of various doses given at 96 hours on the incidence of various malformations.

The compounds with *trans* configuration (isopilocarpine and isopilocarpidine) have a much greater teratogenic effect than the *cis* compounds.

(See also Summary Table 12, under Pilocarpine.)

Thalidomide. Thalidomide exhibits the highest toxicity (mortality) in embryos injected before incubation (*Biagi, 1965*). At 2 and 4 days of age, injection with increasing doses of the substance causes increased embryonic mortality, as is shown in the accompanying table.

Thalidomide Used (mg/egg)	Mortality to 16th Day (%)	
	Treated at 48 Hours	Treated at 96 Hours
0.5	37.0	40.0
1.5	54.0	43.8
2.5	72.4	45.6
5.0	90.3	54.7
10.0	96.0	82.7

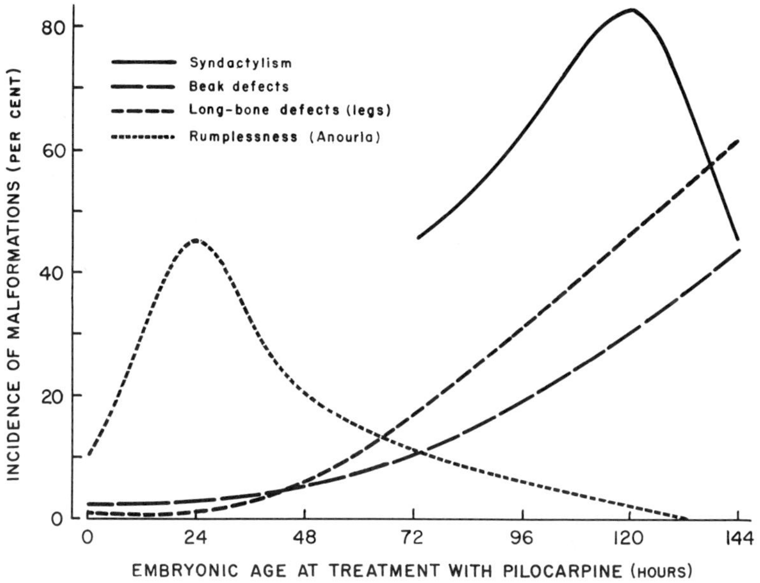

Fig. 130. Effect of 6 mg of pilocarpine on the incidence of various malformations in embryos treated between 0 and 144 hours of incubation, observed in survivors at the age of 13 days. (Plotted from the data of Landauer, 1953b, 1954.)

The teratogenic activity is relatively low when the thalidomide is introduced into the yolk sac. However, its activity is considerably higher when it is locally applied (in powder form), as is demonstrated in the table (*Stanzhevskaya, 1966b*) below.

Time of Treatment (days)	Incidence of Malformations (%)	
	Injected into Yolk Sac	Applied Directly
0	14.2	87.5
1	21.0	30.0
3	5.4	10.0
4	3.9	—

The incidence of malformed embryos produced by injection of various doses of thalidomide at the age of 1 to 3 days was between 20 and 30 per cent. From 15 to 22 per cent showed malformations of the limbs (*Salzgeber and Salaun, 1965*). The limb malformations occurred in the following order of frequency: micromelia, phocomelia, hemimelia, and ectromelia. They were almost always associated with celosomia or anouria.

It has been observed (*Jurand, 1966*) that thalidomide introduced into the albumen at 2½ days causes dilation of the axial limb-bud artery during the 24 hours after treatment.

The incidence of embryos with eyelid defects (encephalocele) was found to vary with the age of incubation at injection of thalidomide into the amniotic cavity (*Williamson, Blattner, and Lutz, 1963*), as follows:

Time of Injection of 1 mg of Thalidomide (days)	Incidence of Encephalocele at 12th Day (%)
4	100
5	75
6	21

For a further list of malformations see Summary Table 12, under Thalidomide.

Summary Table 11 Pathological Effect of Various Inorganic Compounds on the Development of the Avian Embryo

Age (days) Treatment	Age (days) Examination	Malformations (%)	Mortality (%)	Incidence of Typical Syndromes	Reference
Ammonia				(Doses: 0.2 ml of 0.1% solution/egg)	Noto (1967a)
¼–¾	2–3	11	10	Unclosure of neural tube: of tail (8%), of head (2%), of head and tail (1%).	
Barium chloride				(Doses: 20 mg/egg, in yolk sac)	Ridgway and Karnofsky (1952)
4	18	50	50	Inhibition of growth of the toes.	
Boric acid				(Doses: from 0.05 cm^3 of 5% solution to 0.1 cm^3 of 3.5% solution/egg, in yolk sac)	Landauer (1952)
4	18			Retardation of growth, 15/21 grams.	
4	13			Beak defects—all types (69.9%); abnormal toes (14.7%); bent tarsometatarsus, club foot (11.9%); shortened tarsometatarsus (5.6%).	
				(Doses: from 0.05 to 0.1 ml of 5% solution/egg, in yolk sac)	Lane, Darlington and Coursin (1954)
4	13		36–90	Facial anomalies (54.6%); short lower beak (34.4%); abnormal toes (32.5%); short and bent tarsometatarsus (7.9%).	
				(Doses: 0.1 ml of 3.5% solution/egg, in yolk sac)	Lane and Lavy (1958)
4	18	64	17	Short lower beak; abnormal limbs; facial cleft.	
Calcium chloride				(Doses: from 0.01 to 0.04 ml/egg, in yolk sac)	Grabowski (1966)
2–5	9–12	12–17	27–38	Hematomas; defects of brain, eyes, beak, extremities; abdominal hernias.	

Summary Table 11 (*Continued*)

Age (days) Treatment	Age (days) Examination	Malformations (%)	Mortality (%)	Incidence of Typical Syndromes	Reference
Calcium pantothenate				(Doses: 1 ml of 0.02 *M* solution/egg, in yolk sac)	Grabowski (1966)
2	9	17	20	Hematomas (6%); brain defects (4%); anouria (3%); microphthalmia (2%); defects of beak (1.7%) and extremities (1.7%); abdominal hernia (1.4%).	
Cobaltous chloride				(Doses: 0.5 mg/egg, in yolk sac)	Ridgway and Karnofsky (1952)
4–12	18	80–90		Softening and hemorrhaging of the brain; slightly edematous embryos, with enlarged livers.	
Europium				(Doses: from 0.25 to 2.0 mg/egg, in yolk sac)	Zanini (1965)
9	21	12–45		Leg deformities (joint injuries), feather inhibition, edema, subcutaneous blisters.	
Lanthanum chloride				(Doses: from 1 to 100 mg/egg, in albumen)	Machlin, Pearson, and Denton (1952)
1–10	19		100	Hemorrhage (40.7%), shortened beak (7.4%), parrot beak (5.2%).	
Lead acetate				(Doses: from 0.6 to 1.0 μg/egg, in air cell)	Franke, Moxon, Poley, and Tully (1936)
0	14–20	21	26	Ectopia (misplacement of organs).	
				(Doses: from 0.75 to 2 mg/egg, in yolk sac)	Fabre and Girault (1957)
5	19		50	Retardation in growth (15.1/19.4 grams at 19 days); abnormalities, including distended cranial cavity.	
				(Doses: from 10 to 30 mg/egg)	Franciscis and Boccalatte (1962)
8	18	67	100	Retardation in growth (12.5/19.3 grams at 18 days); skeletal abnormalities, especially of the face.	

0	21		100	(Doses: 0.05 ml of 20% aqueous solution/egg, in yolk sac) Failure to hatch, hydrocephaly.	McLaughlin, Marliac, Verrett, Mutchler, and Fitzhugh (1963)
Lead chloride					
0–1	1–2	44	48	(Doses: 0.05 ml/egg, in subgerminal cavity) Abnormalities of the central nervous system.	Catizone and Gray (1941)
Lead nitrate					
0	2–3			(Doses: 0.4 to 0.5 cm^3 of 0.05% solution/egg, in yolk sac) Retardation of growth, especially in the head region.	Hammett and Wallace (1928)
0	13		89	(Doses: from 0.5 to 1.0 mg/egg, in albumen) Meningocele (15%).	Butt, Pearson, and Simonsen (1952)
2–15	2–21			(Doses: from 0.05 to 10 mg/egg, in yolk sac, albumen) Injury to the central nervous system: hemorrhage, necrosis, and hydrocephalus.	Karnofsky and Ridgway (1952)
4	11	74		(Doses: 0.05 cm^3 of 1.6% aqueous solution/egg, in yolk sac) Meningocele, cranioschisis, hydrocephalus, paralytic deformities of the limbs and feet.	Duraiswami (1955)
1–10	7, 19	82	65	(Doses: 0.1 ml of 0.075% solution/egg, in yolk sac) General underdevelopment, hydrocephalus, meningocele, anomaly of legs, microphthalmia.	Puchkov (1963)
Lithium chloride					
0–2	1–4			(Doses: from 0.25 to 4.0 cm^3/egg) Modifications ranging from complete erasure of development to inhibited development of head, somites, heart; or partial inhibition of one or many regions or structures.	Naz and Rulon (1946)

Summary Table 11 (*Continued*)

Age (days) Treat-ment	Examination	Malformations (%)	Mortality (%)	Incidence of Typical Syndromes	Reference
				(Doses: 0.05 or 2.0 ml/egg, in subgerminal cavity)	Noto (1967a)
1	2–3	46	31	Nonclosure of the neural tube: head (29%), trunk (24%), tail (38%), defective eye—cyclopia (21%), enlargement of coelom and aorta (10%), retardation and severe inhibition (9%), tissue transformation (5%).	
Potassium iodide				(Doses: from 156 to 40,000 μg/egg, in albumen)	Wheeler and Hoffmann (1949)
16	21			Significant enlargement of thyroid (10.0 mg/3.1 mg control)	
Potassium selenate				(Doses: from 0 to 2.5 μg/egg, in air cell)	Halverson, Jerde, and Hills (1965)
14	17		83	Depression of growth (12.7/13.4 grams).	
Potassium selenite				(Doses: from 0 to 0.8 μg/egg, in air cell)	Halverson, Jerde, and Hills (1965)
14	17		94	Depression of growth (9.6/16.4 grams).	
Rhodium chloride				(Doses: 10 mg/egg, in yolk sac)	Ridgway and Karnofsky (1952)
8	18	80		Stunted growth, micromelia, edema.	
Selenium				(Doses: 15.15 μg/gram in hen's diet)	Poley, Moxon, and Franke (1937)
Hen		100	100	Edema of neck (77%) and head (35%); shortened upper beak (54%); twisted legs (46%) and neck (19%); absence of both eyes (3.6%); protruding eyes (32%); malposition (17%); absence of one eye (10%) (Fig. 100).	
				(Doses: 17 μg/gram in hen's diet)	Gruenwald (1958)

	1–5			Necrosis (after 2½ to 3 days) in certain areas of the brain, spinal cord, eyes, and limb buds; structural alterations in nearby areas; marked retardation in growth and differentiation of directly or indirectly affected primordia.	
Sodium ortho arsenate				(Doses: 0.3 mg/egg, in yolk sac)	Ridgway and Karnofsky (1952)
4	18	50		Stunted growth; micromelia; abdominal edema.	
Sodium arsenate				(Doses: from 0.01 to 1.0 μg/egg, in air cell)	Franke, Moxon, Poley, and Tully (1936)
0	14	14	93	Ectopic condition.	
Sodium cacodylate				(Doses: 2 mg/egg, in yolk sac)	Landauer (1949)
0	17		58	Anouria (26.4%).	
Sodium chloride				(Doses: 0.02 ml of 0.45 *M* solution/egg, in yolk sac at 2 days, in allantois at 5 days)	Grabowski (1966)
2	4–8	3	24	Microphthalmia (15%); brain defects (12%); abdominal hernia (6%); anouria (3%); eyelid defects (3%).	
5	7–11	14	19	Eyelid defects (15%); abdominal hernia (15%); brain defects (12%); beak defects (6%); microphthalmia (3%).	
				(Doses: 0.1 cm^3/egg, in yolk sac)	Aleksandrov, Bogdanova, and Chernukh (1966)
0	3–14		4–7	Malformations of ventral wall and extremities.	
Sodium fluoride				(Doses: from 0.5 to 2 mg/egg, in yolk sac or in allantoic cavity)	Spira (1956)
5	21		100	Emaciation and short survival; reduction in the amount of bone.	
Sodium selenate				(Doses: from 0.1 to 1.0 μg/egg, in yolk sac)	Franke, Moxon, Poley, and Tully (1936)
0	14–22	86	89	Underdeveloped or missing upper beak and eyes; some encephalocele; ectopia.	

Summary Table 11 (*Continued*)

Age (days) Treatment	Age (days) Examination	Malformations (%)	Mortality (%)	Incidence of Typical Syndromes	Reference
Sodium selenite				(Doses: from 0.01 to 0.9 μg/egg, in yolk sac)	Franke, Moxon, Poley, and Tully (1936)
0	14–22	50	60	Underdeveloped or missing upper beak.	
				(Doses: from 0.0 to 40.0 μg/egg)	Wright and Mraz (1965)
4	21	30	100	Dwarfness; shortened visceral cranium or mandibles; ulna, radius, tarsometatarsus, and tibia; fusion of the third and fourth digits (toes).	
Sodium thiocyanate				(Doses: from 0.3 to 0.7 ml of 0.1 *M* solution/egg)	Nowinski and Pandra (1946)
0	2			Inhibition of growth of various organs; heterauxesis, especially of brain—prosencephalon and mesencephalon; abnormal closing of neural tube; bradyauxesis of heart.	
				(Doses: from 0.4 to 24.2 mg/egg, in yolk sac)	Nowinski and Pandra (1947)
0	2			Omphalocephaly in the region of the forebrain and midbrain; stretching of the notocord.	
				(Doses: from 0.2 to 0.4 *M* solution/egg)	Rossi, Filoni, Bartoli, and Stefanelli (1966)
1–4	19		48–72	Shrunken phalanges of the hind limbs.	

Tantalum hydroxide				(Doses: from 1 to 80 mg/egg, in albumen)	
1, 10	19			Hemorrhage (24.7%); stiff hock joint; soft bones.	Machlin, Pearson, and Denton (1952)
Thallium acetate				(Doses: 0.15 ml of 5% solution/egg, in yolk sac)	
7	21			Stunted growth; micromelia; shortened mandible.	Shipkowitz and Clarke (1955)
				(Doses: 0.5 mg of 1.9 μM solution/egg, in yolk sac)	
4	21		95	Deformities of beak (88%); abnormalities of legs (80%) and of tail (65%).	Landauer (1960a)
Thallium nitrate				(Doses: 0.05 cm³ of 2% aqueous solution/egg)	
4	11			Angulation of one or both tibias (90%); diminutive or absent lower beak with arched upper beak (83%).	Duraiswami (1955)
				(Doses: 0.05 ml of 2% solution/egg)	
3	21	Heavy	Early	Retarded development; embryo choked before hatching; deformation of the beak, extremities, and spine (scoliosis and kyphosis); osteodystrophy.	Cervenansky, Kalman, Micek, Maar, Makovsky, Skrovina (1964)
Thallium sulfate				(Doses: from 0.2 to 2.0 mg/egg, in yolk sac)	
1–12	21	100	80	Achondroplasia (from 0 to 100%); parrot beak; small head; short trunk and extremities; short and curved long bones, especially tibia.	Karnofsky, Ridgway, and Patterson (1950); Ridgway and Karnofsky (1952); Karnofsky and Lacon (1964)
Thorium chloride				(Doses: from 1 to 100 mg/egg, in albumen)	
1, 10	19			Hemorrhage (32.6%); stiff hock joint (22.1%); short bones (7.4%); microphthalmia (4.2%); ectopia (3.2%).	Machlin, Pearson, and Denton (1952)

Summary Table 12 Pathological Effect of Various Organic Compounds on the Developing Avian Embryo

Age (days) Treatment	Age (days) Examination	Malformations (%)	Mortality (%)	Incidence of Typical Syndromes	Reference
Acetazolamide				(Doses: at 1 day—from 2.5 to 10 mg, at 4 days—from 1 to 20 mg, in yolk sac)	Landauer and Wakasugi (1967)
1	18	35–73	23–49	Short or abnormal upper beak (52%), anouria (21%).	
4	18	0–74	9–32	Short or abnormal upper beak (74%), muscular hypoplasia, bending of tibiotarsus shaft.	
Acetone				(Doses: from 5 to 45 mg/egg)	Caujolle, Caujolle, Cros, Dao-Huy-Giao, Moulas, Tollon, and Caylus (1965)
3	11	20–54	0–50	Defects of limbs (25%); celosomia (22%); twisting of axial skeleton (14%); defects of eye (13%) and beak (1.4%); anouria (0.8%).	
4	11	4–26	0–88	Defects of limbs (5.9%) and eye (5.7%); celosomia (3.3%); defects of axial skeleton (1.2%) and beak (0.8%); anouria (0.6%) (Fig. 115).	
Acetylcholine				(Doses: from 1.2 to 2.85 mg/egg, in yolk sac, air cell)	Bueker and Platner (1956)
4–12	7–13			Decreased weight (8/12 grams), visceral ectopia.	
Acriflavine				(Doses: from 0.042 to 0.125 cm^3/egg, injected into head)	Ancel (1955)
1	15–21	58	28	Acleiencephalia (37%), cleiencephalia (11%), facial malformations (11%).	

Aflatoxin B				(Doses: from 0.08 μg to 3.0 μg/egg, into air cell)	Shibko, Arnold, Morningstar, and Friedman (1968)
3–18	19	0	2–98	Severe growth retardation (after twelfth day); in the same ratio, reduction in weight of the liver.	
Alcohol, anhydrous				(Doses: from 5 to 15 parts/100 of water, on embryo)	Patry and Ferrier (1934)
2			Rapid	Acceleration of heartbeat.	
				(Doses: from 0.05 to 0.3 ml/egg, in yolk sac)	McLaughlin, Marliac, Verrett, Mutchler, and Fitzhugh (1963)
0			5–100	Intermediate toxic effect.	
				(Doses: from 0.4 to 0.6 ml/egg, in air cell)	Sandor and Elias (1968)
0	8		49	General malformations (61%), deformed brain vesicle (15%), caudal malformation (24%), malformation of heart (9%), abnormal flexure (3%) (Fig. 116).	
1	8		95	General malformations (89%), deformed brain vesicle (24%), caudal malformations (24%) (Fig. 116).	
Allopurinol				(Doses: from 0.001 to 10 mg/egg, in yolk sac)	Krakoff and Meyer (1965)
8	14		0–50	Little toxic effect.	
Alloxan				(Doses: concentration $M/50$, applied directly to embryo)	Diwan (1967)
½				*In vitro:* abnormalities mainly in the brain region and neural tube.	
o-Aminoazotoluene				(Doses: 0.3 ml of suspension in albumen/egg, in blastoderm)	Noto (1967c)
1	2–3	22	25	Inhibition of closure of the caudal neural tube (56%); nonclosure of neural tube: of tail (33%), of head (11%).	

Summary Table 12 (*Continued*)

Age (days) Treatment	Age (days) Examination	Malformations (%)	Mortality (%)	Incidence of Typical Syndromes	Reference
Aminobutyric acid				(Doses: 1 mg in 1 ml physiological solution/egg, in yolk sac)	Aleksandrov, Bogdanova, and Chernukh (1966)
0	3–14		12	Defects of brain (13%), of body (8%), of eye (2%), and of extremities (2%).	
Aminoguanidine				(Doses: from 2 to 60 mg/egg, in yolk sac)	Neuman and McCoy (1955)
4	14		25–100	Severe inhibition of liver growth (45/173 mg) (Fig. 117).	
				(Doses: 20 mg)	Nosal and Watterson (1959)
4	13–20		17–94	Reduction in size of liver (25/133 mg); below 26 mg., gallbladder seldom formed (Fig. 117).	
				(Doses: from 0.5 to 50.0 mg/egg)	Stoll, Bodit, and Maraud (1968)
3	14	100		Brachymelia (7%), ankylosis (2%).	
β-Aminopropionitrile				(Doses: from 0.1 to 20.0 μl/egg)	Chang, Witschi, and Ponseti (1955)
			25–100	Lesions of legs, bending of femurs.	
				(Doses: fed to hen, from 0.01 to 0.06%)	Barnett, Richey, and Morgan (1957)
Hen's diet		40–80	4–100	Hemorrhages of dead embryo from 40 to 80% (Fig. 118).	
				(Doses: from 0.05 to 0.4 ml of 25% solution/egg, in yolk sac)	Rosenberg (1957)
4–9	10–12	41–100	28–81	Among living: bowing of both tibias, marked buckling and bowing of mandibles.	

			Among dead: pronounced changes in the beak, agenesis of the body wall.	
			(Doses: from 0.5 to 20.0 mg/egg, on allantoic membrane)	Levene and Gross (1959)
4, 14	17		Increase in fragility, thickening of femur and tibia with dislocation of joints; often, nearly complete rotation of tibia (Fig. 119).	
			(Doses: from 0.001 to 1.0 mg/egg, in albumen)	Naber and Blackwood (1962)
0–15	3–21	10–100	Malformed legs, twisted and shortened lower beak.	
			(Doses: from 0.001 to 0.1 mg/egg, in albumen)	Champion and Schnabel (1966)
0	21	95	Embryonic stunting, hemorrhages in the vessels of the allantois and amniotic sac.	
3½	21	93	Stunting and gross malformation of embryos, retardation in development of blood-vessel network.	
4½	10	100	Stunting and malformation of embryos.	
8		77–96	A stereotyped pattern of stunting of embryos at hatching; short upper and/or short lower mandibles and twisting/bowing of the mandibles.	
Aminopterin			(Doses: from 0.001 to 1.0 mg/egg, in yolk sac)	Wagley and Morgan (1948)
7	14	30–100	Marked changes in hematopoietic tissue.	
			(Doses: from 2 to 1000 μg/egg, in albumen)	Snell and Cravens (1950)
0	21	42–100	With 10 μg—100% mortality; with 100 μg or more—complete inhibition of embryonic development.	
			(Doses: 0.025 and 0.05 mg/egg, over blastoderm)	Blattner, Williamson, Simonsen, and Robertson (1960)
2	5	100	Affected: neural tube, auditory vesicle, lens vesicle, visceral arches, somites, mesodermal structures; overall general degeneration.	

Summary Table 12 (*Continued*)

Age (days)					
Treatment	Examination	Malformations (%)	Mortality (%)	Incidence of Typical Syndromes	Reference
Aminopterin				(Doses: 0.3 ml of suspension in albumen/egg)	Noto (1967c)
1	2–3	34	41	Nonclosure of neural tube—head and tail (25%); retardation and inhibition (12%); tissue transformation–hyperplasia, double embryo (9%).	
				(Doses: 0.032% solution on paper 1 mm^3, on embryo)	Puchkov (1967b)
1	6	67		Microphthalmia (57%); defects of body and tail (52%) and extremities (41%); microcephaly (33%).	
2	6	43		Defects of body and tail (30%) and extremities (29%); microphthalmia (27%); microcephaly (10%).	
Antimycin-A				(Doses: from 0.1 to 1.0 μg/egg, in albumen or yolk sac)	McKenzie and Ebert (1960)
1	2			Prominent anomalies of the nervous system, some inhibition of growth.	
L-Arginine				(Doses: 3 mg/egg, in side of blastoderm)	Noto (1967a)
¼–¾	3	16	10	Nonclosure of neural tube: of tail (10%), of head (5%).	
Atropine				(Doses: from 0.6 to 1.5 mg/egg, in yolk sac, on embryo)	Bueker and Platner (1956)
4–12	13		59	Slight visceral ectopia, decreased weight (8.6/12.0 grams).	

Azaserine					
0–12	12–16	0–92	8–100	(Doses: from 0.075 to 0.3 mg/egg, in yolk sac) Growth inhibition (7.5/12.4 grams); beak defects: facial coloboma, parrot and short beak, short lower beak; leg defects: micromelia, ectrodactyly, short or bent tarsometatarsus, missing metatarsus; wing defects; feather growth inhibition; edema (Fig. 120).	Dagg and Karnofsky (1955)
4	18			(Doses: 0.1 mg/egg, in yolk sac) Abnormal lower beak, small eyes, defective upper eyelid, defect in growth of appendicular skeleton, micromelia, feather growth inhibition.	Karnofsky (1955)
4	18	Many	High	(Doses: from 0.15 to 0.2 mg/egg, in yolk sac) Inhibition of growth, paleness, edema and impaired feather formation; gross abnormalities of head and extremities: facial coloboma with cleft palate, shortened mandible and tarsometatarsus with ectrodactyly, hypophalangy.	Murphy, Dagg, and Karnofsky (1957)
2	3	0–100	8–50	(Doses: from 0.007 to 0.2 mg/egg, in vitelline membrane) Growth inhibition and developmental retardation in specific organs: forebrain and midbrain, optic cup and lens primordia.	Blattner, Williamson, and Simonsen (1958)
4	16			(Doses: from 0.15 to 0.25 mg/egg, in yolk sac) Extremely short beak, facial coloboma with cleft palate, shortened mandibles, micromelia, short and bent tarsometatarsus, ectrodactyly, hypophalangy, hemimelia of wings and legs.	Karnofsky and Lacon (1964)
6-Azauridine					
4, 8	14–20	13–68	45	(Doses: from 1 to 10 mg/egg, in yolk sac) Most frequent: frontal encephalocele, microcephalia, anophthalmia, and spastic paraplegia.	Grafnetterova, Grosi, Fumagalli, Morganti, and Grafnetter (1966)

Summary Table 12 (*Continued*)

Age (days) Treatment	Age (days) Examination	Malformations (%)	Mortality (%)	Incidence of Typical Syndromes	Reference
Benzidine				(Doses: 0.3 ml. of suspension in albumen—7 mg/egg, in blastoderm)	Noto (1967c)
1	2–3	24	31	Inhibitory effect on closure of the cephalic neural tube: retardation and inhibition (69%), nonclosure of head (31%).	
Benzimidazole				(Doses: from 0.1 to 0.7 mg/egg, in albumen)	Blackwood (1961)
0–7	18		18–100	(See Fig. 121.)	
				(Doses: from 0.1 to 0.7 mg/egg, in albumen)	Blackwood (1962)
0–3		31–100	3–100	Malformations of head (9 to 24%), heart (4 to 6%), trunk (8 to 43%), whole embryo (3 to 22%).	
Benzyl alcohol				(Doses: 0.01 and 0.02 cm^3/egg, in yolk sac)	Duraiswami (1954)
0–7	20			Limb deformities (62%); meningocele (40%); beak defects (27%); localized blebs and/or generalized edema (15%).	
Busulfan				(Doses: 0.4 μg/egg, in yolk sac)	Aleksandrov and Stanzhevskaya (1969)
0–4	7–9	29	50	Retardation in growth (21% at 9 days); malformations of limbs (15.5%), eyes (10.8%), brain (10.0%), tail (4.6%), beak (1.9%), axial skeleton (0.8%).	
Carbutamide				(Doses: from 0.05 to 0.4 ml of 4.8% solution/egg, in yolk sac)	Lane and Slepecky (1960)
0–5	17		35	Decrease in body size (22.9/26.5 grams at 17 days); micromelia (39.9%); beak abnormalities—cross beak (8.6%); abnormal down (2.7%).	

Chloramphenicol					
0	4			(Doses: 0.5 and 1.0 mg/egg) Marked inhibition of growth and development without gross malformations.	Blackwood (1962)
p-Chloroacetophenone					
17	38			(Doses: 0.0005 *M* solution to entire embryo, *in vitro*) Abnormalities of brain region (81%), neural tube (44%).	Lakshmi (1962)
Chromomycin A_3					
3	9	15–59	15–59	(Doses: 1 and 3 μg/egg) Various types of malformations of the body, limbs and, most of all, the head (deformities of beak, of eyelids).	Maraud, Coulard, and Stoll (1963)
Colchicine					
1	11–20		14–70	(Doses: from 0.01 to 0.025 mg/egg, injected into head) Various forms of exencephaly, micrognathia superior, absence of upper eyelids.	Ancel (1955)
Cyclophosphamide					
0	7–16	34–47	60–77	(Doses: 0.1 mg in 0.5 ml Tyrode solution/egg, in albumen) Retardation in growth (6/14 grams at 16 days); brachymelia, parrot beak, absence of eyelids, twisting of internal organs.	Gerlinger, Ruch, and Clavert (1962)
3	7, 11	38		(Doses: 0.1 mg/egg, in albumen) Damage of bulbotruncal region (46%), of communication between small arteries or anterior branches (46%); malposition of bulbus aortae (8%).	Ruch, Kremer, Rumpler, and Gerlinger (1965)
4	7, 11	43	30	(Doses: 0.1 mg/egg, in albumen) Gross anomalies: micromelia, hypotrophy, parrot beak, ablepharia. (Anomalies of heart: cardiac malposition (48%), arterial intervention (48%), cardiac malformation (15%), bulboventricular communication (7%).	Ruch, Zahnd, Kremer, Rumpler, and Gerlinger (1965)

Summary Table 12 (*Continued*)

Age (days) Treat-ment	Exami-nation	Malfor-mations (%)	Mor-tality (%)	Incidence of Typical Syndromes	Reference
Cyclophosphamide				(Doses: 40, 66, and 100 μg/egg, intraependymal injection)	Miclea and Arcan (1967)
4	5–7		100	Retarded growth, hemorrhagic areas at limbs and along the embryonic axis, severe degenerative pyknotic processes.	
Cytosine (arabinoside)				(Doses: from 0.01 to 1.0 mg/egg, in allantois, in yolk sac)	Karnofsky and Lacon (1966)
3–12	18	30	30–89	Inhibition of development (3 to12/18 grams at 18 days); beak defects (33 to 93%): twisted, parrot and short upper beak, short lower; facial defects (27 to 39%): facial coloboma, cleft palate; limb defects (17 to 100%): micromelia, missing or short toes; eye defects (7 to 25%): corneal cysts, eyelid defects, and small eyes (Fig. 122).	
Dactinomycin				(Doses: from 0.0625 to 0.25 μg/egg, in yolk sac)	Pierro (1961)
1–4	17	4–53	14–80	Rumplessness (53%) and trunklessness (28%), some edema in the posterior trunk region, hypoplasia of the lower extremities.	
				(Doses: from 0.01 to 0.02 μg/egg)	Cury and Montenegro (1969)
1	4			Alterations in nervous system, especially in the region of forebrain (prosencephalon), and the neural tube at the caudal region.	

2-Deoxyguanosine				(Doses: from 0.25 to 1.0 mg/egg, in yolk sac)	Karnofsky and Lacon (1964)
4	18			Facial coloboma and cleft palate, short lower beak, angulation of femur, fusion of femur and tibia, fusion of humerus and ulna, missing metatarsals and metacarpals, malformation of vertebrae, hemimelia.	
				(Doses: from 25 to 200 μg of aqueous solution/egg, in yolk sac)	Pierro and Spiggle (1966)
2–5	17	17–66	50–80	Skeletal defects: axial (0 to 43%), facial (2 to 41%); phocomelia (0 to 5%), micromelia (0 to 49%), foot defects (1 to 63%) (Fig. 123).	
Desoxypyridoxine				(Doses: from 0.1 to 1.0 mg/egg)	Conti and Milio (1965)
0–2½		2–15	11–48	Inhibits development; no specific teratogenic effects.	
Diazinon				(Doses: 1.0 mg/egg, in yolk sac)	Khera and Bedok (1967)
4	28	88		Effects similar to those of treatment with parathion (see below).	
Dimethyl sulfoxide				(Doses: from 10.3 to 12.2 mg/embryo, intravenous injection)	Caujolle, Caujolle, Cros, and Calvet (1967)
3	10–16	19		Anophthalmia and other eye malformations (16.9%); malformations of the beak (11.4%).	
4	10–16	26		Malformations of limbs—from ectromelia to brachymelia (21.8%); of beak (4.6%); of eyes (3.9%); anourous embryos (3.7%); celosomia (5.5%).	
2,4-Dinitrophenol				(Doses: from 1 to 750 μg/egg, in air cell)	Feldman, Ferguson, Rigdon, Reid, Cross, and Couch (1958)
0, 5, 8, 15	21		11–80	Cataract in embryos and chicks (0 to 20%).	

Summary Table 12 (*Continued*)

Age (days) Treatment	Age (days) Examination	Malformations (%)	Mortality (%)	Incidence of Typical Syndromes	Reference
2,4-Dinitropheno				(Doses: from 200 to 1000 μg/egg, in air cell)	Feldman, Ferguson, and Couch (1959)
6–15	21		7–100	Cataract in embryos and chicks (5 to 75%), edema (35 to 69%), with hemorrhage (23 to 40%) at the back of the neck.	
				(Doses: *in vitro*—from 10^{-4} to 2×10^{-3} *M*/embryo)	Bowman (1967)
1	2			Degeneration and/or complete absence of neural tissue accompanied by reduction in number and inhibition of somites.	
Ethionine				(Doses: LD_{50} = 2 to 5 mg/egg)	Karnofsky (1955)
16		LD_{50}		Growth inhibition, thin legs; enlarged, fatty liver; severe edema.	
				(Doses: 19 to 20 mg/egg)	Dessi, Rizzoli, and Cestari (1959)
4	14			Inhibition of embryonic growth (1.9/2.3 grams at 9 days; 1.2/4.3 grams at 12 days).	
				(Doses: from 10 to 20 mg/egg, in yolk sac)	Rizzoli, Dessi, and Cestari (1959)
4	14		35–70	Decrease in growth, damage of the ocular cutaneous tissues and liver, reduction in respiratory activity (60% at 10 days)	
N-Ethylnicotinamide				(Doses: from 2.5 to 10 mg/egg, in yolk sac)	Landauer and Wakasugi (1967)
1	18	5–56	32–86	Short or abnormal upper beak (50%), rumplessness (6%).	
4	18	10–100	10–25	Short or abnormal upper beak (63%), short and abnormal tibiotarsus-fibula (74%).	

Floxuridine				(Doses: 0.08 μg/egg, in yolk sac)	Karnofsky and Lacon (1964)
4	18			Short lower beak, short fused metatarsals, fusion of femur and tibia, micromelia.	
				(Doses: from 0.005 to 0.08 μg/egg, in yolk sac)	Kury and Craig (1966)
1–4	21	7–29	64–78	Gross malformations (23%), defective eyelids (13%), micromelia (8%), corneal cysts (5%).	
				(Doses: from 0.5 to 3.1 μg/egg, in yolk sac)	Kury and Crosby (1967)
1–4	19–20			Severe retardation in growth (28%); skeletal abnormalities (58%) affecting upper and lower extremities, axial skeleton, palate and calvarium. Other gross anomalies: bilateral hypoplasia of kidneys (13%), omphalocele (19%), enlargement and discoloration of liver (9%), exteriorization of viscera (6%).	
5-Fluorouracil				(Doses: 0.2 mg/egg, in yolk sac)	Karnofsky and Lacon (1964)
4	18			Facial coloboma and cleft palate, short lower beak, changes in long bones and toes, micromelia.	
				(Doses: from 2 to 1000 μg/egg, in albumen)	Conti and Aita (1966)
0–8	21	4–100	2–81	Micromelia, microphthalmia, ectrosomia (Fig. 124).	
				(Doses: from 0.045 to 0.1 mg on third day, 0.2 mg on fourth day, in yolk sac)	Kury and Craig (1966)
3–4	20	11	90	Decrease in weight, micromelia of metatarsus (3%), brain hemorrhage (3%).	
				(Doses: 1 cm^2 of filter paper saturated with 0.1% solution, applied to embryo for 10 min *in vivo*)	Puchkov (1966)
1	7	69		Microcephaly (23%); microphthalmia (45%); malformations of body, tail and wings (53%), of legs (31%).	

Summary Table 12 (*Continued*)

Age (days) Treat-ment	Exami-nation	Malfor-mations (%)	Mor-tality (%)	Incidence of Typical Syndromes	Reference
5-Fluorouracil				(Doses: 1 cm^2 of filter paper saturated with 0.1% solution, applied to embryo for 10 min *in vivo*)	Puchkov (1967b)
2	7		52	Microcephalia (12%); microphthalmia (50%); absence of part of trunk and tail (23%), of wings (10%), of limbs (10%), or of allantois (15%).	
Guanazolo				(Doses: from 0.125 to 0.5 mg/egg, in yolk sac)	Youngner, Ward, and Salk (1950)
4	16	3–30	38–80	Growth inhibition (10/13 grams at 16 days), developmental abnormalities (Fig. 126).	
				(Doses: from 0.1 to 10 mg/egg)	Karnofsky (1955)
4	15–18			Growth inhibition; abnormalities in most surviving embryos, including parrot beak, short lower beak, thin stiff legs with curled toes, edema.	
Hydroxyurea				(Doses: from 0.25 to 1.0 mg/egg, in yolk sac)	Bodit, Stoll, and Maraud (1966)
6	14	27–100	56–93	Brachymelia (32%), bending of limbs (14%), fusion of joints (4%), and dislocation of tibia-tarsometatarsus; malformation of beak: upper (24%), lower (10%); atrophy of upper eyelids (16%).	
Hypoglycine A				(Doses: 20 and 200 μg/egg, injected under area vasculosa)	Persaud and Kaplan (1970)
1, 2	4, 6		14–40	No significant increase in abnormalities.	

Isoniazid				(Doses: 0.02 cm³/egg, in yolk sac)	Duraiswami (1961)
3–5	18		71–74 at 11 days	Limb defects (61 to 64%), beak defects (15 to 25%), visceral ectopia (4 to 6%).	
Janus Green				(Doses: from 3.5 to 16.2 μg/egg, in yolk sac)	Braun (1954)
0–3	3–15	12–93		Microphthalmia (24%), phocomelia (17%), dysplasia cerebri (14%), cardiacus amorphus (13%), hydrocephalus (8%), hemorrhagia cerebralis (6%), dystopia extremitatum (5%), ectopia cordis (4%), and many others (26 in all).	
				(Doses: from 2 to 50 μg/egg, in amnion)	Menkes and Deleanu (1964)
2	6–10		10–100	Retardation in development (80% of control), syndactylia (0 to 100%) (Fig. 112).	
				(Doses: from 6 to 12 μg/egg *in vivo*, intraependymal injection)	Menkes and Miclea (1964)
4	8		90	Complete or incomplete destruction of the spinal cord, shortened tail.	
				(Doses: 1 μg/egg, injected into pharynx)	Menkes, Pavkov, and Mircov (1967)
3	10	7–70	70	Defects of cardiovascular system, trunk, main arterial vessels.	
Leucine				(Doses: 0.2 ml of 1% solution/egg, in amnion)	Bergström, Erilä, and Pirskanen (1967)
9	21	86	96	Abnormalities of legs (92%): inward flexion of toes.	
Mechlorethamine				(Doses: 0.01 mg/egg, in yolk sac)	Murphy, Dagg, and Karnofsky (1957)
4	18		Early	Occasional cross beak, thin legs.	
				(Doses: from 0.025 to 0.045 mg/egg, inoculated over blastoderm)	Blattner, Williamson, Simonsen, and Robertson (1960)
2	3			Growth inhibition; abnormalities of neural tube (100%), somites (89%), visceral arches (78%), lens vesicles (56%), auditory vesicles (45%) (Fig. 127).	

Summary Table 12 (*Continued*)

Age (days) Treatment	Age (days) Examination	Malformations (%)	Mortality (%)	Incidence of Typical Syndromes	Reference
Mechlorethamine				(Doses: from 6 to 9 μg of aqueous solution/egg, on embryo)	Salzgeber (1962)
3–4	13	94	40	Reduction in size of anterior and posterior extremities (legs, wings, toes), ectrodactyly, syndactyly, phocomelia, brachymelia—all unilateral or bilateral (Fig. 127).	
				(Doses: from 0.6 to 2.0 μg of aqueous solution/egg, on embryo)	Salzgeber (1966)
2–4	12, 20	90		Phocomelia (49 to 62%), reduction in size of legs and wings, disappearance of fibula in leg and of ulna and radius in wing (Fig. 127).	
				(Doses: 1.26 μg/egg, in lumen of the neural tube canal)	Miclea and Arcan (1968)
4	5–18			Neuroepithelial alterations by necrotic and necrobiotic degeneration.	
				(Doses: 50 μg/ml of Tyrode solution, *in vitro*, 2 to 8 min)	Salzgeber (1968)
3				Wings: phocomelia (46 to 52%); micromelia (23%); ectromelia (43%). Legs: phocomelia (78%), ectromelia (22%).	
Methacycline				(Doses: from 0.3 to 7.5 mg/egg, in albumen)	Genazzani, Pelagalli, Pagnini, and DiCarlo (1965)
4	15		40–95	At high doses: somatic and skeletal modifications. (See Fig. 108 on mortality.)	

Methenolone				(Doses: from 0.05 to 0.4 mg/0.05 ml sesame oil/egg, in yolk sac)	Lakshmi (1967b)
3	13	28	46	Edema (34%), chin pouch (31%), ectopia (23%), micromelia (9%), beak abnormalities (3%).	
Methotrexate				(Doses: 0.005 mg/egg, in yolk sac)	Karnofsky (1955)
4	18			Growth inhibition, microphthalmia (cyst over cornea, small eyes), short lower beak, evisceration, edema, increase in amniotic fluid, maldevelopment of allantois.	
3-Methylcholanthrene				(Doses: saturated aqueous solution on 0.5 cm^2/egg, on albumen)	Beskrovni (1941)
0–8		9–29	30–100	Retardation of growth; greatest effect on nervous tissue, especially brain.	
				(Doses: a few crystals/egg, dropped on allantois)	Beohar, Goldhaber, and Hertig (1964)
4–10	18		30–100	Keratogenic metaplasia of the chorionic ectoderm. (See text table on mortality.)	
Methylene Blue				(Doses: from 1.4 to 14 μg/egg, in albumen, in cephalic region)	Kirrmann (1961)
2	8	44	53	Partial or complete absence of roof of skull and malformation of brain (50%); facial coloboma (33%); micromelia (33%); absence of eyelids (11%); celosomia (11%).	
Methyl pyridyl ketone				(Doses: from 400 to 600 μg/egg, in yolk sac)	Ackermann and Taylor (1948)
4	5		0–100	Lethal in 24 hours.	
				(Doses: 0.5 mg/egg, in yolk sac)	Duraiswami (1955)
4	11		53	Varying degrees of myodystrophy of spinal muscle, giving rise to scoliosis, thin limbs, short upper beak.	

Summary Table 12 (*Continued*)

Age (days) Treatment	Age (days) Examination	Malformations (%)	Mortality (%)	Incidence of Typical Syndromes	Reference
Methyl pyridyl ketone				(Doses: 0.75 mg of 6.2 μM solution/egg, in yolk sac)	Landauer (1957)
4	13, 22		57, 98	Muscular hypotrophy (97%), short upper beak (24%).	
				(Doses: from 375 to 500 μg/egg, in yolk sac)	Landauer and Clark (1962b)
4	13, 22	24	59, 75	Muscular hypotrophy (32%), short upper beak (3%).	
				(Doses: 0.5 mg/egg)	Landauer and Clark (1964c)
3	21	67	94	Muscular hypoplasia (61%), short upper beak (22%), twisted neck (9%).	
4	21	48	87	Muscular hypoplasia (47%), short upper beak (2%), twisted neck (2%).	
5	21	40	71	Muscular hypoplasia (36%), short upper beak (5%), twisted neck (6%).	
Mitomycin C				(Doses: from 2 to 28 μg/egg, in yolk sac)	Kury and Craig (1967)
2–4	6–20	38–42	59–70	Stunting (83%), bilateral micromelia (78%), malformation of lower extremities (70%), abnormalities of beak (44%), defects of abdominal wall (23%), cerebral hemorrhage (17%).	
Neostigmine				(Doses: 3 injections, 5 mg/egg, in yolk sac)	Bueker and Platner (1956)
4, 5, 8	13			Shortened beak, micromelia, retarded development of eyelids; with larger doses (15 mg), severe midline bending of vertebral column.	

4	8			(Doses: 2 mg/egg, in yolk sac) Fused cervical vertebrae, twisted spinal column, angulation of femur.	Karnofsky and Lacon (1964)
Nicotinamide					
				(Doses: at 1 day, 2.5 μg of 0.018 μM, and at 4 days, 10 μg of 0.07 μM solution/egg, in yolk sac)	Landauer (1957)
1	13		56	Rumplessness (51%), ectromelia, hypodactyly, etc. (15%).	
4	13		30	Micromelia (97%), parrot or cross beak (78%), miscellaneous defects (0.7%).	
4	18			(Doses: 0.01 mg/egg, in yolk sac) Gross skeletal abnormalities: parrot beak, micromelia, bending of tibiotarsus and tarsometatarsus.	Murphy, Dagg, and Karnofsky (1957)
2	3		20–100	(Doses: from 1 to 10 μg/egg) Strong antimetabolic and antimitotic effects.	Taglioretti and Eridani (1958)
4	22	15–81	23–93	(Doses: 2.5, 4, 5, or 10 μg/egg, in yolk sac) Micromelia (9 to 98%), parrot and/or short lower beak (0 to 76%), short upper beak (0 to 2%).	Landauer and Clark (1962b)
4	14–18			(Doses: from 0.1 to 0.2 cm^3, in 5% solution of carboxymethylcellulose in saline/egg, in yolk sac) Micromelia, curved upper beak, poorly ossified toes.	Karnofsky and Lacon (1964)
4	13	1	1	(Doses: 1.5 μg/egg) Malformations (1%).	Landauer and Clark (1964c)
3–5	13			(Doses: 4 μg/egg) Resistant strain: micromelia (10–40–13%), parrot beak (0–6–4%). Susceptible strain: micromelia (35–65–13%), parrot beak (8–13–3%).	Landauer (1965a)

Summary Table 12 (*Continued*)

Age (days) Treatment	Age (days) Examination	Malformations (%)	Mortality (%)	Incidence of Typical Syndromes	Reference
Nicotine					
0	21		24–97	(Doses: from 0.1 to 5.0 mg/egg) Reduction in body weight, torticollis, edema (anasarca), some changes in heart and kidney.	Kiwaki (1956)
0–5	13, 20	5–100	90–100	(Doses: from 2.5 to 5.0 mg/egg, in yolk sac) Abnormal cervical vertebrae, shortening and twisting of the neck (63 to 100%); muscular dystrophy (19%); shortening of the upper beak (5%); edema (anasarca).	Landauer (1960b)
Noprylsulfamide					
4	20			(Doses: 0.05 cm³ of 10% aqueous solution/egg) Angulation of one or both tibias (57.6%); typical parrot-beak defects (30.1%).	Duraiswami (1955)
Oxytetracycline					
4	15		25–65	(Doses: from 0.3 to 7.5 mg/egg, in albumen) Somatic and skeletal malformations. (See Fig. 108 on mortality.)	Genazzani, Pelagalli, Pagnini, and DiCarlo (1965)
Oxythiamine					
5, 10, 15	21		5–70	(Doses: from 0.5 to 4.0 mg/egg) Tissue edema.	Naber, Cravens, Baumann, and Bird (1954)
0–18	21			(Doses: from 0.5 to 1.0 ml/egg) No significant effect.	Kawamura (1967)
Pantethine					
0	4–28		63	(Doses: 43 mg/turkey egg, in yolk sac) Inhibition of growth; malformations during the first week: retarded development of brain and neural cord, heterotaxia, with anemia and hemorrhage; during the final week: micromelia, curled	Goetinck, Abbott, and Kratzer (1957)

			toes, rumplessness, bilateral facial coloboma, polydactyly, hypertrophy of gizzard.	
Parathion			(Doses: 1.0 mg/chicken egg, in yolk sac)	Khera and Bedok (1967)
4	19	85–90	Skeletal defects: brachymorphic neck (at 7 days); inhibited vertebral column, with axial length about one half of normal (at 19 days); compression of head anteroposteriorly; short neck and trunk, micromelia (Fig. 114).	
			In duck—more or less the same anomalies at corresponding developmental ages.	
			(Doses: from 1.17 to 29.2 mg/egg, in yolk sac)	Greenberg and LaHam (1969)
4–12	15–16	100	Growth retardation (two-thirds normal); abnormalities of legs and beak, micromelia.	
			(Doses: from 0.05 to 2.0 ml/egg)	Lutz-Ostertag, Meiniel, and Lutz (1969)
1, 2, 4	7, 14	73	Lordosis of neck (100%), anouria (100%), achondroplasia (98%), shortening of posterior part (90%), lordosis of body (86%), maldevelopment of feathers (75%), edema (68%), nanism (58%), celosomia (30%), spina bifida (11%).	
			(Doses: immersion of quail eggs in 10% solution for 30 sec)	Meiniel and Lutz-Ostertag (1970)
0	12–14		Teratogenesis of skeleton: retardation of growth, abnormalities of long bones, malformation of axial skeleton—lordosis of the neck and numerour flexions in the sagittal plane.	
Penicillin			(Doses: from 0.05 to 0.5 ml/egg, in air cell, allantois, yolk sac)	Gentry (1958)
10			Edema and hemorrhage in wings, legs, and head.	

Summary Table 12 (*Continued*)

Age (days) Treatment	Examination	Malformations (%)	Mortality (%)	Incidence of Typical Syndromes	Reference
Phthalic acid					
3	21		50–92	(Doses: from 0.025 to 0.10 ml/egg, in yolk sac) Cranium bifidum and anophthalmia; marked exophthalmia (from absence of bone tissue of skull) and blindness (due to failure of the cornea to develop).	Bower, Haberman, and Minton (1970)
Physostigmine salicylate					
0	9–16		90	(Doses: from 0.1 to 1.0 mg/egg) Subcutaneous edema of face, neck, etc. (90%).	Ancel and Lallemand (1943b)
4–8	13			(Doses: 0.5 mg/egg-day, in yolk sac) Shortened beak, micromelia, retarded development of eyelids.	Bueker and Platner (1956)
4	18			(Doses: 1 mg/egg, in yolk sac) Vertebral fusion, micromelia, parrot beak.	Karnofsky and Lacon (1964)
Physostigmine sulfate					
4	17	80–100		(Doses: from 0.25 to 1.0 mg/egg, in yolk sac) (Micromelia (25 to 64%), syndactyly (12 to 38%), deformities of beak (25 to 66%).	Landauer (1949)
2–6	10+	53	16	(Doses: from 25 to 50 μg in 0.25 cm^3 aqueous solution/egg, in air cell) Skeletal deformities of hind limbs (32%), short, parrot or cross beak (6.7%), anophthalmia (10%), edema around lower jaw (13%), ectopia of viscera (3%).	Agarwal (1956)
4	21		63	(Doses: 0.5 mg/egg, in yolk sac) Micromelia (96%), parrot or short lower beak (95%), short and/or crooked neck (93%).	Landauer (1957)

				(Doses: from 25 to 75 μg in 0.25 cm^3 aqueous solution/egg, in air cell)	Agarwal and Monga (1958b)
2–6	10+	36–71	16–55	Quantitative variations in response in three breeds of chicken.	
				(Doses: from 0.25 to 2.0 μg in 0.25 ml aqueous solution/egg, in air cell)	Agarwal, Monga, and Agarwal (1959)
3¾	10	14–88	27–90	Teratogenic and toxic responses (Fig. 128).	
Pilocarpine				(Doses: 3, 6, and 12 mg in saline solution/egg, in yolk sac)	Landauer (1953a)
1–5	13	85	13–70	Syndactyly (17 to 79%), rumplessness (0 to 43%), short first toes (0 to 75%), tarsometatarsal defects and micromelia (30 to 84%). (See Fig. 130.)	
				(Doses: 3, 6, and 12 mg/egg, in yolk sac)	Landauer (1953b)
4	18	16–84	21–60	Syndactyly (79%), beak defects (42%), micromelia (38%), short and bent tarsometatarsus (35%).	
				(Doses: 3, 6, and 12 mg in saline solution/egg, in yolk sac)	Landauer (1954)
0–6	13		16–100	Syndactyly (58 to 94%), defects of long bones (0.6 to 84%), beak defects (0.9 to 42%) (Figs. 129, 130).	
				(Doses: 3 and 6 mg/egg, in yolk sac)	Landauer (1956e)
4	13	9	14–23	Syndactyly (7 to 58%), shortened leg bones (24%), abnormal beak (19%), short first toes (11%).	
				(Doses: 10 mg/egg, in yolk sac)	Karnofsky and Lacon (1964)
4	18			Curved upper beak, shortened lower beak, moderate micromelia, bent tibia, fibula, and metatarsals.	

Summary Table 12 (*Continued*)

Age (days) Treat-ment	Age (days) Exami-nation	Malfor-mations (%)	Mor-tality (%)	Incidence of Typical Syndromes	Reference
Pyrimethamine				(Doses: 0.1 to 1.0 mg/egg, in yolk sac)	Stanzhevskaya (1966a)
0–4	10	35–58	72–98	General underdevelopment (29%); malformations of extremities (42%), of eye (32%); ectopia cordis (19%); reduction of lower jaw (17%); malformation of brain (15%); deformation of body axis (8%); reduction of tail (8%); hydrops (7%).	
Pyrithiamine				(Doses: from 0.25 to 2.0 mg/egg)	Naber, Cravens, Baumann, and Bird (1954)
5, 10, 15	21		5–100	Polyneuritis in chicks at hatching, growth inhibition.	
Quinine				(Doses: from 0.1 to 5.0 mg of 10% aqueous solution/egg)	Kitamura (1957)
0				With 1.0 mg—accelerated development, with increase in embryo's weight; with 3 mg—retarded development, with decreased weight; with 5 mg—retarded development and toxic symptoms.	
Reserpine				(Doses: from 10 to 250 μg/duck egg, in yolk sac)	Limborgh and Meenen (1965)
4, 11, 18	8, 15, 22	None	12–96	Noxious effects on growth, mortality.	
Ricin				(Doses: from 0.125 to 0.25 mm^3/egg, injected into head)	Ancel (1955)
1	17	70	31	Exencephalia (48%), acephalia (13%), acleiencephalia (5%), facial malformations (4%).	

Saponin(s)					
1	16	65	22	(Doses: 0.125 mm^3/egg, injected into head) Acleiencephalia (27%), facial malformations (22%), cleiencephalia (16%).	Ancel (1955)
Semicarbazide hydrochloride					
4–6	14–16		0–100	(Doses: from 1.0 to 5.0 mg/egg, in yolk sac) Shortened and malformed lower beak, bent tarsometatarsal and tibiotarsal bones.	Neuman, Maxwell, and McCoy (1956)
2	4			(Doses: 2 mg/egg, in yolk sac) Slight retardation in growth, gross bending of tibia and fibula, maldevelopment of beak—downward bending of upper beak.	Cameron (1962)
3	14	18–84	49–59	(Doses: from 0.5 to 1.0 mg/egg, in yolk sac) Malformations of upper and lower beak (29%); of lower extremities: bending (26%), shortening (16%), dislocation (13%); atrophy of upper eyelids (16%).	Bodit, Stoll, and Maraud (1966)
Sucrose					
0	3–5	45		(Doses: 1 ml of 2 *M* solution/egg, in albumen) Various abnormalities of the central nervous system (14.6%); heterotaxia (10.9%), rumplessness (9.0%).	Hunt (1951)
Sulfamide					
2	20	100		(Doses: from 0.25 to 5.0 mg/egg) Brachymelia, beak defects, syndactyly.	Ancel and Lallemand (1943a)
2				(Dose: 1 mg/egg) Brachymelia, parrot beak, syndactyly.	Ancel (1945)
2	9–18			(Doses: from 0.1 to 30 mg/egg) Somewhat retarded growth; changes in liver and kidneys (nephrosis-like changes in urinary tubules).	Matsumoto (1957)

Summary Table 12 (*Continued*)

Age (days) Treatment	Age (days) Examination	Malformations (%)	Mortality (%)	Incidence of Typical Syndromes	Reference
Sulfanilamide				(Doses: from 0.5 to 2.0 mg/egg, in yolk sac)	Zwilling and DeBell (1950)
1–9	14	87	42	Inhibition of growth; micromelia (94%); anouria (8%).	
				(Doses: 1 mg/egg, in yolk sac)	Landauer and Clark (1964c)
4	21	60	66	Micromelia (46%), parrot beak (38%).	
Sulfapyridine				(Doses: 20 mg of 20% solution/egg)	Borsuk (1964)
2	14, 21		70	Disproportionately large head; inhibited ossification of cranial bones; deformed beak, malformed eyes; celosomia without eventration.	
Sulfathiazole				(Doses: from 0.1 to 30 mg/egg)	Matsumoto (1957)
0	9–18			Some malformations, regressive changes in the liver cells, granular degeneration of urinary tubules of the kidney.	
				(Doses: 20 mg of 20% solution/egg)	Borsuk (1964)
2	21		40	Inhibited growth; markedly enlarged head, micro- and macrophthalmia, hypoplasia of the beak.	
Sulfisoxazole				(Doses: from 1 to 50 mg of 10% solution/egg)	Ajitomi (1959)
8–18				Ankylosis of knee joint and toes, short legs.	
Tetracycline				(Doses: from 0.1 to 5 mg/egg, in yolk sac)	Bevelander, Nakahara, and Rolle (1960)
8	12			Reduction in size of the embryo and of its bones, with abnormally low calcification.	

4	6–20		45–90	(Doses: from 0.3 to 7.5 mg/egg) Somatic and skeletal modifications. (See Fig. 108 on mortality.)	Genazzani, Pelagalli, Pagnini, and DiCarlo (1965)
2–13	10–17			(Doses: 2.5 mg/egg, in yolk sac) Retardation in growth, inhibition of mineralization of long-bone cartilage. (See Fig. 108.)	Rolle and Bevelander (1966)
Tetrahydro-3-furanol					
3	7, 11	25		(Doses: 0.5 mg/egg, in albumen) Damaged bulbothoracic aorta (80%), faulty interventricular communication (20%).	Ruch, Kremer, Rumpler, and Gerlinger (1965)
4	11			(Doses: 0.5 mg/egg, in albumen) Same as above.	Ruch, Zahnd, Kremer, Rumpler, and Gerlinger (1965)
Thalidomide					
3–5	16–18	21–100		(Doses: 2 mg/egg, in amnion) Encephalocele (7%); eye defects (bilateral 20%, right 10%); crossed and/or short upper beak (7%).	Williamson, Blattner, and Lutz (1963)
3	11	18–34	20–30	(Doses: from 0.3 to 0.5 mg/egg) Malformed limbs (15 to 20%): micromelia, phocomelia, hemimelia, and ectromelia; always associated with celosomia or anouria.	Salzgeber and Salaün (1965)
0–4	10	21–87	74	(Doses: 4 mg/egg, in yolk sac, and applied locally) Reduction of tail (26%), micromelia (17%), microphthalmia (14%), deformity of brain (9%) (see p. 240).	Stanzhevskaya (1966b)
Thiadiazole					
4	18			(Doses: 13.5 mg/egg, in yolk sac) Occasional shortening of upper beak.	Murphy, Dagg, and Karnofsky (1957)
Thiocolchicine					
3	9	0–17	0–45	(Doses: from 0.15 to 0.35 μg/egg) Retardation in growth, bilateral brachymelia, celosomia, excencephaly, malformations of beak.	Faucounau, Stoll, and Maraud (1963a)

Summary Table 12 (*Continued*)

Age (days) Treat-ment	Exami-nation	Malfor-mations (%)	Mor-tality (%)	Incidence of Typical Syndromes	Reference
Thiopental sodium					
1	6	52		(Doses: from 0.25 to 1.5 mg/egg) Reduction in size; malformations of cephalic portions, including platyneuria, micro- and anophthalmia, cyclopy, diastematomyelia, with rosette cell structures, hemorrhages, parietal adherences, edema, etc.	Garcia (1965)
2	6	87		(Doses: 1.5 mg/egg) Undeveloped ureter (10%), uronephrosis (24%), displaced neural tube (9%), duplication of neural tube (18%), morbid development (25%).	Ojeda, Bosque, and Barbosa (1970)
Thiosemicarbazide					
3	14	33–66	65–80	(Doses: from 0.5 to 1.0 mg/egg, in area vasculosa) Malformations of beak and wing (similar to those caused by semicarbazide).	Bodit, Stoll, and Maraud (1966)
Triazene					
4	18	20–100	35–93	(Doses: from 0.25 to 2.0 mm^3/egg, in yolk sac) Growth inhibition (14/11 grams), skeletal deformities of beak (parrot), and appendages (micromelia, bending of tibiotarsus and, occasionally, of tarsometatarsus), some edema and inhibition of feather growth.	Dagg, Karnofsky, Stock, Lacon, and Roddy (1955)
4	18			(Doses: from 0.5 to 1.0 mg/egg, in yolk sac) General effects: weight inhibition, edema, and short clubbed feathers. Specific effects on appendicular skeleton, with micromelia, bending of tibio-	Murphy, Dagg, and Karnofsky (1957)

				tarsus and tarsometatarsus. Beak abnormalities, from parrot beak to shortening of upper and lower beak.	
2,2′,2″-Trichlorotriethylamine					
3	10	56–65	11–74	(Doses: from 0.6 to 2.0 μg/egg) Ectromelia, ectrodactyly, brachydactyly, syndactyly, celosomia.	Faucounau, Stoll, and Maraud (1963b)
Triethylenemelamine					
2	3			(Doses: 0.012 mg/egg, injected over blastoderm) Developmental retardation, growth inhibition; specific defects in neural tube (100%), visceral arches (100%), somites (100%), auditory vesicles (67%), lens vesicles (67%).	Blattner, Williamson, Simonsen, and Robertson (1960)
3–4	7–8		50	(Doses: from 5 to 10 μg/egg, in albumen or yolk sac) Retardation in growth; teratisms in beak, extremities (micromelia), and skeleton.	Kocher (1963, 1965)
4	¼–3			(Doses: from 6 to 12 μg/egg, injected into neural tube) Diminishing mitoses, followed by necrosis, with various malformations.	Miclea and Arcan (1965)
4	17			(Doses: from 4.4 to 5.4 μg/egg, injected into neural tube) Delay in growth and differentiation, reshuffled structure of medulla, incomplete differentiation of white and gray matter.	Miclea and Arcan (1966)
3–4	9–11		50	(Doses: from 5 to 10 μg/egg, in air cell, yolk sac) Marked reduction in body weight and dimensions of various parts; edema; umbilical hernia; bending and torsion of extremities; bending downward of the upper beak; reduction in size of the lower beak; cleft palate.	Kocher (1969, 1970)

Summary Table 12 (*Continued*)

Age (days) Treatment	Age (days) Examination	Malformations (%)	Mortality (%)	Incidence of Typical Syndromes	Reference
Trimethylene glycol				(Doses: 0.05 ml/egg, in amnion, in yolk sac)	Gebhardt (1968)
1–9	14–21	19–88	0–20	Micromelia (88 to 19%).	
Trypan Blue				(Doses: 1 mg/egg, injected subgerminally)	Beaudoin and Wilson (1958)
1	10–12	13–75	11–74	Anouria (69%), beak defects (18%), anophthalmia (4%), microphthalmia (4%). (See Fig. 113.)	
				(Doses: from 0.01 to 0.05 mg/egg, in yolk sac)	Stephan and Sutter (1961)
1–3		47	53	Dose on *second day:* (1) edema, (2) anomalies of somites, (3) agenesis of caudal extremity; dose on *third day:* (1) celosomy, (2) scoliosis, lordosis and strophosomy, (3) anomalies of great arteries.	
				(Doses: 0.1 mg/egg, in yolk sac)	Kaplan and Grabowski (1967)
2		58	25	Rumplessness (28%), hematomas of limb buds (27%), edema (21%), hematomas in brain or eyes (17%), microphthalmia (5%), microcephaly (5%).	
				(Doses: 1 μl of 1% solution in saline or serum/egg, injected intravenously)	Kaplan and Johnson (1970)
3	4	17–58	42–83	Dissolved in saline: hematomas of head, caudal region, or both (58%), mortality (42%); in serum: abnormalities (17%), mortality (83%).	

Urea				(Doses: from 0.05 to 0.9 gram/egg, in albumen)	Dollander, Marconnet, and Autelin (1962)
¼–¾	2–5½			Retarded development (9%), and hypomorphy (5%), anomalies in curvature of spine (lordosis), and in cephalic and caudal regions.	
				(Doses: 0.2 ml of 2% agar suspension/egg, in side of blastoderm)	Noto (1967b)
¼–¾	3	18	2	Nonclosure of neural tube: of head (5%), of tail (4%), of head and tail (2%).	
Urethan				(Doses: from 30 to 40 mg/egg, in yolk sac)	Landauer (1956d)
0–6	13	5–39		Abnormalities of beak: short upper beak and cross beak.	
				(Doses: 0.5%/egg, in albumen)	Adhikari (1967)
0	4			Shortened embryonic axis; chaotic, cytotoxic neural tube; many degenerate cells.	
Uric acid				(Doses: 0.2 ml of 2% agar solution/egg, in blastoderm)	Noto (1967a)
¼–¾	3	19	7	Nonclosure of neural tube: of head (5%), of tail (4%), and of head and tail.	
Vinblastine				(Doses: from 0.3 to 2.0 μg/egg, on allantois)	Cros, Tollon, and Oreglia (1965)
4	13	24–45	3–100	Malformations of beak (22%), celosomia (10%), strophosomia (9%), malformations of eyes (8%).	

VIII

Hormonal Derangement

Hormones have potency to play
A highly dominating role:
Divert from embryonic way
To Nature's veiled morbid goal.

Hormones are specific organic products of living cells. Functional activity of hormones in the avian embryo starts during the later stages of development. From that time, the reaction of the embryo to hormonal substances is essentially the same as that of the adult.

One of the principal functions of hormones is the control of certain phases of morphogenesis. An excess or lack of these substances will profoundly influence development and maintenance of some structures.

All hormonal preparations, naturally occurring or synthetic, are capable of producing specific teratologic effects and death of the embryo.

Hormonal deficiencies, especially of the thyroid, can be produced by antihormonal substances such as thiourea, thiouracil, and others.

ADRENAL CORTICAL EXTRACT

Growth. After injection of adrenal cortical extract, retardation in growth of the chicken embryo without increase in the incidence of developmental defects was noted by Landauer (*1947c*).

BETAMETHASONE

Growth. The inhibitive effect of the corticosteroid betamethasone on the dry weight of the developing chicken embryo (*Graziani and Biagi, 1966*) is shown in the tabulation below.

Dose of Betamethasone (mg)	Dry Weight of 18-Day Embryo (grams)
none	3.01*
0.0003	1.26
0.001	0.81
0.005	0.15

* See Romanoff, 1967, p. 24.

Malformations. (See Summary Table 13, under Betamethasone.)

CORTISONE

Cortisone exerts both lethal and teratogenic effects when injected into the developing egg.

Growth of Whole Embryo

The principal effect of cortisone is marked dwarfism of the embryo.

Dose. It appears that there is a negative correlation between the dose and the growth of the whole embryo. This is illustrated (*Karnofsky, Ridgway, and Patterson, 1951*) in the table below.

Dose of Cortisone at 4 Days (mg, in yolk sac)	Weight of Embryo at 18 Days (grams)
0	16.4
1	13.3
2	11.5
3	9.0
4	6.5

Time of Injection. The effects of cortisone on the growth of the embryo depend on the stage of embryonic development at the time of treatment. Since the cortisone acts continuously (*Karnofsky, Ridgway, and Patterson, 1951*), its effect on the growth of the embryo is undiminished until the later stages of development. (This response of embryonic growth to cortisone treatment is well illustrated in Fig. 131*B*.)

There is a wide distribution in the weights of 18-day-old cortisone-treated embryos (Fig. 132). On the average, they weigh nearly 50 per cent less than normal embryos.

The relative weights of various organs of the cortisone-treated embryos are shown in Fig. 133. Although cortisone inhibits the growth of all parts of the embryo, not all organs of the dwarfed embryos are affected proportionally. The weight of the liver is reduced to about 70 per cent of normal; the weight of the heart is unaffected; and the gonads—both ovaries and testes—and the eyes are relatively heavier.

In vitro, cortisone inhibits the growth of explanted femur anlage (Fig. 134). This effect is evident 24 hours after explantation; but increases in length continue until the fifteenth day and beyond.

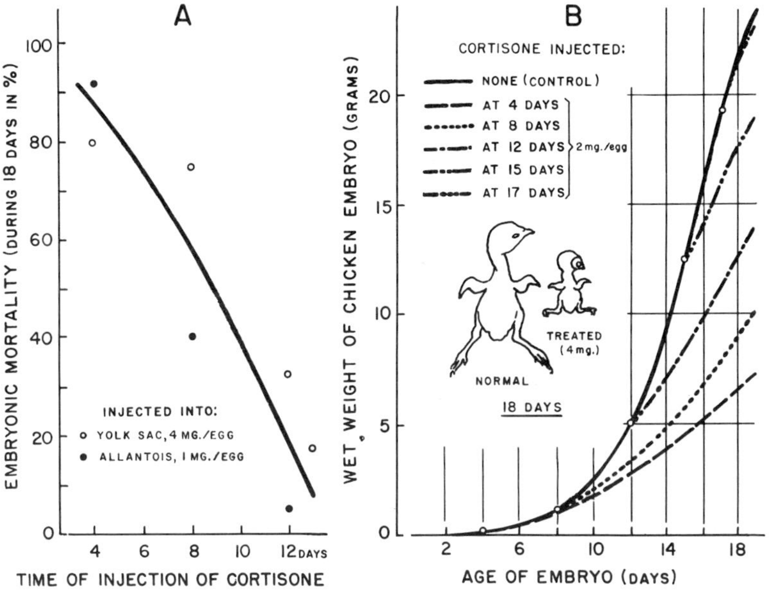

Fig. 131. Effects of cortisone on the growth and mortality of the chicken embryo. *A,* embryonic mortality in relation to the route of administration of cortisone at various stages of development (from the data of Karnofsky, Ridgway, and Patterson, 1951); *B,* the effects of 1.5 to 2.0 mg/egg of cortisone acetate, injected at various developmental stages, on the growth of the embryo (reconstructed after Moog and Richardson, 1955; and Moscona and Karnofsky, 1960), with an *insert* showing gross appearance of normal (16.4 grams) and cortisone-treated (4 grams) embryos (after Karnofsky, Ridgway, and Patterson, 1951.)

Mortality

Cortisone is toxic to the avian embryo. Embryonic mortality varies with dose, age of the embryo at injection, and site or route of injection.

Dose. It has been shown (*Piotrowski, 1966*) that embryonic mortality rises with increases in dose, as illustrated below.

Dose of Cortisone Injected at 2 Days (μg)	Incidence of Mortality at Hatching (%)
0	14
1	24
5	48
10	52
100	72
500	96

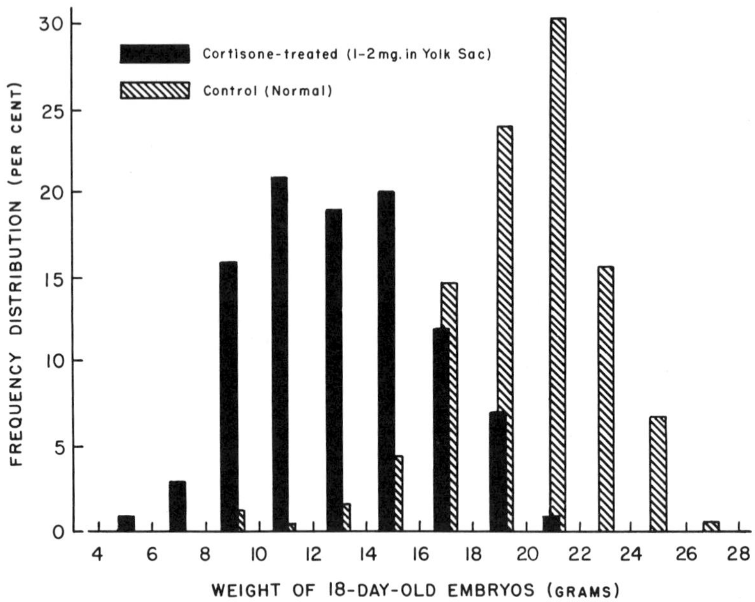

Fig. 132. Effect of cortisone on the weight of the 18-day-old embryo. (Recalculated after the experimental data of Karnofsky, Ridgway, and Patterson, 1951; and the control after Romanoff, unpublished.) *Note:* cortisone, in doses of 1 to 2 mg, was injected into the yolk sac at 4 to 12 days, and the surviving embryos were weighed on the eighteenth day.

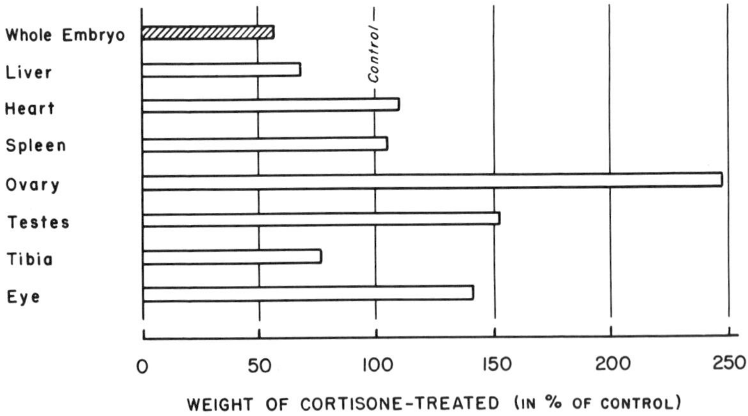

Fig. 133. Relative weights of the body and organs of cortisone-treated embryos. (Plotted from recalculated data of Evans, 1953.) *Note:* chicken embryos 8 days of age were treated with a single dose of 1 mg of cortisone injected into the allantoic membrane.

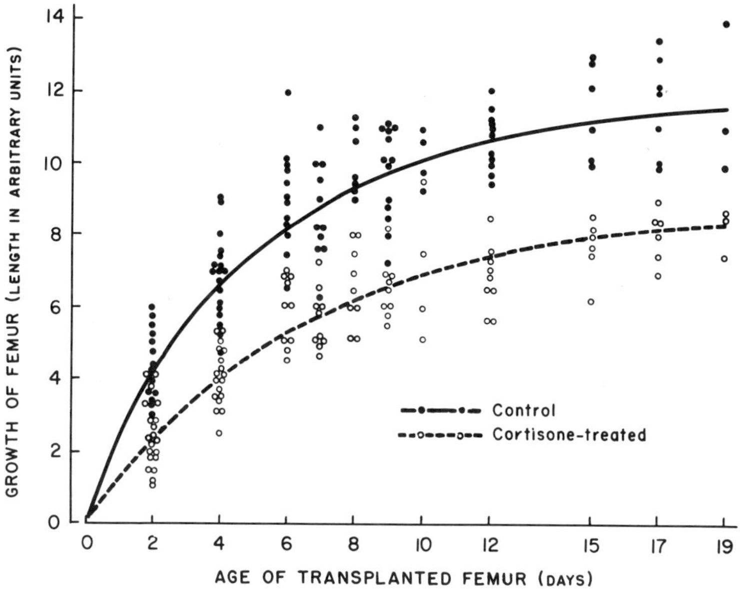

Fig. 134. Effect of cortisone on growth of the femur of the chicken embryo *in vitro*. (Redrawn after Buno and Goyena, 1955.) *Note:* the femurs were from 7-day-old embryos and transplantations were made every 2 to 3 days. The cortisone was introduced in a concentration of 0.025 mg/ml.

Time of Injection. The effect of cortisone (1.0 to 1.5 mg) on mortality decreases later in embryonic development (*Moscona and Karnofsky, 1960*), as is shown below (see also Fig. 136).

Age at Injection of Cortisone (days)	Incidence of Mortality at 14 Days (%)
3	88
4	78
5	74
6	44
7	18
8	15

Site of Injection. The toxicity of cortisone is markedly influenced by its route of introduction into the egg. The embryonic mortality rate is about three times greater when the dose is injected into the allantois than when it is injected into the yolk sac, at all stages (from 4 to 12 days) of development (Fig. 135; see also Fig. 131A).

Injections of cortisone into the albumen (*Agarwal and Monga, 1958a*) presumably have effects similar to those of injections into the yolk sac.

The calculated 50 per cent lethal dose (LD_{50}) for injection by various routes is shown below.

Route of Injection	LD_{50} (mg cortisone)
Yolk sac	2.0
Albumen	1.5
Allantois	0.25

Malformations

Cortisone produces characteristic anomalies which are distinct morphologically from those caused by other chemicals.

The total number of abnormalities resulting from treatment of embryos with cortisone varies with the age of the embryo at the time of treatment (Fig. 136). Furthermore, there are various syndromes that

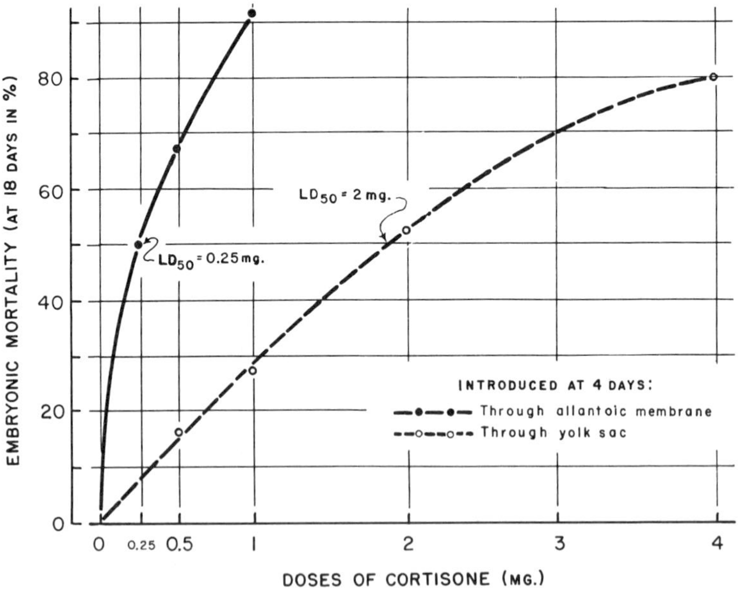

Fig. 135. Effect of various doses of cortisone on embryonic mortality. (Based chiefly on the data of Karnofsky, Ridgway, and Patterson, 1951.)

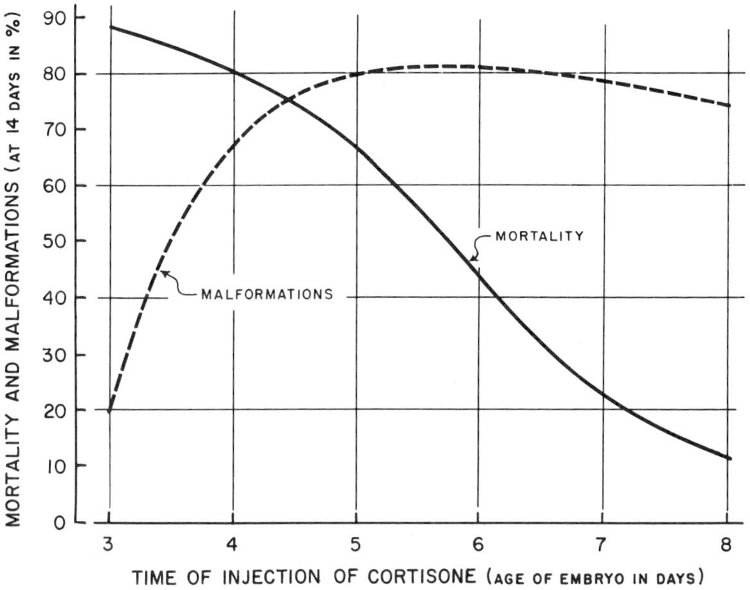

Fig. 136. Effect of the time of injection of cortisone on the incidence of mortality and malformations among survivors at 14 days of incubation. (Plotted from the data of Moscona and Karnofsky, 1960.) *Note:* a single dose of cortisone acetate in aqueous suspension, 1 mg/egg, was either dropped on the allantoic vesicle (on the third and fourth days) or, at later stages, applied to the allantoic membrane.

are characteristic of embryos treated by cortisone at particular stages (Fig. 137). In spite of overlapping in the appearance of these syndromes, they could be grouped into distinct categories (*Karnofsky, Ridgway, and Patterson, 1951; Moscona and Karnofsky, 1960*):

Category I, treated at 3 to 5 days: shortening and bending of femur and humerus, facial coloboma with cleft palate.

Category II, treated at 6 to 7 days: exteriorized viscera.

Category III, treated at 8 to 9 days: shortened beak, bent upper beak, delayed ossification of bones of the roof of the skull.

(See also Summary Table 13, under Cortisone, for detailed information on toxicity and typical syndromes.)

DEOXYCORTICOSTERONE

Mortality. Experimental observations (*Lakshmi, 1967b*) show that the incidence of embryonic mortality tends to increase with increase in the

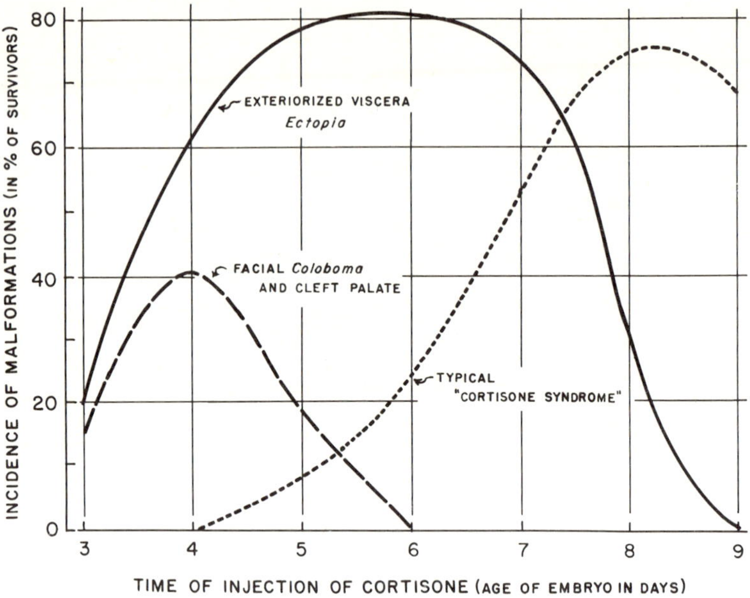

Fig. 137. The incidence of malformations among embryos surviving 14 days after single doses of cortisone, 1 mg/egg, applied to the allantois during the period between the third and ninth days of embryonic development. (After Moscona and Karnovsky, 1960.)

dose of deoxycorticosterone. The results are summarized below.

Deoxycorticosterone Injected at 3 Days (mg)	Embryonic Mortality at 10 Days (%)
0.1	50
0.2	83
0.4	86

Malformations. For a detailed description see Summary Table 13, under Deoxycorticosterone.

DEXAMETHASONE

The effect of dexamethasone on the mitotic index of the fibroblasts of the chicken embryo's heart *in vitro* (*Ruch and Rumpler, 1964*) is shown in Fig. 138.

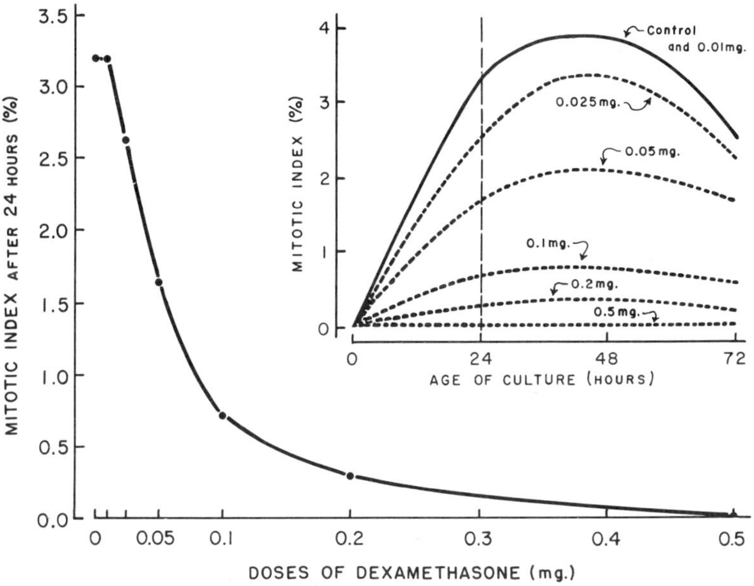

Fig. 138. Effect of dexamethasone on the mitotic index of heart fibroblasts from the 10-day-old chicken embryo. (After Ruch and Rumpler, 1964.) *Insert:* mitotic index of the heart fibroblasts as influenced by the age of the culture at various doses of dexamethasone.

With increase in dose, the mitotic index rapidly decreases. At all dose levels, the index rises during the first 24 hours (see *insert* of Fig. 138), then falls with further aging of the culture.

DIETHYLSTILBESTROL

Malformations. Dipping chicken eggs into a solution of diethylstilbestrol (40 mg/100 cm^3 of isopropyl alcohol) for 10 sec caused modifications in gonads (*Pincus and Erickson, 1962*), as shown by the number of ovotestes, given in the table below.

Age at Dipping (days)	Ovotestis in Left Testis (%)
0	31
1	30
2	43
3	48
4	86
5	100

For a detailed description of all malformations see Summary Table 13, under Diethylstilbestrol.

ESTRADIOL

Mortality. After administration of estradiol, embryonic death occurred throughout the 10 days of post-injection observation (*Lakshmi, 1967b*), and no specific period of susceptibility was observed.

Malformations. There are more malformations in dead estradiol-treated embryos than in live ones (*Lakshmi, 1967b*).

Embryos	Malformations (%)
Dead	84.1
Alive	15.9

The increased death rate is obviously due to abnormalities (between 3 and 13 days).

For more detailed information about types of abnormalities, see Summary Table 13, under Estradiol.

INSULIN

The effect of insulin on the development of the embryo has been well investigated in relation to growth, mortality, and incidence of malformations.

Growth

Whole Body. It has been shown (*Parhon, Laurian, Balaceanu, and Albu-Aderca, 1956*) that relatively small but repeated doses of insulin, injected into the allantois, bring about a noticeable retardation in embryonic development (Fig. 139). The average decrease in weight of the insulin-treated embryos on the twenty-first day was about 38 per cent. This was caused primarily by a delay of about 3 days in development; treated embryos can reach normal weight on the twenty-fourth day.

Organs. Various organs tend to be heavier in proportion to body weight in the insulin-treated embryos than in control embryos (Fig. 140). It

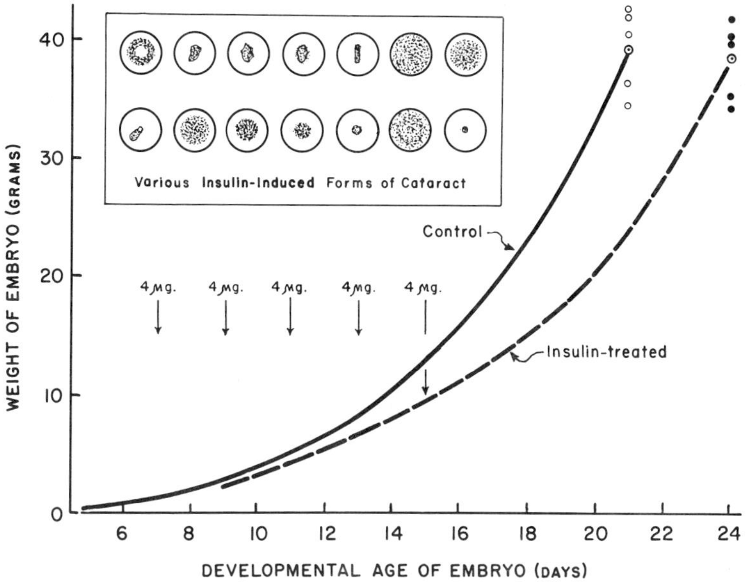

Fig. 139. Effect of five successive injections of insulin (in total 20 μg) on the growth of the developing embryo. (Plotted from the data of Parhon, Laurian, Balaceanu, and Albu-Aderca, 1956.) *Insert:* various insulin-induced forms of cataract in the chicken embryo. (Redrawn after Parhon, Laurian, Balaceanu, and Albu-Aderca, 1956.)

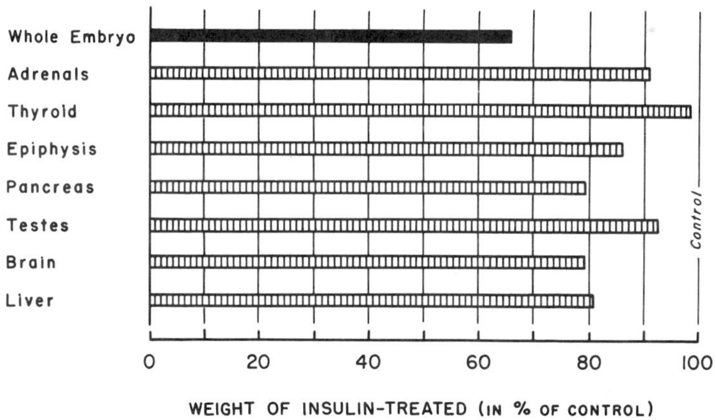

Fig. 140. Weight of the whole insulin-treated embryo (given 20 μg in five equal doses) and its organs relative to that of control groups. (Calculated from the data of Parhon, Laurian, Balaceanu, and Albu-Aderca, 1956. See effect of cortisone on embryo and its organs by Evans, 1953.)

is important to note that all relative weights are thus distorted (e.g., large eyes, small liver; see *Evans, 1953; Siegel, Smith, and Gerstl, 1957*). Some organs grow disproportionately large whereas other body tissues are reduced in proportional weight.

Tissues. Limb bones cultivated *in vitro* with 4 μg of insulin per 1 ml of medium (*Chen, 1954*) show retardation in the length of the shaft (with great enlargement and distortion of the bone ends).

Mortality

The toxicity of insulin to the embryo, as measured by embryonic mortality, is shown in Fig. 141. The relation of mortality both to the dose of insulin injected, and to the time of injection with a constant dose, has been studied.

Dose. The dosage–mortality curve (*insert* of Fig. 141) shows a rapid increase in mortality, from 5 to 32 per cent, with an increase in dosage from 3 to 125 μg.

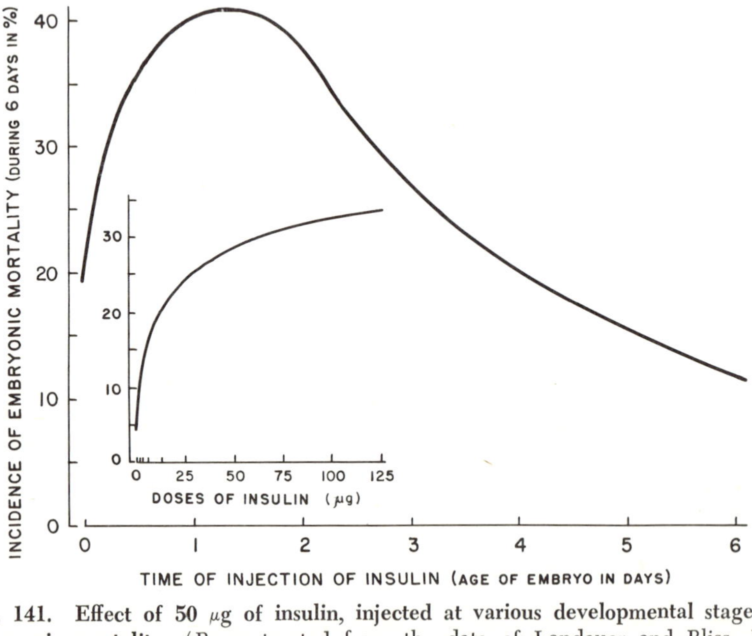

Fig. 141. Effect of 50 μg of insulin, injected at various developmental stages, on embryonic mortality. (Reconstructed from the data of Landauer and Bliss, 1946; Landauer, 1947a.) *Insert:* effect of various doses (from 0 to 125 μg) of insulin, injected into yolk sac before incubation, on mortality of the avian embryo during the first 6 days.

Time of Infection. With varying times of injection and a constant dose (Fig. 141), mortality increases rapidly with age to a peak of about 40 per cent between the first and second day, then declines gradually to about 10 per cent with treatment on the sixth day.

Distribution of Dead Embryos. A dose of 50 μg of insulin, injected on the fourth day, affects the distribution of dead embryos throughout the period of development (*Landauer and Clark, 1962a*), as is shown below.

Stage of Development (days)	Embryonic Mortality (%)	
	Control	Insulin-Treated at 4 Days
1–6	3.8	26.9
7–13	1.9	5.9
14–22	21.1	26.9
Total	26.8	59.7

Repeated tests with slightly different doses and times of injection (*Landauer and Clark, 1964b*) produced almost identical results with respect to the quantitative distribution of mortality (see also Fig. 13, or normal distribution of embryonic mortality, in Chapter II).

Species Difference. There is some difference in the immediate toxic effects of insulin on chicken and duck embryos (see Fig. 146*D*). In the earliest developmental stages, chicken embryos show a higher mortality after insulin treatment than do duck embryos. The mortality of the chick embryo rises slowly and declines thereafter. For the duck embryo, the toxicity of insulin rises steeply to a maximum of over 80 per cent at 3 days, and declines rapidly during subsequent stages of development.

Malformations

Insulin has well-defined teratogenic effects on the developing avian embryo. These effects are chiefly skeletal (*Landauer, 1946; 1947a*). In order of severity, the abnormalities produced are shortening of long bones (micromelia), rumplessness (taillessness or anouria), various beak deformities, and eye abnormalities (microphthalmia in chickens, and buphthalmia in ducks—*Landauer, 1951*).

The extent and sequence of occurrence of major insulin-induced abnormalities depends wholly upon the dose and the time of treatment in relation to the age of the embryo.

Dose. The relation between the dose of insulin and the incidence and severity of the resulting deformity is well demonstrated in the accompanying table (*Duraiswami, 1952*).

Dose of Insulin (μg)	Incidence of Deformities (%) Vertebral Column (Injected at 1 Day)	Limbs (Injected at 4 days)
0	1	2
25	18	26
38	21	32
63	25	46
125	31	85
150	36	90

There is a definite relationship between dose of insulin and frequency of rumplessness (anouria) among embryos which survive the seventeenth day of incubation (Fig. 142). In spite of the heterogenicity

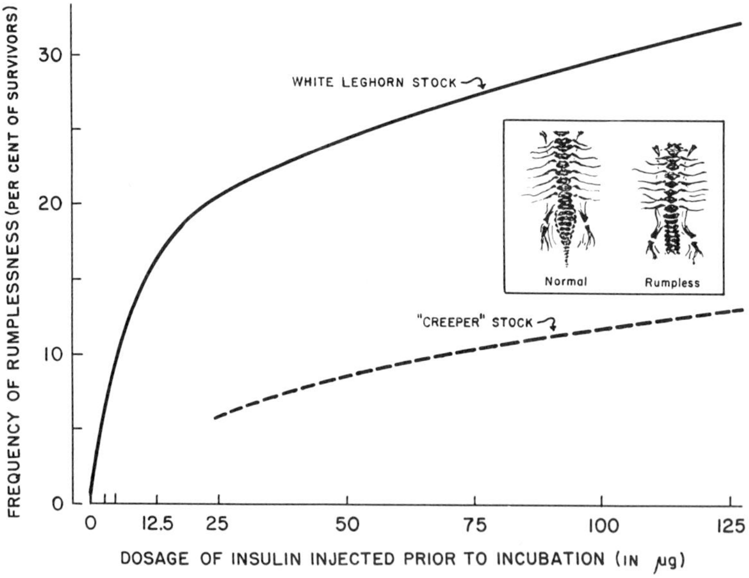

Fig. 142. Relationship between the dosage of insulin injected into yolks of fresh eggs, and the frequency of rumplessness (anouria) in White Leghorn and Creeper stocks, based on the survivors at the seventeenth day of incubation. (From the data of Landauer and Bliss, 1946.) *Insert:* appearance of pelvic regions of normal and rumpless—lacking the coccygeal vertebrae—embryos. (Redrawn from Landauer, 1947b.)

of general stocks of White Leghorn and Creeper chickens, the observations agree, with parallel curves. To obtain an equal incidence of anouria, however, the dosage of insulin for eggs from Creeper fowl would have to be increased to 13 times that required for eggs of Leghorns (*Landauer and Bliss, 1946*).

The difference in response to 50 μg insulin by three species: White Leghorn, Creeper, and Pekin duck (*Landauer, 1945; Landauer and Bliss, 1946; and Landauer, 1951;* respectively), is shown below.

	Incidence of Anouria (%)	
Species	Control	Insulin-Induced
Leghorn	1.5	41.6
Creeper	1.0	17.1
Duck	0.0	7.4

Time of Injection. Observations show (*Duraiśwami, 1952*) that insulin affects primarily the part of the body or tissues which is in the most active stage of growth and differentiation at the time of injection. This is exemplified in the tabulation below.

Insulin-Induced Abnormalities

Time of Injection (days)	Part of the Body Affected
1–2	Vertebral column
3	Feet
4–5	Limbs; beak
6	Hip (dislocations)

The effect of 50 μg of insulin, injected between 0 and 7 days of incubation, on the incidence of maldeveloped tail, beak, legs, and eyes of embryos, is shown in Fig. 143.

The injection of insulin before incubation or during the early hours of development induces rumplessness (anouria). In later stages, insulin treatment results in defects of the beak, micromelia, buphthalmia and, occasionally, polydactyly. Among embryos at the end of incubation or newly hatched chicks, abnormalities of the axial skeleton are more severe than in adult animals (*Landauer, 1947b*).

The curves for incidence of micromelia after injection of either 50 or 125 μg of insulin at various developmental stages of the embryo are closely parallel (Fig. 144). The maximum incidence of micromelia occurs after treatment at about 5½ days of development.

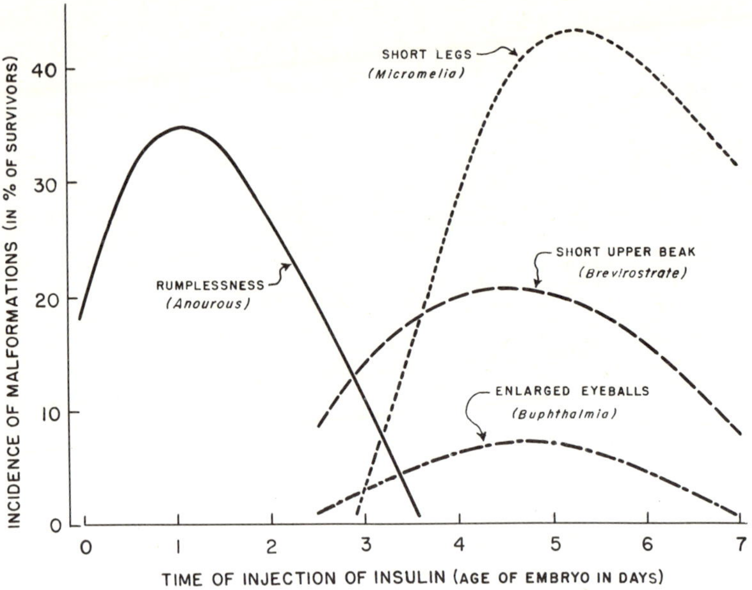

Fig. 143. Effect of insulin on the incidence of major malformations after injection of a single dose of 50 μg into the yolk sac at various stages of incubation. (After the data of Landauer and Bliss, 1946; and Landauer, 1947a.)

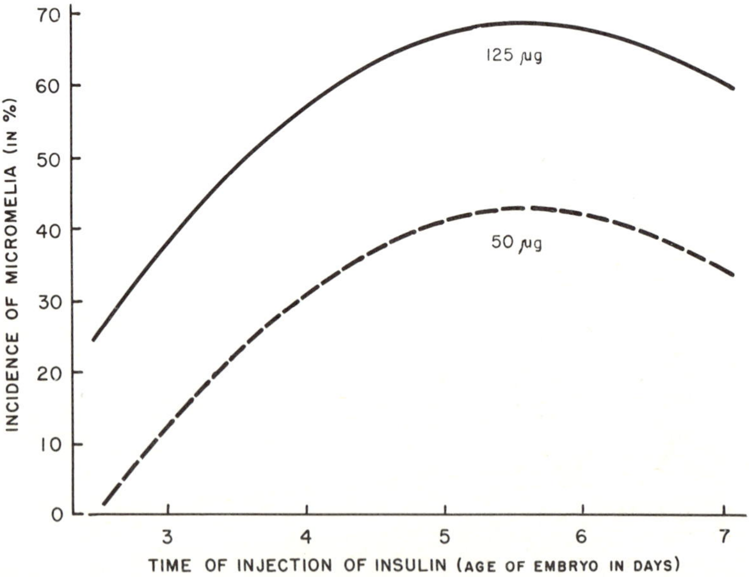

Fig. 144. Incidence of micromelia produced by different doses of insulin at various stages of development. (Plotted from the data of Landauer, 1947a.)

At certain developmental stages, the injection of insulin leads to polydactyly, as is shown in the accompanying table.

Time of Treatment (hours)	Incidence of Polydactyly (%)	
	With 50 μg	With 125 μg
48	0	0
72	0	10.5
96	4.3	18.8
120	0	0

Species Difference. The response of the duck embryo to 50 μg of insulin at various stages of development is generally the same as that of the chicken embryo (Fig. 145). The difference is only in the degree of response (see Fig. 146).

The relative occurrence of various malformations in chicken and duck embryos after injection of 50 μg of insulin at various stages of development is shown in Fig. 146. From this chart, it is evident that the duck embryos respond more readily with micromelia (Fig. 146*B*) and less with beak abnormalities (Fig. 146*C*) and rumplessness (anouria) (Fig. 146*A*) than the chick embryo.

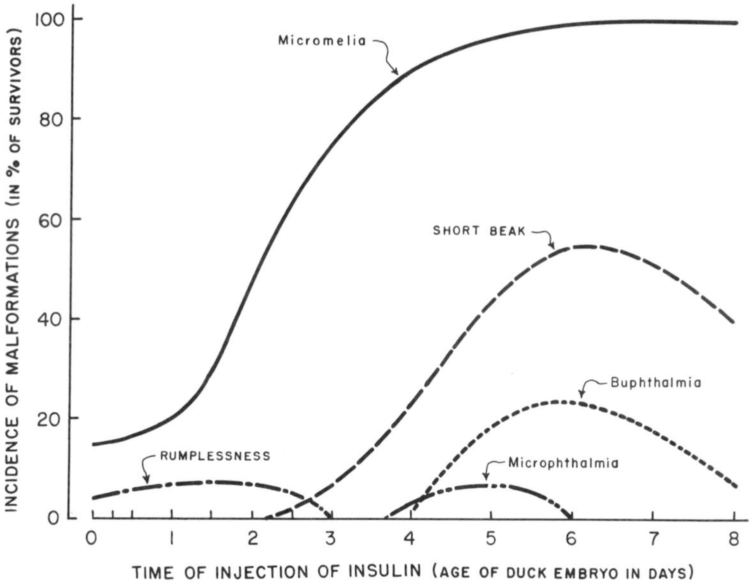

Fig. 145. Effects of 50 μg of insulin, injected into the yolk sac of duck eggs between 0 and 8 days, on the occurrence of various abnormalities among survivors at the thirteenth day of incubation. (Plotted from the data of Landauer, 1951.)

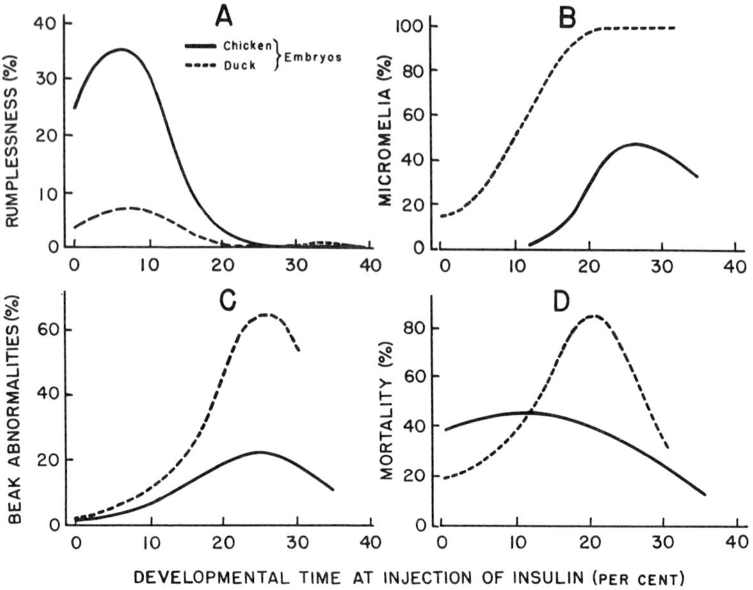

Fig. 146. Comparative effects of insulin (50 μg), injected between 0 and 8 days of incubation, on various malformations and toxicity (mortality) in Leghorn chicken and Pekin duck embryos, examined at the age of 13 days. (Recalculated from the data of Landauer, 1951.) *A*, the incidence of rumplessness; *B*, the incidence of micromelia; *C*, the incidence of beak abnormalities; and *D*, the incidence of embryonic mortality (toxic effects.)

Comparative general effects of insulin treatment on chicken and duck embryos are tabulated below.

Malformation (at a Peak)	Incidence (%)	
	In Chicken	In Duck
Rumplessness (at 1 day)	35	6
Micromelia (at 4 days)	60	100
Polydactyly (at 4 days)	0	19
Syndactyly (at 4 days)	5	0
Beak defects (at 5 days)	26	57
Microphthalmia (at 5 days)	0	11
Abnormal eyes (at 5 days)	6	0
Buphthalmia (at about 6 days)	0	24

A small number of other abnormalities have been observed. (Cerebral hernia occurred only after experimentation in duck and chicken material, and was absent in the control groups.)

See also Summary Table 13, under Insulin, for detailed information on toxicity and typical syndromes.

THYROXINE

It has been observed that the avian embryo is very responsive to a lack of thyroid hormone (see Antithyroid Drugs), as well as to an excess of thyroxine.

Experimental observations indicate that both the dose of thyroxine and the day of incubation on which it is injected influence the response of the embryo.

Growth

It has been shown (*Beyer, 1952*) that thyroxine accelerates the growth of the embryo. The effect becomes noticeable at approximately the fourth day of incubation (Fig. 147*A*). The time of hatching is delayed by 1 to 1½ days.

The accelerated growth is accompanied by an increase in metabolism, demonstrated by higher consumption of oxygen (Fig. 147*B*). It has also been observed (*Hanan, 1928*) that administration of thyroxine raises carbon dioxide production from the seventh day.

Hypothyroid Symptoms. Indications are that a hypothyroid condition is produced by higher doses of thyroxine administered in the later stages of development. This is shown in the table below.

Dose of Thyroxine (μg/egg)	Weight of Thyroid at Hatching (% of control)	Reference
0.1	100	Romijn, Fung, and Lokhorst (1952)
1.0	77	Booker and Sturkie (1949)
5.0	38	Vidal (1953b)
10.0	29	Adams (1958)

Thyroxine produces an obvious accelerating effect on the appearance of ossification centers in the vertebrae (Fig. 148), and, at the same time, a retardation in the differentiation of the thyroid, as is illustrated in the *insert* of Fig. 148.

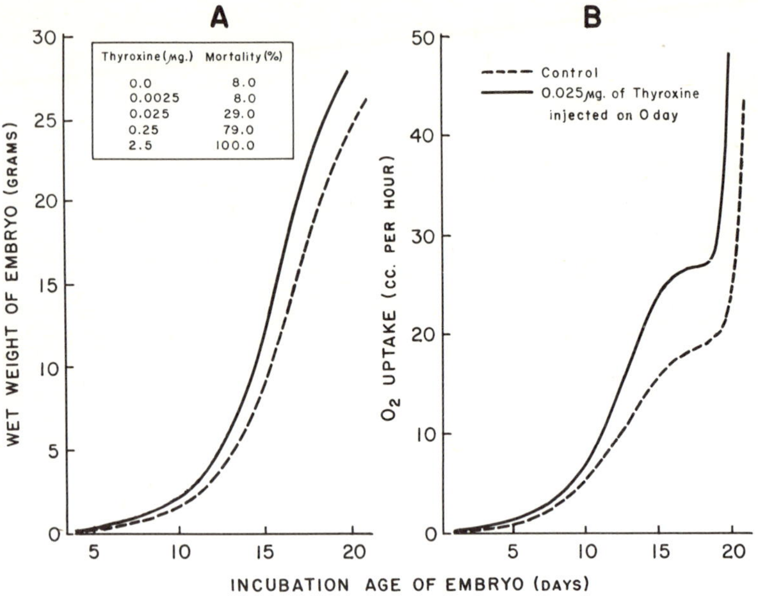

Fig. 147. The effect of thyroxine (0.025 μg injected into the air cell before incubation) on growth and metabolism (measured by O_2 consumption) of the developing embryo. (Plotted with modifications from the data of Beyer, 1952. See also Romijn, Fung, and Lokhorst, 1952.) *A,* the weight of the embryo, with *insert* showing the effect of various doses of thyroxine on embryonic mortality; *B,* oxygen (O_2) consumption by the embryo. (See also Romanoff, 1941, and, for the correction of O_2 consumption by extraembryonic membranes, Romanoff, 1967, pp. 278–279, and Table 28, p. 287.)

Mortality

Dose. Groups of eggs injected prior to incubation with various doses of thyroxine (*Beyer, 1952*) display a remarkable response, as shown by embryonic mortality on the fifth day of incubation (*insert* of Fig. 147A).

Time of Injection. The effect of time of injection of thyroxine, at a dose of 3 μg/egg (*Stoll, Coulaud, Faucounau, and Maraud, 1966*), is shown below.

Age at Injection (days)	Incidence of Embryonic Mortality (%)
2	70
3	34
5	30
6	0

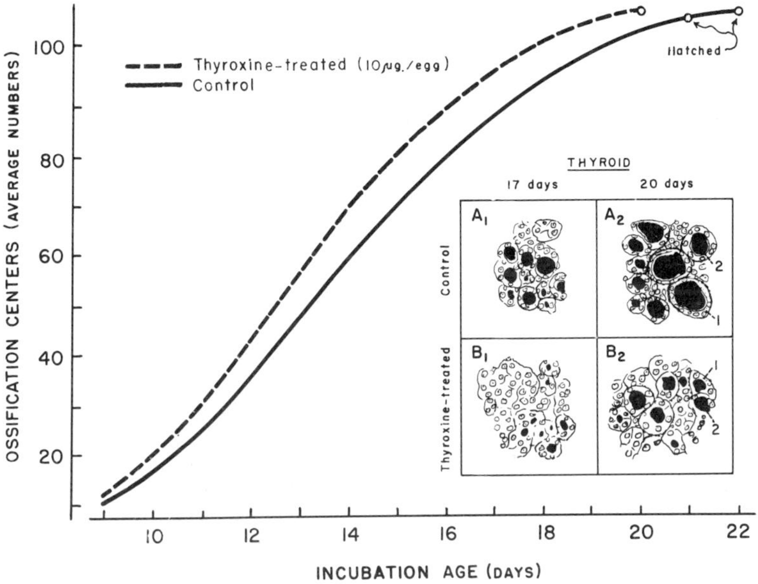

Fig. 148. **The effect of thyroxine on the time of appearance of ossification centers in the vertebrae—ossa extremitatis thoracicae, costal (ribs), sternum, and ossa extremitatis pelvinae.** (Plotted from the data of Honjo and Kuth, 1959.) *Insert:* retardation in thyroid differentiation. (Redrawn from Vidal, 1953b.) *1,* follicle; *2,* chromophile colloid.

A representative curve of embryonic mortality is also presented in Fig. 149*A*.

Malformations

Thyroxine applied to the avian embryo during its early development disturbs morphogenesis and induces an important malformation of the body form, namely, strophosomy (see sketches in the *insert* of Fig. 149*A*).

Characteristic effects of various forms of thyroxine (L, DL, and D forms) on the incidence of strophosomy are tabulated in the *insert* of Fig. 149*B*.

Dose. The incidence of strophosomy induced by various doses of thyroxine is graphically shown in Fig. 149*B*.

Time of Injection. The effect of time of injection with 3 µg/egg of thyroxine on the incidence of strophosomy is illustrated in Fig. 149*A*.

(See also Summary Table 13, under Thyroxine, for detailed information on toxicity and typical syndromes.)

COMBINED EFFECT OF HORMONES

Tests have been made on the combined effects of cortisone and insulin injected into the yolk sac at 4 and 8 days of incubation (*Franciscis and Landauer, 1959*).

Mortality. Among embryos treated at 4 days with both cortisone and insulin, the rate of mortality was considerably higher than in groups treated with either hormone alone (Fig. 150*A*). At 8 days, on the other hand, the combined treatment did not produce a mortality rate significantly different from the rate for either insulin or cortisone treatment alone.

Malformations. The incidence of short legs (Fig. 150*B*) increased with age at injection of insulin, but insulin in combination with cortisone showed slight effect. There was no significant difference in incidence of beak abnormalities (Fig. 150*C*).

(See also Synergism, in Appendix.)

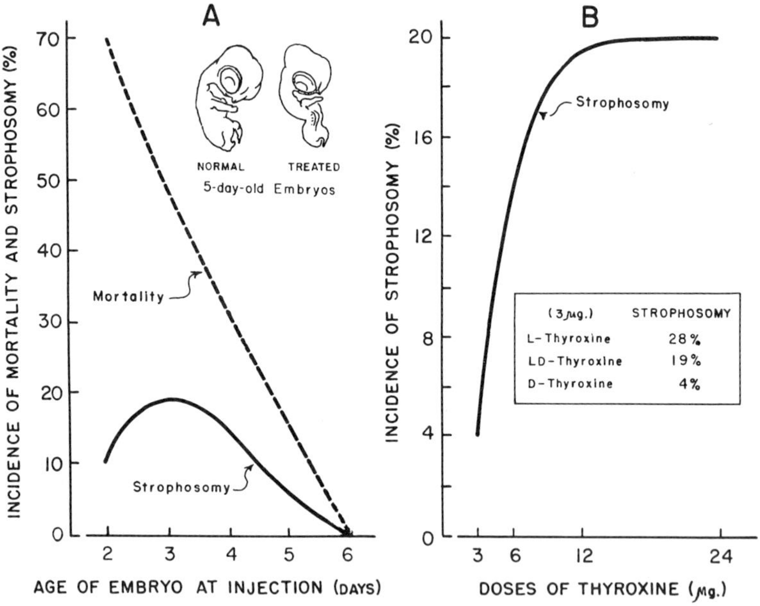

Fig. 149. Effect of thyroxine on the incidence of embryonic mortality and strophosomy. (Rearranged and plotted from the data of Stoll, Coulaud, Faucounau, and Maraud, 1966.) *A*, incidence of mortality and strophosomy as influenced by the time of injection of 3 μg of thyroxine (DL form), with *insert* showing the gross appearance of a strophosomic embryo; *B*, incidence of strophosomy as influenced by various doses of thyroxine (DL form), with *insert* showing the characteristic effects of various forms of thyroxine.

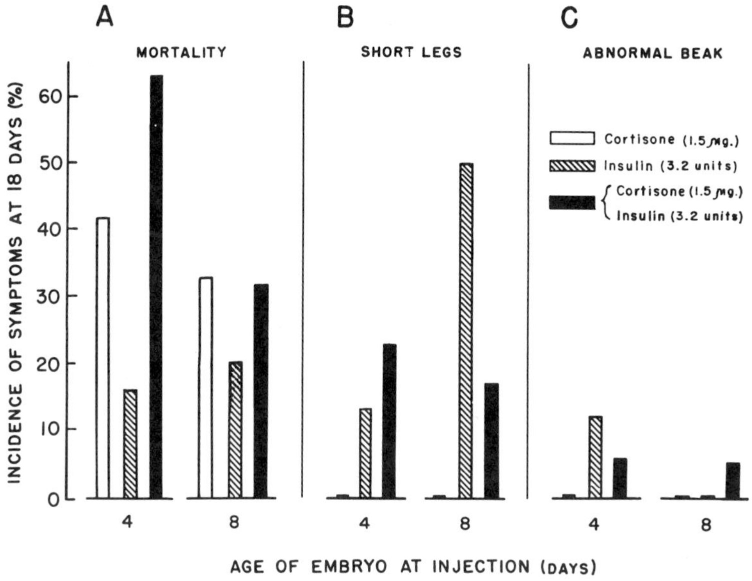

Fig. 150. Effects of cortisone and insulin, alone or combined, on the incidence of several symptoms. (Plotted from the recalculated data of Franciscis and Landauer, 1959.) *A*, embryonic mortality; *B*, short legs; *C*, abnormal beak.

ANTITHYROID DRUGS

Antithyroid (goiterogenic) drugs prevent the synthesis of thyroid hormone and so produce a state of hypothyroidism in the avian embryo.

Among the depressants of hormonal function in the avian embryo are thiourea, thiouracil, and their derivatives, such as methimazole, amphenone B, and aminothiazole. (See also under each name in Summary Table 13, for a detailed account of toxicity and typical syndromes.)

Thiourea is more effective than thiouracil, thiouracil more effective than methimazole, etc.

Thiourea

The presence of a goitrogen in the developing egg, whether deposited there from maternal dietary sources (*Andrews and Schnetzler, 1945*) or injected directly, results in retarded growth, delayed hatching, increased mortality, thyroid hypertrophy, and other anomalies (*Romanoff and Laufer, 1956*).

Growth of Embryo. The effect of various doses of thiourea, injected into the albumen of fertile eggs before incubation, on the weight of the embryo at 21 days is shown in the *insert* of Fig. 151. There is an obvious delay in development, which is also demonstrated by the increase in length of the incubation period (*Romanoff and Laufer, 1956*).

Dose of Thiourea (mg/egg)	Length of Incubation Period (days)
0	21
3	24–28
5	26–30
10	29–31

This prolongation of embryonic development 3 to 10 days beyond the normal period of 21 days is again demonstrated by the cumulative weight of the thiourea-treated embryo in comparison with the control (Fig. 151; and *insert* of Fig. 152).

If the dose of thiourea is repeated at a later stage of embryonic

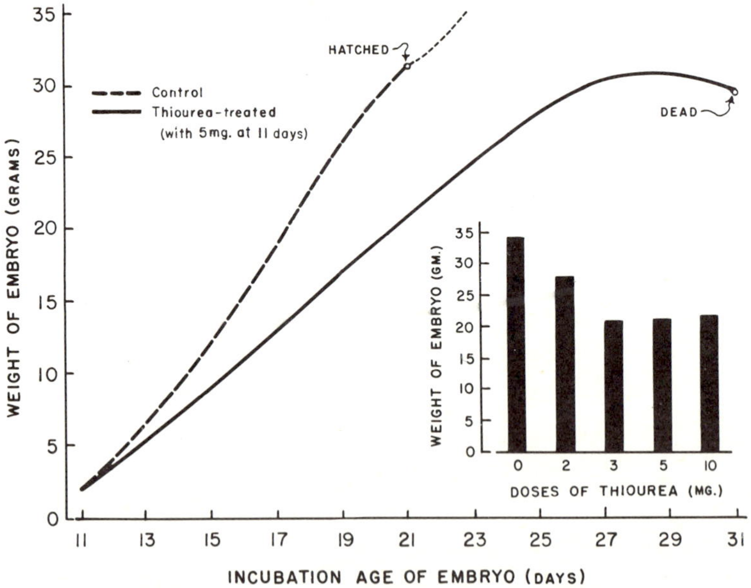

Fig. 151. Effect of thiourea (5 mg) on the growth of the developing embryo. (With modifications after Romanoff and Laufer, 1956. See also Vidal, 1952, 1953a.) *Insert:* effect of various doses of thiourea on the average weight of the embryo on the twenty-first day of incubation.

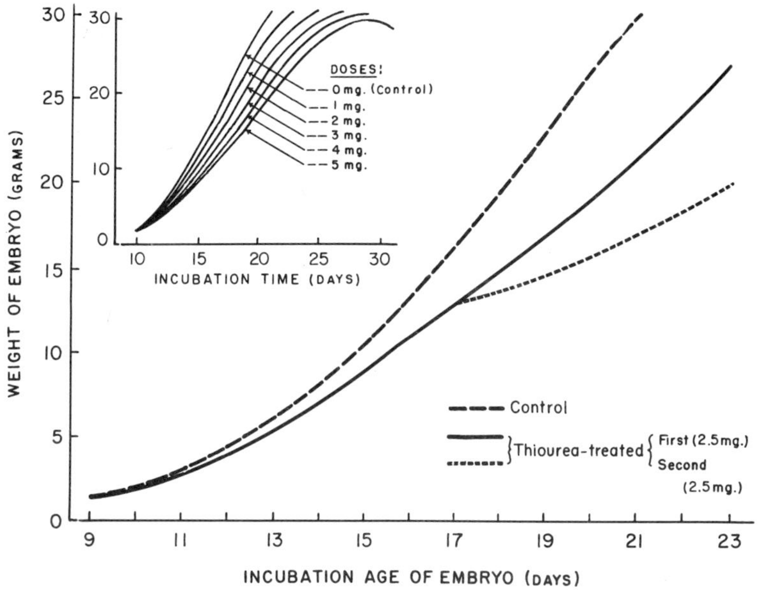

Fig. 152. Effect of repeated injections of thiourea (on the ninth and seventeenth days) on the growth of the chicken embryo. (After Vidal, 1953a.) *Insert:* delay in embryonic growth caused by various doses of thiourea injected on the ninth day of incubation. (Reconstructed from the data of Grossowicz, 1946; Adams and Bull, 1949; Yushok, 1950; Vidal, 1953a; Romanoff and Laufer, 1956; Moog, 1962.)

development (17 days) a further retardation of growth ensues (Fig. 152).

Growth of Organs. Relative weights (to normal) of several organs from 20-day-old embryos after injection of 2 and 5 mg of thiourea into the albumen on the eleventh day of incubation (*Yushok, 1950*) are shown in the accompanying table.

Organ (mg)	Relative Weight of Thiourea-Treated (%)	
	With 2 mg/egg	With 5 mg/egg
Liver	61	61
Heart	74	77
Kidney	89	83
Brain	98	91

The growth of the liver of thiourea-treated embryos deviates considerably from normal. The rate of increase in weight of the whole brain is least affected by thiourea.

Deviations in the weights of thyroid, adrenals, and gonads from the experimental embryo at the time of death, at about 28 days (*Romanoff and Laufer, 1956*), are shown below.

Endocrine Glands	Deviation in Weight from Control (in %)
Thyroid	+319
Adrenals	+103
Testes	+43
Left ovary	−5

The effect of time of injection of 5 mg of thiourea on the weight of the whole embryo and some of its organs on the twenty-first day of development is illustrated in Fig. 153.

Hyperthyroid Symptoms. When various doses of thiourea were injected into the albumen on the eleventh day, the smallest dose (2 mg) caused the largest increase in the wet weight of the thyroid (*insert* of Fig. 154).

The curve for thyroid weight in relation to time of treatment with

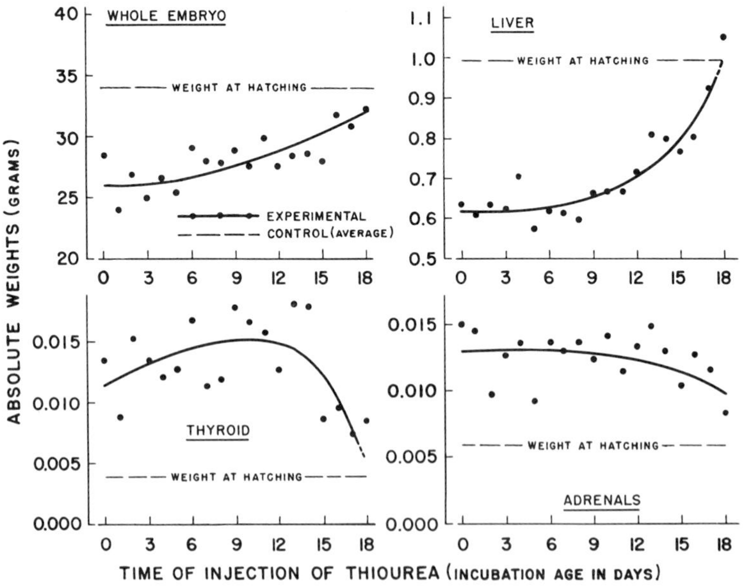

Fig. 153. The influence of the time of injection of 5 mg of thiourea (from 0 to 18 days) on the weight of the chicken embryo and some of its organs on the twenty-first day of development. (After Romanoff and Laufer, 1956.) *Note:* the weights of the body and its organs under experimental conditions are compared with those under normal (control) conditions.

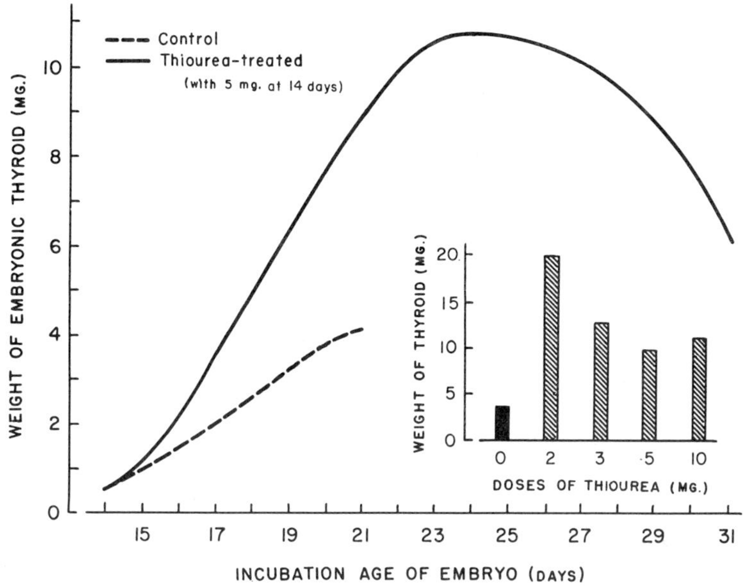

Fig. 154. Effect of thiourea injected into the egg at the fourteenth day on the growth of the thyroid of the developing embryo. (Plotted from the recalculated data of Adams and Buss, 1952; and Romanoff and Laufer, 1956.) *Insert:* effect of various doses of thiourea, injected into the egg albumen at 11 days, on the size of the embryonic thyroid on the twenty-first day of incubation. (After Romanoff and Laufer, 1956.)

5 mg of thiourea (Fig. 153) indicates that the maximum thyroid enlargement is produced when the drug is injected at 9 to 12 days of incubation.

It has been observed (*Adams and Bull, 1949*) that the increase in weight of the thyroid after thiourea treatment is due not only to hypertrophy of the follicular cells present in the gland when treatment begins, but also to a marked increase in the number of cells, by mitosis, and their subsequent hypertrophy.

Mortality. The influence of various doses of thiourea injected on the eleventh day of incubation (*Romanoff and Laufer, 1956*) is shown in the following tabulation.

Thiourea Injected (mg)	Total Mortality at Hatching (%)
0	7
2	25
3	85
5	100

Malformations. The main effects of thiourea injections (*Adams and Bull, 1949*) are connected with either (1) retardation of growth and lowered body weight (shortened limbs), or (2) enlargement of the thyroid (hyperemia, hyperplasia, hypertrophy).

See also Summary Table 14, under Thiourea, for a detailed account of toxicity and typical syndromes.

Thiouracil

Growth of Embryo. The effect of injection at 2 days of various doses of thiouracil on the growth of the embryo up to 21 days of age is shown, by continuous curves of body weight, in Fig. 155. With increased doses of thiouracil, the average weights of the embryos were gradually diminished, especially at hatching time.

Growth of Organs. It has been observed (*Yushok, 1950*) that injecting 1 to 2 mg/egg of thiouracil into the albumen has an insignificant effect

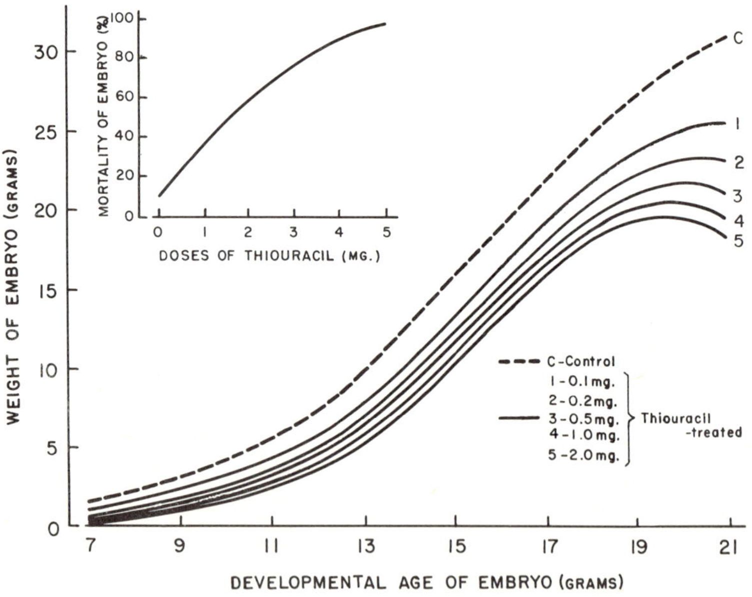

Fig. 155. Effect of various doses of thiouracil (injected at 48 hours) on subsequent growth of the developing embryo. (Plotted from the data of Sinha, Ringer, Coleman, and Zindel, 1959. See also Honjo and Kuth, 1959.) *Insert:* effect of various doses of thiouracil on the total embryonic mortality at hatching. (Same reference.)

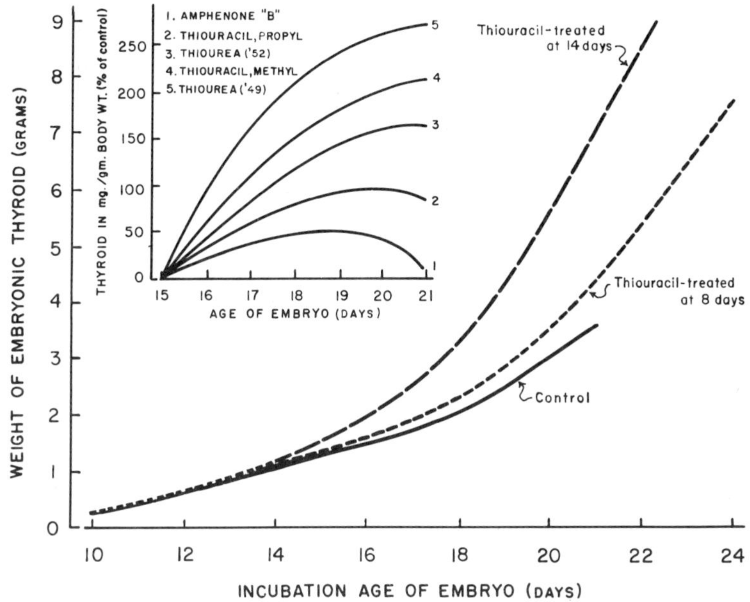

Fig. 156. Effect of time of injection of thiouracil (1 to 2 mg) on the enlargement of the embryonic thyroid. (Reconstructed from the data of Adams and Bull, 1949.) *Insert:* relative effect of antithyroid drugs on the weight of the embryonic thyroid. (After Adams and Bull, 1949; Adams and Buss, 1952; Adams and Bennett, 1953.)

on weight of the liver, brain, heart, and kidneys (see the effect of thiourea).

Hyperthyroid Symptoms. When thiouracil is fed to hens at a level of 2 mg/gram, it appears in both the albumen (0.11 mg/gram) and in the yolks (0.10 mg/gram) of their eggs in 3 weeks. The chicks hatched from these eggs show an increase in thyroid weight from 5.8 mg (control) to 10.5 mg, that is, an increase of over 50 per cent (*Andrews and Schnetzler, 1945*).

With an increase in the level of thiouracil in the diet of the hen (*McCartney and Shaffner, 1949*), there is a further increase in the weight of the thyroid of the chick at hatching, as is shown below.

Dietary Thiouracil (mg/gram)	Weight of Thyroid at Hatching (mg)
0	5.4
1	7.9
3	15.7

The effect of time of injection of 1 to 2 mg of thiouracil on the growth of the thyroid is presented graphically in Fig. 156. The thyroids of embryos treated with a single dose of thiouracil on the fourteenth day are almost twice as heavy as those of the control group at hatching time. There is less significant weight gain if the thiouracil is injected on the eighth day of incubation, because the thyroid in the chicken embryo does not become completely functional until the eleventh day (*Romanoff, 1960, p. 871*).

The continuous growth of thyroid (in mg/g body weight) under the influence of various antithyroid drugs is shown in the *insert* of Fig. 156.

Mortality. Injection of 5 mg of thiouracil into eggs containing 2-day-old embryos was uniformly fatal. Successively reduced levels of thiouracil gradually decreased the embryonic mortality, as shown in the *insert* of Fig. 155.

Malformations. The same malformations are caused by thiouracil as by thiourea.

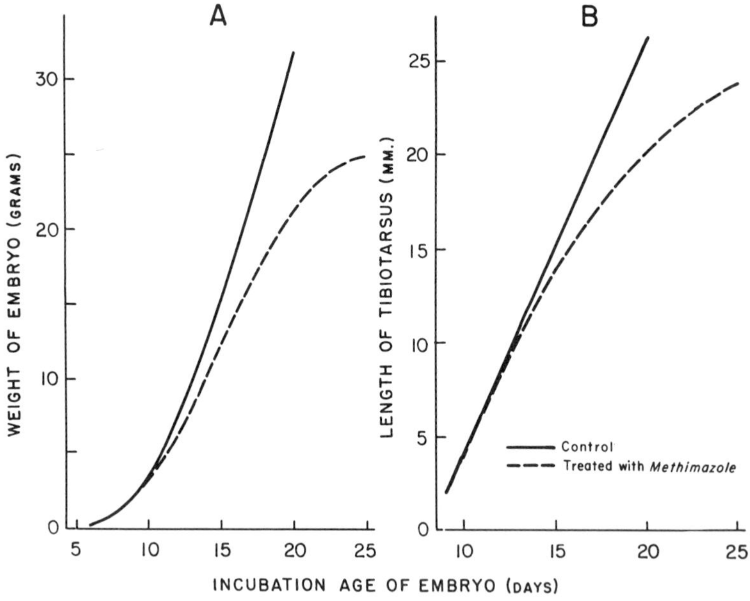

Fig. 157. Effect of injected methimazole on the growth of the chicken embryo. *A,* wet weight of the embryo, under control conditions and after the injection of 2.5 mg of the drug on the seventh day of incubation (after Zahnd and Rumpler, 1961a); *B,* length of tibiotarsus under control and experimental conditions (after Zahnd and Rumpler, 1961b.)

See also Summary Table 14, under Thiouracil, for a detailed account of toxicity and typical syndromes.

Methimazole

The effects of 2.5 mg of methimazole, injected on the seventh day, on the growth of the embryo, are shown in Fig. 157*A*, and the effects on length of the tibiotarsus in Fig. 157*B*.

Summary Table 13 Effect of Various Hormonal Preparations on the Development of the Avian Embryo

Age (days) Treatment	Age (days) Examination	Malformations (%)	Mortality (%)	Incidence of Typical Syndromes	Reference
Adrenocorticotropin					
10–17	18			(Doses: 17.5 mg/egg) Retardation of growth.	Biener, Balaceanu, and Albu (1960)
4, 8	18	10–64	7–40	(Doses: from 1 to 2 mg/egg in yolk sac; at 4 days, LD_{50} = 2 mg) Retardation of growth (4 to 18 days—15/19 grams; 8 to 18 days—12/17 grams)	Franciscis, Finelli, Gargiulo, and Renda (1960)
Androsterone					
4–5	21			(Doses: from 0.3 to 2.0 mg/egg) With doses over 0.5 mg.—intersexuality.	Wolff, Et. and Wolff, Em. (1936)
4–5	21			(Doses: from 0.1 to 1.0 mg/egg) With doses from 0.7 to 1.0 mg—feminization of the male embryo and masculinization of the female embryo. (See Romanoff, 1960, pp. 844, 847.)	Wolff, Et. and Wolff, Em. (1937)
Betamethazone				(Doses: from 0.001 to 0.08 mg/egg, in yolk sac; LD_{50} = 0.01 mg)	Biagi and Graziani (1966)
9	10–18		50–100	Mortality relative to the dose and the age of the embryo.	
10	16			Inhibitive effect on the dry weight of embryo (see text table).	Graziani and Biagi (1966)
Cortisone				(Doses: from 0.2 to 4.0 mg/egg, in allantois or yolk sac)	Karnofsky, Ridgway, and Patterson (1951)
0–16	3–18		17–100	Inhibition of growth (6.5/16.4 grams) (Figs. 131, 132, 134).	

				(Doses: 2 mg/egg, in yolk sac)	Duraiswami (1955)
4	11			Marked inhibition of growth (35%); visceral ectopia, including gut, liver, and heart (73%); inhibition of extraembryonic .membranes (30%); general edema (12%).	
				(Doses: from 1.0 to 1.5 mg/egg, in allantoic vesicle).	Moscona and Karnofsky (1960)
3–8	14–18	10–88	17–94	Growth inhibition (18 days—5.5/21.6 grams), defects in facial bones and tibia, exteriorized viscera (celosomia). (See Fig. 137.)	
				(Doses: 1.25 mg/egg, in allantois)	Pantíc, Kosanovíc, and Pavlovíc (1970)
8–15	11–18		17–100	Decrease in body weight (31 to 41%); malformation of sternum; atrophic eyes, and occasionally, cataract; deformed extremities; hemorrhage, edema, and cysts of the skin and various organs.	
Deoxycorticosterone				(Doses: from 0.25 to 0.50 mg/egg, in yolk sac)	Sames and Leathem (1951)
6	18			Retardation of growth (17/26 grams).	
				(Doses: 0.4 mg/egg, in allantois)	Hirayama (1960)
8	13–21			Inhibition of ossification.	
				(Doses: from 0.05 to 0.4 mg/0.05 ml sesame oil/egg, in yolk sac)	Lakshmi (1967b)
3	10	15–42	40–86	Chin pouch (43%), abnormal beak (33%), general edema (25%), ectopia (23%), bent neck (15%), curved digits (15%), microphthalmia (8%), pericardial edema (5%).	
				(Doses: from 0.5 to 2.0 mg/sesame oil/egg, in albumen)	Vannini (1947)
2				Inhibition of development of the medullary tissue of the gonads.	

Summary Table 13 (*Continued*)

Age (days) Treatment	Age (days) Examination	Malformations (%)	Mortality (%)	Incidence of Typical Syndromes	Reference
Dexamethasone					
0	3–4			(Dose: 0.01 mg/egg, in yolk sac) Retardation of growth, decrease in number of blood vessels of allantois, decrease in mitotic index of neural tube.	Starr and Weston (1962)
6–12	7–15			(Dose: 0.1 mg/egg, in yolk sac) Accumulation of bile pigments and accentuation of green color in the liver and intestines.	Ruch (1964)
0–3	1–3			(Doses: from 0.01 to 0.5 mg/ml.) Inhibition of growth, diminution of mitotic index in culture of fibroblasts of heart. (See Fig. 138.)	Ruch and Rumpler (1964)
4	11			(Doses: 0.01 mg/egg, in albumen) Malposition of bulbus, disturbance in interventricular communication.	Ruch, Zahnd, Kremer, Rumpler, and Gerlinger (1965)
Diethylstilbestrol					
0–5	21		5–42	(Doses: eggs dipped for 10 sec in solution of 40 to 100 mg/50–100% isopropyl alcohol) Feminization of portions of the male reproductive system, especially in early stages.	Pincus and Erickson (1962)
3	7, 11	21		(Doses: 2 mg/egg, in albumen) Malformations of heart: defects in thoracic aorta (43%), defects in communication between the small arteries and arterial branches (57%).	Ruch, Zahnd, Kremer, Rumpler, and Gerlinger (1965)

				(Doses: eggs dipped for 10 to 20 sec in solution of 40 to 50 mg/100 ml 100% isopropyl alcohol)	Miyamori (1966)
4	11–20		25–50	Stimulation of ovarian differentiation in the female; feminizing effects on the developing left gonad in the male.	
				(Doses: eggs dipped in solution of 50 mg/100 ml isopropyl alcohol)	Asayama and Maruoka (1967)
4	11			Feminized left gonad; when cultured, abnormally hypertrophied epithelium and developing testicular cords.	
Epinephrine				(Doses: 0.1 cm^3/egg, every third day, in allantois)	Dalton, Sergeyeva, and Selye (1940)
6	12–18			Edema, liver necrosis, gastroschisis with ectopia of viscera, protrusion of cornea.	
Estradiol				(Doses: 500 μg of 25 mg/aqueous solution 1 ml/egg in allantoic vesicle)	Asayama (1965)
3	11		31	No external abnormalities, but a conspicuous feminizing influence upon the developing gonads.	
				(Doses: from 1 to 5 mg in olive oil/egg)	Piotrowski (1966)
2	21		63–72	No teratogenic effect.	
				(Doses: from 0.05 to 0.4 mg/0.05 ml oil/egg, in yolk sac)	Lakshmi (1967b)
3	13	20	35–68	General edema (32%), ectopia (21%), micromelia (21%), beak abnormalities (15%), chin pouch (11%).	
Estrone				(Doses: from 10 to 500 μg/olive oil 0.03 to 0.05 ml/egg)	Piotrowski (1966)
2⅙	21	none	33–40	No anomalies.	
				(Doses: 1 mg/egg)	Franciscis, Gargiulo, Manzo, and Renda (1960)
4, 8	18		100–76		

Summary Table 13 (*Continued*)

Age (days) Treatment	Age (days) Examination	Malformations (%)	Mortality (%)	Incidence of Typical Syndromes	Reference
Ethylestrenol				(Doses: from 0.015 to 0.4 mg/0.05 ml of sesame oil/egg, in yolk sac)	Lakshmi (1967a)
3	13		19–69	LD_{50} = 0.2 mg.	
Follicle-stimulating Hormone				(Doses: 0.005 ml of antiserum in 0.2 ml saline/explanted embryo)	Sherbet and Lakshmi (1968)
¾		46	4–98	Arrested development (16%), degeneration of neural tube and somites.	
Insulin				(Doses: from 1 to 6 units/egg, in yolk sac)	Duraiswami (1952)
0–6	21	90		Abnormalities of vertebral column (18 to 36%); deformities of limbs (29 to 90%), including spina bifida, scoliosis, talipes, tibial kyphosis, pseudarthrosis, micromelia, osteogenesis imperfecta, achondroplasia.	
				(Doses: 2 units/duck egg, in yolk sac)	Landauer (1951)
0–9	13	15–100	56	Micromelia (11 to 100%); beak abnormalities (60%), including short upper beak (15%); eye abnormalities: buphthalmia (24%), microphthalmia (11%); rumplessness (6%). (See Fig. 145.)	
				(Doses: 75 μg/egg, in yolk sac)	Duraiswami (1955)
4	11			Angulations of tibias (48%); beak defects (12%); including "parrot beak."	
				(Doses: 50 μg/egg, in yolk sac)	Landauer and Rhodes (1952)
4	17		31	Micromelia (60%); beak abnormalities (26%); syndactyly (5%).	

0–4	17	5–33	13–35	(Doses: from 1.25 to 125 μg/egg, in yolk sac) Rumplessness (anouria), varying with dose and time. (See Figs. 142, 143.)	Landauer and Bliss (1946)
0–7	5–7		8–38	(Doses: from 1.25 to 125 μg/egg, in yolk sac) Maximum incidences—micromelia (87.3%), beak abnormalities (55%), rumplessness (32%), abnormalities of eyes (10%). (See Figs. 144, 145.)	Landauer (1947a)
0–7	17			(Doses: from 50 to 125 μg/egg) Taillessness, beak defects, short limbs, deformity of eye.	Ancel (1950)
3–5	18	39–73	56	(Doses: from 50 to 100 μg/egg, in yolk sac) Reduction in body weight (16/21 grams at 18 days); reduction in length of femur (13/17 mm), tibiotarsus (17/25 mm), and tarsometatarsus (14/19 mm).	Landauer (1953a)
7–9	9–21			(Doses: from 4 to 72 μg cumulatively/egg, in allantois) Delayed development (microsomia), achondroplasia, bilateral cataract. (See Figs. 139, 140.)	Parhon, Laurian, Balaceanu, and Albu-Aderca (1956)
Nandrolone					
5	20			(Doses: 0.63 mg. in 0.1 cm^3 corn oil/egg, at pointed end) Complete inhibition of the bursa of Fabricius in both sexes.	Meyer, Rao, and Aspinall (1959)
5–17	19			(Doses: from 0.1 to 6 mg in 0.1 cm^3 corn oil/egg, in albumen) Profound effects on the epithelial component of the bursa of Fabricius; sometimes complete inhibition.	Rao, Aspinall, and Meyer (1962)

Summary Table 13 (*Continued*)

Age (days): Treatment	Age (days): Examination	Malformations (%)	Mortality (%)	Incidence of Typical Syndromes	Reference
Parathyroid hormone				(Doses: 0.1 cm^3 of physiological solution/egg)	Pelagalli and D'Angelo (1965)
5	12, 14			Retarded skeletal development, inhibition of nuclei of ossification.	
				(Doses: 0.1 ml/egg, in allantois)	Jones (1970)
4, 6	5–15		29–68	Inhibition of ossification, retardation of development of cartilage, chondrolysis of capsular matrix in zone of cellular hypertrophy.	
Pituitrin				(Doses: 20 mm^3/egg, in albumen)	Cunningham and Stanfield (1928)
7	21	53			
PMS–gonadotropin				(Doses: from 0.5 to 1.0 cm^3/egg, in albumen or allantois)	Venzke (1941)
2, 14	20			Increase in size of gonads of both sexes; hyperplasia of testicular and ovarian tissue.	
Prednisolone				(Doses: 0.1 ml solution/egg, in air cell and intravenously)	Warner and Burnet (1961b)
12	16	88 and 0	47 and 7	Inhibition of growth (10.4/13.9 and 8.7/10.4 grams), edema, anomalies of feather growth.	
Progesterone				(Doses: from 10 to 1000 μg/0.03 to 0.05 ml olive oil/egg)	Piotrowski (1966)
2⅙	21	2–16	10–94	Malformations of hind limbs (74%), abdominal wall (17%), central nervous system (7%), sense organs (2%). LD_{50} = 300 μg.	
Testosterone				(Doses: from 1 to 2 mg/egg, in yolk sac, allantois)	Warner and Burnet (1961a)
4–17	15–21			Complete inhibition of the bursa of Fabricius.	

			(Doses: from 0.015 to 0.4 mg/0.05 ml of sesame oil/egg, in yolk sac)	Lakshmi (1967a)
3	13	23–51	Increase in dry weight of the bursa of Fabricius, no malformations. $LD_{50} = 0.4$ mg.	
			(Doses: dipped for 5 sec in solution of 0 to 1280 mg/100 ml ethyl alcohol)	May and Glick (1964)
0	21		Reduced size of bursa of Fabricius at hatching.	
			(Doses: dipped—from 640 to 1200 mg/egg; injected—from 0.5 to 1.0 mg/egg, into albumen)	Nakamura (1967)
3 dipped 5 injected		58–86 17–21	Reduced size of bursa of Fabricius at hatching; abnormalities of brain development, of optic vesicle, of beak, and of legs.	
Thyroxine			(Doses: 10 μg to 0.5 cm^3 of 0.002% thyroxine solution/egg, in yolk sac)	Honjo and Kuth (1959)
5	6–21		Increase in body weight, early hatch (1 day), significant increase in number of ossification centers (in cervical, thoracic, lumbosacral, and caudal vertebra) after 10 days. (See Fig. 148.)	
			(Doses: from 3 to 24 μg/egg)	Stoll, Coulaud, Faucounau, and Maraud (1966)
2–6	5–17	70–30	Strophosomy (9 to 19%), susceptibility to lordosis, effects on thoracic and abdominal organs. (See Fig. 149.)	
			(Doses: 1 μg/egg, in albumen)	Booker and Sturkie (1949)
0–6	21		Reduction in size of the thyroid (7.8/10.2 mg)	
			(Doses: from 0.013 to 0.1 μg in 0.25 ml alkaline aqueous solution/egg, in air cell)	Romijn, Fung, and Lokhorst (1952)
8	9–21		Some reduction in dry and wet body weight, but not in thyroid weight; increased oxygen consumption and carbon dioxide elimination, indicating higher metabolic rate, with early hatching (about 1 day).	

Summary Table 13 ***(Continued)***

Age (days) Treatment	Age (days) Examination	Malformations (%)	Mortality (%)	Incidence of Typical Syndromes	Reference
Thyroxine				(Doses: from 0.0025 to 2.5 μg/egg, in aqueous solution, in yolk sac)	Beyer (1952)
0	1–21		8–100	With 0.025 μg dose—increase in both wet and dry weights of the embryo and elevated oxygen uptake at all stages of development; shortening of the incubation period by more than one day. (See Fig. 147.)	
				(Doses: 5 μg in one or three doses/egg, in yolk sac)	Vidal (1953a)
7–17	10–21			Reduction in size of the growing embryo, accelerated hatching (about 1 day).	
				(Doses: 5 μg/egg, in yolk sac)	Vidal (1953b)
7–17	11–21			Retarded differentiation and growth of thyroid: at 19 days, 2.8/5.5 mg; at 20 days, 2.6/6.9 mg.	
				(Doses: from 2.5 to 10 μg/egg, in yolk sac)	Adams (1958)
10–14	21			Retarded growth and differentiation of thyroid gland.	
Triamcinolone				(Doses: from 0.3 to 300 μg/egg)	Guzman-Garcia and Laguna (1965)
9	13–21		20–100	Inhibition of growth, increase in uric acid secretion.	

Summary Table 14 Effect of Antihormonal Compounds on the Development of the Avian Embryo

Age (days)		Malformations (%)	Mortality (%)	Incidence of Typical Syndromes	Reference
Treatment	Examination				
2-Aminothiazole					
8–18	10–21			(Doses: 2 cm^3 of 5% aqueous solution/egg, in air cell) Effects on thyroid: hypertrophy, hyperplasia, and hyperemia.	Mangione (1957)
Amphenone B					
14–19	15–21			(Doses: 1 mg [0.2 mg daily for 5 consecutive days]/egg, in yolk sac) Increase in thyroid weight, with hypertrophy and hyperplasia of its cells, increased mitosis.	Adams and Bennett (1953)
Methimazole					
7	10–13		Up to 100	(Doses: from 0.1 to 2.0 mg/egg, in yolk sac) Growth retardation (7.5/9.3 grams). (See Fig. 157*A*.)	Zahnd and Rumpler (1961a)
7	20			(Doses: 2.5 mg of 10% aqueous solution/egg, in yolk sac) Retardation in bone growth (tibiotarsus), delay in hatching. (See Fig. 157*B*.)	Zahnd and Rumpler (1961b)
5, 7	16			(Doses: 2.5 mg of 1/40 aqueous solution/egg) Retarded growth (9/11 grams and 10/12 grams).	Rumpler (1966)
2-Thiouracil					
8–14	21			(Doses: 0.5 cm^3 of 0.1 to 0.2% solution/egg, in yolk sac) Retarded growth and hatching, enlargement of thyroid with associated hyperplasia and hypertrophy. (See Fig. 156.)	Adams and Bull (1949)
14	15–21			(Doses: 0.5 cm^3 of 0.1 to 0.2% solution/egg, in yolk sac) Histological changes of enlarged thyroid: hyperemia, colloid loss, cell hypertrophy, and hyperplasia.	Adams and Buss (1952)

Summary Table 14 (*Continued*)

Age (days) Treatment	Age (days) Examination	Malformations (%)	Mortality (%)	Incidence of Typical Syndromes	Reference
8	5–21			(Doses: 0.25 mg/egg, in air cell) Lowered rate of metabolism (O_2 consumption and CO_2 production), increase in size of thyroid.	Romijn, Fung, and Lokhorst (1952)
5	7–26			(Doses: 1 mg/egg, in yolk sac) Loss of body weight, decrease in ossification centers.	Honjo and Kuth (1959)
0	4–21			(Doses: 2 mg in 0.3 ml saline solution/egg, in albumen) From the tenth day: reduction in size of embryo; increase in size of thyroid (18/3 mg), with increase in iodine content (8.4/3.8 μg).	Rogler, Parker, Andrews, and Carrick (1959)
2	7–21		8–100	(Doses: from 0.1 to 5.0 mg/egg, in albumen) Significant decrease in body weight after eleventh day. (See Fig. 155.)	Sinha, Ringer, Coleman, and Zindel (1959)
Thiourea					
10–14	21+			(Doses: from 0.3 to 3.0 mg/egg, in yolk sac) Delayed development (up to 9 days).	Grossowicz (1946)
8, 14	10–24			(Doses: 0.5 cm³ of 0.4% solution/egg, in yolk sac) Delayed hatching, decreased growth of embryo, enlargement of thyroid with typical hyperemia, hyperplasia, and hypertrophy, marked increase in height of follicular epithelium, vacuolization of colloid and colloid release.	Adams and Bull (1949)
				(Doses: 2 mg/egg, in albumen)	Yushok (1950)

11	12–24		Decrease in body weight (at 20 days—19.4/29.8 grams) decrease in weight of liver (0.5/0.8 grams), decrease in weight of heart (166/223 mg).	
			(Doses: 0.5 cm^3 of 0.4% solution/egg, in yolk sac)	Adams and Buss (1952)
14	15–21		Decrease in body weight, increase in size of thyroid; elevated mitotic activity, with subsequent hypertrophy of these cells. (See Fig. 154.)	
			(Doses: 2.5 mg/0.5 cm^3 saline solution/egg, in yolk sac)	Vidal (1952), Vidal (1953a)
7	9–23		Retarded growth. (See Fig. 152.)	
			(Doses: from 2 to 10 mg/egg, in albumen)	Romanoff and Laufer (1956)
0–18	0–31	7–100	Retarded growth, delay in development (from 1 to 14 days), hypertrophy of thyroid (over 300 per cent), reduction in weight of liver and enlargement of adrenals. (See Figs. 151, 153, 154.)	
			(Doses: 3.8 mg/egg)	Moog (1962)
11	28		Inhibition of growth, increase in weight of thyroid, enlargement of adrenals.	
			(Doses: 0.25 ml of 0.9% saline/egg, in allantois)	Stocum (1966)
11	12–24		Subnormal growth in length and diameter of duodenum.	

IX

Pathogenic Afflictions

Microbial invasion can
Upset the embryo's true course,
Of its whole life reduce the span—
Suppress the work of vital force.

Many infectious agents may produce certain developmental anomalies. Among known pathogens of the avian embryo are viruses, bacteria, protozoa, and fungi, as well as neoplasms and some diseases of unknown origin.

The more specific effects of these pathogens take place during a definite stage of embryonic development. The chicken embryo is most susceptible at the age of 10 to 15 days. During this period, certain organs, tissues, body fluids, and special sites of the developing embryo, as well as the nutritional and environmental conditions, are most favorable for proliferation of pathogenic microorganisms.

Beyond the 15-day stage to hatching, susceptibility of embryonic cells becomes more restricted, especially to those more specific lesions which characterize infection of well-formed embryos.

VIRUSES

Bronchitis

It has been noted (*Beaudette and Hudson, 1937*) that during propagation of infectious bronchitis virus in the chicken embryo, the virus becomes more pathogenic for the embryos with each successive passage, and less pathogenic for adult birds.

Growth. Ten-day-old embryos, injected with 0.2 ml of inoculum into the allantoic sac, are about one-half normal size on the nineteenth day (*Loomis, Cunningham, Gray, and Thorp, 1950*).

Weight of Organs. Among the embryonic organs, some deviations in weight have been noted (*Loomis, Cunningham, Gray, and Thorp, 1950*). Development of the lung is markedly retarded; on the nineteenth day, the lungs are about one-half normal size.

In all infected embryos, the heart is noticeably decreased in size, and the spleen is enlarged to twice normal size.

Mortality. No outstanding factors characterize embryonic deaths during the first five passages of the bronchitis virus. During the fifth to tenth passages, some embryos die on the fourth and fifth days following inoculation. More embryos begin to die between the tenth and fifteenth passages. Between the twenty-fifth and thirtieth passages, there is a marked increase in the number of embryonic deaths, occurring at the end of the fifth day following inoculation. With each succeeding passage, the virus becomes more virulent for the embryo and continues to kill at about the same rate until the ninetieth passage (*Delaplane and Stuart, 1941*). Then, the ability of the virus to kill embryos apparently becomes fixed or delineated.

Malformations. For a description of all syndromes see Summary Table 15, under Bronchitis virus.

Coxsackie

Mortality. The virus is not immediately lethal to the embryo (*Peers, Ransom, and Huebner, 1952*). However, all embryos die at hatching or a few days after.

Malformations. Skeletal muscle lesions characteristic of infection with group A Coxsackie virus are produced in the immature embryos (6 days of age), with extensive necrosis of epidermis and feathers, visceral lesions, and extensive liver necrosis (*Peers, Ransom, and Huebner, 1952*).

A sharp reduction in severity of the lesions becomes apparent in embryos inoculated on the twelfth day of incubation.

Distemper

Mortality. Among embryos inoculated with various concentrations of distemper virus at 2 days of age, the incidences of death observed during

Time After Inoculation (days)	Mortality of Embryos (in %)		
	(Virus 20%)	(Virus 2%)	(Virus 0.2%)
5	0	0	0
6	2	0	0
7	21	19	0
8	74	43	13
9–13	98	67	38
14–18	100	81	38

successive days after inoculation are summarized in the accompanying table.

Malformations. Four or 5 days after injection of a 20 per cent suspension of virus, small spot- or threadlike lesions begin to appear on the allantoic membrane (*Heath, Shear, Imagawa, Jones, and Adams, 1956*). During the next few days, these lesions grow in size and coalesce to form large, ulcerative areas on the surface of the membrane.

The incidence of such lesions after injection of various concentrations of virus is shown in the table below.

Time After Inoculation (days)	Incidence of Lesions (%)		
	Virus 20%	Virus 2%	Virus 0.2%
3	0	0	0
4	2	10	0
5	37	19	0
6	60	29	13
7	98	67	38
8	100	81	38
9–13	—	81	38
14–18	—	81	38

In a few cases, similar, though smaller, lesions could also be detected on the amnion.

Encephalitis

Mortality. The effect of time of injection of encephalitis virus on the distribution of embryonic mortality is illustrated in Fig. 158. The peak of mortality gradually tapers off and appears later with increasing age of the embryo at inoculation.

Encephalomyelitis

Mortality. Chicken embryos infected with equine encephalomyelitis virus die sooner than embryos of any other experimental animal studied (*Crawley, 1948*).

Malformations. Allantoic membranes infected with the virus soon show vascular collapse, thrombosis, and hemorrhage (*Bang, 1943*).

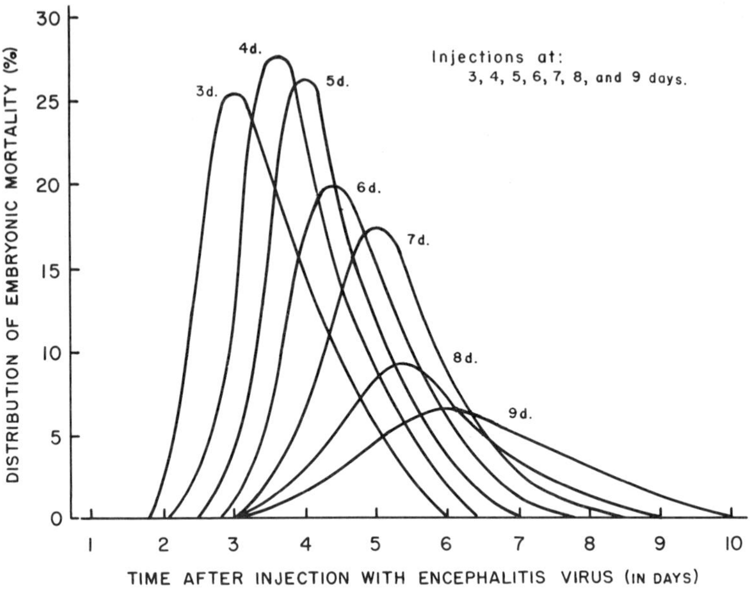

Fig. 158. Effect of the time of injection with tick-borne encephalitis virus on the time of appearance and the distribution of early embryonic mortality—during the first critical period. (Reconstructed from the data and charts of Slonim and Roslerova, 1965.)

The embryo itself initially reveals scattered hemorrhages and thromboses, indicating that the virus first destroys vascular endothelium.

With progressive development, the allantoic membrane and embryo become more resistant to the virus, especially at hatching time.

Hepatitis

Mortality. Mortality of 10-day-old chicken embryos, after inoculation of duck hepatitis virus into the allantoic cavity at different passages (*Hwang, 1965*), is shown in the tabulation below.

Passages (number)	Embryonic Mortality (%)
25	78
40	86
55	93
60	100

Malformations. The most characteristic lesions produced by hepatitis virus are observed in the liver and spleen—degenerative, infiltrative, and proliferative lesions (*Paunescu, 1964*).

Herpes Simplex

Growth. The total weight of embryos infected with herpes simplex is decidedly less than normal, Weight of 15-day-old embryos 3 days after inoculation was 10.2 grams, as compared to a normal weight for that age of 11.5 grams (*Ackermann and Francis, 1950*).

Weight of Organs. Three days after injection of herpes virus into the allantoic membrane, the wet weights of the liver and heart are 42 and 34 per cent greater than those in normal embryos (*Ackermann and Francis, 1950;* see also *Anderson, 1940*), as shown below.

	Weight of Organ at 15 Days (mg)	
Organ	Control	Viral
Heart	88	118
Liver	172	245

This accelerated growth of organs occurs despite the fact that the total weight of the infected embryo is less than that of a normal embryo. The infected organ tissues appear to develop more rapidly at the expense of the rest of the tissues. (See disproportionate growth of organs under Hormonal Derangement, Chapter VIII, Figs. 133 and 140.)

Mortality. The mortality rate of embryos during successive days after inoculation with various dilutions of herpes simplex virus (*Heath, Shear, Imagawa, Jones, and Adams, 1956*), is illustrated in Fig. 159*B*, and data for 2 days after injection (2-day embryos) are given in the accompanying table.

Concentration of Herpes Simplex (dilution %)	Incidence of Mortality (%)
10^{-1}	100
10^{-2}	97
10^{-3}	79
10^{-4}	57
10^{-5}	9
10^{-6}	0

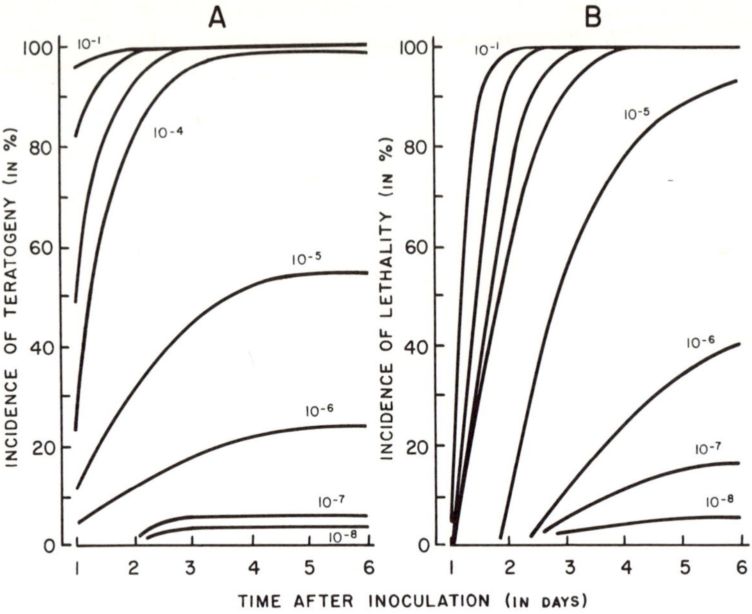

Fig. 159. The general pattern of the effect of inoculation, at the age of 2 days, with either herpes simplex, influenza A, or vaccinia virus, in concentrations from 10^{-1} to 10^{-8}, on the incidence of embryonic teratogeny and mortality. (Plotted as composed values recalculated from the data of Heath, Shear, Imagawa, Jones, and Adams, 1956.) *A,* effect of the viruses on teratogeny; *B,* effect of the viruses in terms of lethality.

Malformations. The incidence of anomalous development in 2-day-old embryos (2 days after infection with various doses of inoculum) is illustrated in Fig. 159*A* and listed in the table below.

Concentration of Herpes Simplex (dilution %)	Incidence of Malformations (%)
10^{-2}	100
10^{-3}	96
10^{-4}	93
10^{-5}	27
10^{-6}	10
10^{-7}	0

The primary effects of this virus are micrencephaly and axial flexion. In addition to the above anomalies, it was observed that the eyes tend to become compressed along their anteroposterior axis in a high percentage (*Heath, Shear, Imagawa, Jones, and Adams, 1956*).

Observations also show that at all concentrations of the herpes virus, growth of the embryo after anomalies become apparent proceeds at a slower rate until death. Infected embryos that fail to show anomalies by the sixth day may appear normal at death.

Influenza A

Growth. In developing embryos inoculated at 11 days with 0.1 ml of a 10^{-3} dilution of influenza A virus in a saline solution, body weights were significantly lower than normal (*Parodi, Lajmanovich, Pennimpede, and Mittelman, 1948*), as is shown in the accompanying table.

	Weight of Embryo	
Age of Embryo (days)	Control (grams)	Infected (grams)
12	4.5	4.6
13	6.4	6.0
14	8.7	7.6
15	11.5	9.4
16	14.3	11.1
17	17.3	13.1

Concurrently, there is an increase in the weight of the allantoic membrane and in the volume of the allantoic fluid.

Also, there is a marked decrease in oxygen consumption (and in carbon dioxide production) in the virus-infected embryonated eggs (Fig. 160).

Mortality. Inoculation with active influenza A virus produces alterations which are incompatible with embryonic survival, leading to a very high mortality during the first few days after infection. Embryos showing no abnormalities and surviving the third day seem to have remained uninfected (*Hamburger and Habel, 1947*).

Consequently, the mortality curve is quantitatively abnormal, because of the appearance of a great peak shortly after the beginning of incubation (Fig. 161). The difference between the mortality of the influenza A-infected group and that of the control group is highly significant.

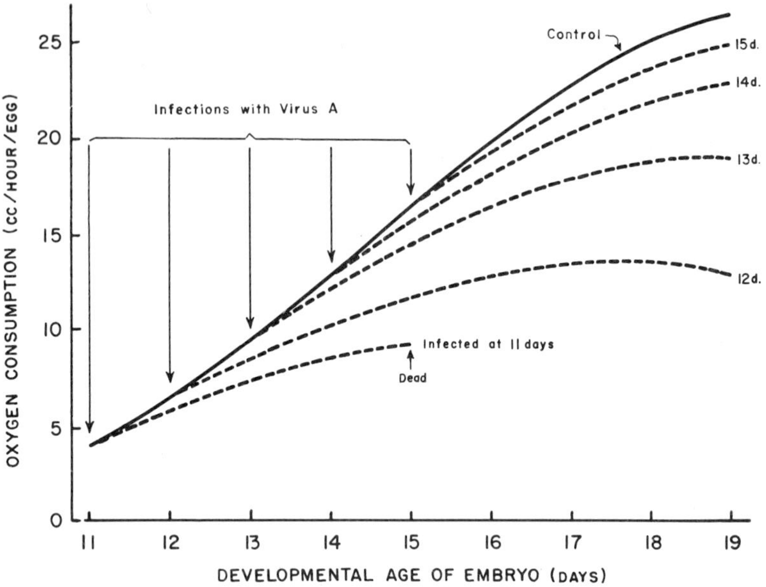

Fig. 160. Composite curves showing the effect of influenza A virus injections on the oxygen consumption at various stages of embryonic development. (Rearranged from the data of Reczko, 1955; see also Parodi, Lajmanovich, Pennimpede, and Mittelman, 1948.)

The total embryonic mortality during incubation, in control eggs and eggs inoculated with influenza A virus, allantoic fluid, or saline is shown below.

Substance Injected	Embryonic Mortality (%)
Control (none)	28
Active virus	99
Heat-inactivated virus	87
Allantoic fluid	70
Buffered saline	80

Incidence of lethal effects produced by various dilutions of influenza A virus upon the early embryo (*Heath, Shear, Imagawa, Jones, and Adams, 1956*) is shown in Fig. 159*B*.

Malformations. Results of influenza A infection in the chicken embryo confirm the observations on rubella in humans, in that only injections of early embryos result in abnormalities (*Hamburger and Habel, 1947*).

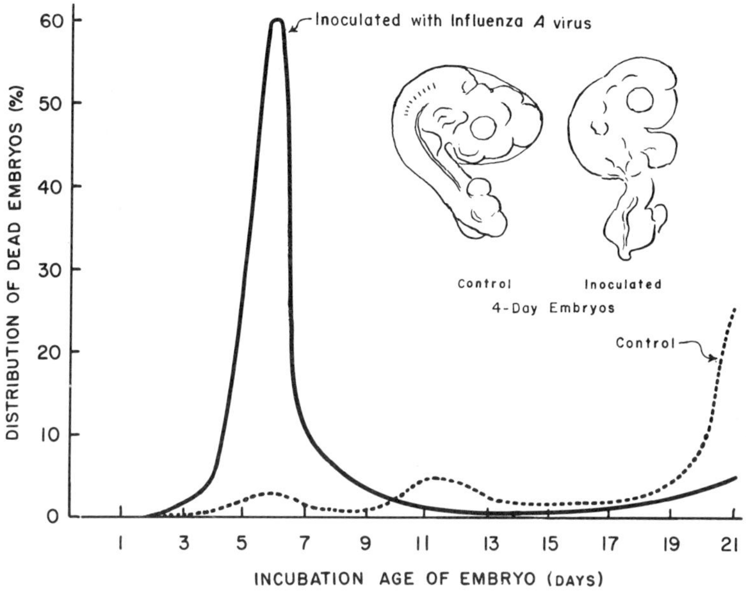

Fig. 161. Effect of inoculation with active influenza A virus on the distribution of embryonic mortality. (After Cruz, Campillo-Sainz, Munoz-Armas, Pascual-Chavez, and Zamudio-Bandera, 1963.) *Insert:* an influenza A virus-injected embryo shows a swollen auditory vesicle, micrencephaly, and axial flexion. (After Heath, Shear, Imagawa, Jones, and Adams, 1956.)

The effect of various dilutions of influenza A virus (injected at the age of 18 hours) on the total incidence of embryonic anomalies on the fifth day (*Collado, 1965*) is shown in the accompanying table (see also Fig. 159*A*).

Dilution of Influenza A (concentration in %)	Incidence of Malformations (%)
10^{-3}	100
10^{-4}	83
10^{-5}	40
10^{-6}	14
10^{-7}	0

Different anomalies may be classified according to the developmental stage of the embryo at the time of inoculation (*Collado, 1965*), as follows:

Time of Inoculation (age in hours)	Regional Modifications
19–29	Cephalic pole
29–45	Caudal pole
45–72	Apparently normal

The primary symptoms of infection with influenza A virus (*Heath, Shear, Imagawa, Jones, and Adams, 1956*) are micrencephaly and axial flexion, a form of lordosis.

It has been observed (*Robertson, Williamson, and Blattner, 1960*), that there is a relationship between the time of inoculation of the embryo with the virus and the incidence of myeloschisis, as is shown in the tabulation below.

Time of Inoculation (hours)	Incidence of Myeloschisis (%)
43–44	47
48–50	33
60	6

The cephalocaudal extent of the defect generally is greater in the younger embryos.

For a more detailed enumeration of anomalies, see Summary Table 15, under Influenza A virus.

Mumps

Growth. It has been observed (*Williamson, Blattner, and Simonsen, 1957*) that 2-day-old embryos inoculated with 0.05 ml of mumps virus under the vitelline membrane or over the blastoderm show general growth inhibition and retarded development of the amnion.

Mortality. The rate of embryonic death by the fifth day after inoculation increases with the concentration (50% infectious dose) of virus in the inoculum, as is shown below.

Amount of Virus (concentration)	Embryonic Mortality (%)
$10^{5.8}$ ID_{50}	58
$10^{3.8}$ ID_{50}	31
$10^{1.8}$ ID_{50}	18

Two-day-old embryos are more sensitive to the virus than 8-day-old embryos (*Hamburger and Habel, 1947*). Virus injection is invariably lethal to the young embryos.

Malformations. Mumps virus inoculated into developing eggs at 48 hours of incubation produces a high incidence of cataracts in embryos which survive from 3 to 15 days after inoculation (*Williamson, Blattner, and Simonsen, 1957*). Figure 162 shows that the frequency of cataract bears a definite relationship both to the dilution of inoculum (Fig. 162*A*) and to the age of the embryo at the time of inoculation (Fig. 162*B*).

Other defects are listed in Summary Table 15, under Mumps virus.

Newcastle Disease

Growth. Newcastle disease virus causes general retardation in growth of the embryo and, in varying degrees, retardation in growth of the amnion (*Blattner and Williamson, 1951*).

Mortality. The rate of embryonic mortality in eggs inoculated with the virus (concentration of 10^7 per 0.05 ml of inoculum) differs somewhat for

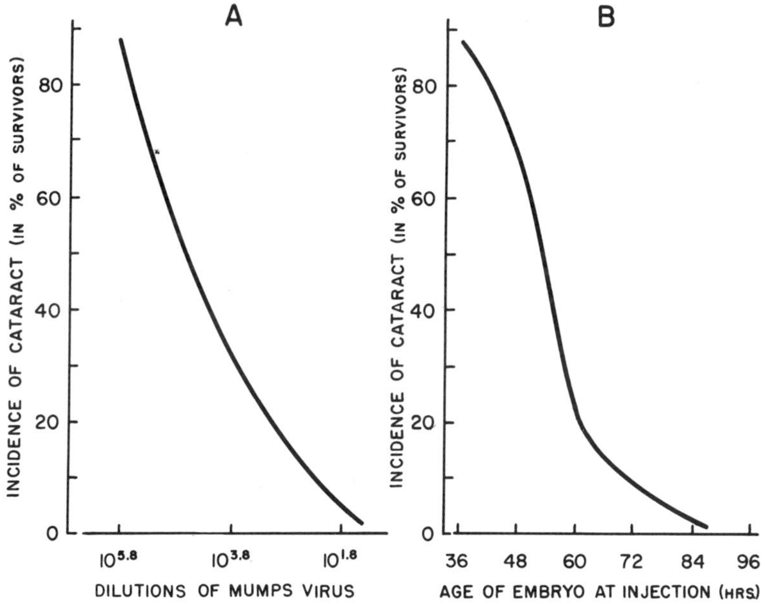

Fig. 162. Effect of injecting mumps virus on the incidence of cataracts. *A,* effect of various dilutions of virus injected into the allantoic cavity of the 2-day-old embryo (plotted from the data of Williamson, Blattner, and Simonsen, 1957); *B,* effect of time of injection of a 10^5 dilution (after the data of Robertson, Williamson, and Blattner, 1964.)

various age groups (*Robertson, Williamson, and Blattner, 1955*), as shown below.

Age of Embryo at Injection (hours)	Embryonic Mortality After 2 Days (%)
36	83
48	60
60	22
72	38

It is apparent that the toxic effects of this virus are more pronounced in the younger embryos, the older embryos showing more resistance.

Malformations. The effectiveness of different doses of virus in the inoculum in producing defects in 11-day-old chicken embryos (*Williamson, Blattner, and Robertson, 1953*) is shown in the accompanying tabulation.

Virus Titers (0.05 ml of inoculum)	Defective Embryos (% of survivors after 24 hours)
$10^{7.0}$	98
$10^{6.5}$	85
$10^{4.5}$	40
$10^{2.5}$	10
$10^{0.5}$	7
$10^{0.0}$	0

The susceptibility of different developmental stages of the embryo to production of defects by Newcastle disease virus is demonstrated in the table below.

Age of Embryo at Inoculation (hours)	Defective Embryos Among Survivors (%)
36	88
48	75
60	77
72	16
84	0

These two tables distinctly show that the amount of inoculum and the age of the embryo at inoculation time definitely affect the occurrence of anomalies.

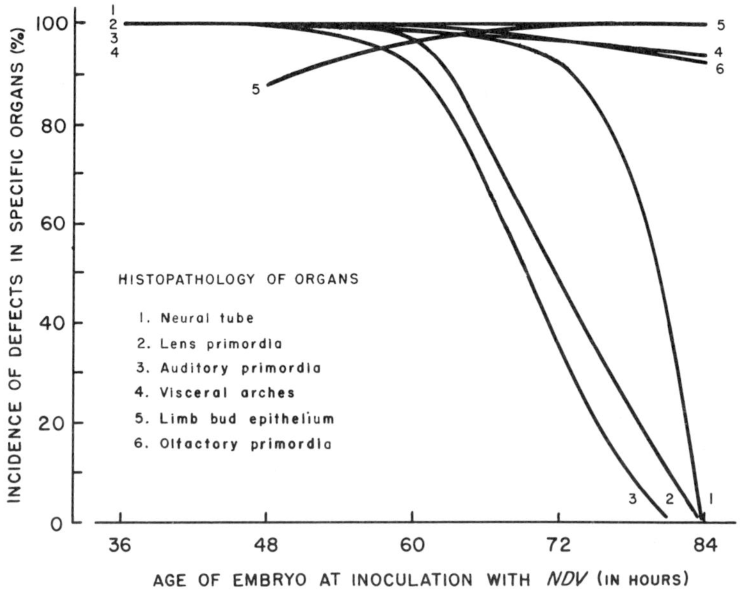

Fig. 163. Effect of Newcastle disease virus (NVD) on the histopathology of various organs of the chicken embryo, in relation to the time of administration of 0.05 ml of inoculum. (Plotted from the data of Robertson, Williamson, and Blattner, 1955.) *Note:* all observations were made 24 hours after inoculation.

In the inoculated embryos, a high incidence of characteristic lesions is observed in the neural tube, lens primordia, auditory primordia, visceral arches, limb buds, and olfactory primordia (Fig. 163).

For further details see Summary Table 15, under Newcastle disease virus.

Pseudorabies

Malformations. The chicken embryo responds to the virus of pseudorabies by development of lesions on the allantoic membrane, followed by infection of all tissues of the central nervous system (*Bang, 1942*). The reaction of the embryo to the virus varies with the age of the embryo.

Vaccinia

Growth. Injection with vaccinia virus in a dose of 1 ml of 10^{-5} dilution on the ninth day of incubation induces delay in embryonic development

(*Hoffstadt and Tripi, 1946*) as shown in the table below.

Incubation Age (days)	Weight of Embryo (grams)	
	Control	Vaccinia-injected
10	2.4	2.2
12	5.2	3.8
14	9.8	7.8
16	15.3	11.9
18	22.5	16.0

The final average weight for infected embryos is 6.5 grams less than for normal ones.

Mortality. The effect of vaccinia on embryonic mortality is directly related to the dosage of inoculum (see Fig. 159*B*).

Malformations. The concentration of vaccinia virus inoculum affects the incidence of malformations in the avian embryo to a degree that is highly significant (see Fig. 159*A*).

Yellow Fever

The course of infection in the chicken embryo indicates early multiplication of yellow fever virus at the site of inoculation (the allantoic membrane), with rapid invasion of the blood and infection of the embryo (Fig. 164), where multiplication occurs in all the tissues (*Fox and Laemmert, 1947*).

The concentration of virus is greatest on the third day after inoculation in embryos 6 to 8 days of age; the lowest concentrations are observed in 12- and 14-day-old embryos. This is shown in the accompanying table.

Age of Embryo at Inoculation (days)	Virus Content	
	After 3 Days	After 4 Days
6	7,300	4096
7	6,800	2400
8	15,600	2450
10	19,500	1600
12	570	897
14	256	800

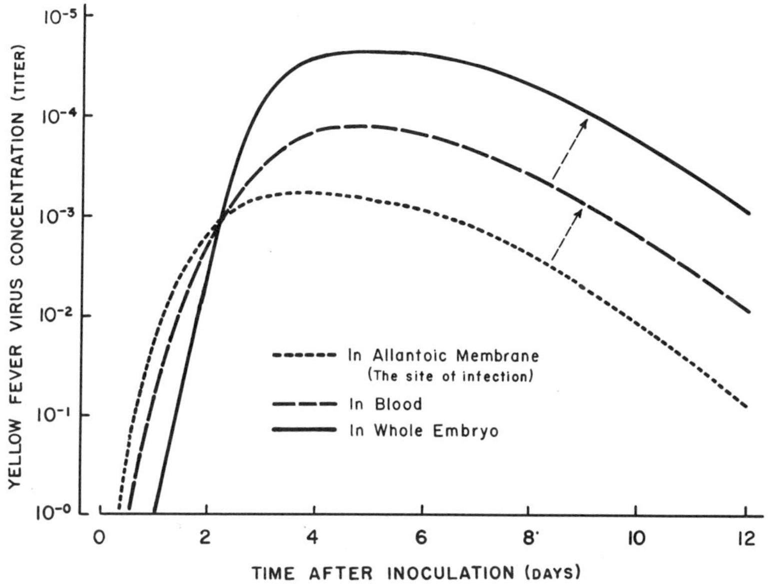

Fig. 164. Course of yellow fever virus injection in the embryo, from the allantoic membrane (the site of injection), to the blood, then to the whole embryo. (Curves based on the data of Fox and Laemmert, 1947.)

Growth. Among gross findings due to yellow fever virus is retarded development, with hatching of surviving embryos delayed from 2 to 8 days (*Fox and Laemmert, 1947*).

Mortality. The incidence of mortality in infected embryos is closely related to their age at the time of inoculation (*Fox and Laemmert, 1947*). This is well illustrated by the data in the accompanying table.

Age of Embryo (One Day after Inoculation)	Total Mortality (% inoculated)
8	83
10	25
12	13

The data indicate that, with age, embryos acquire resistance to the infective agent.

Susceptibility to yellow fever virus of 7-day-old embryos inoculated by different routes is shown in the accompanying table.

The largest dose is noninfective when injected into the air cell, and embryonic susceptibility is low to virus inoculated into the albumen

Site of Inoculation	Embryonic Infection (%)				
	Concentration of Inoculum (0.06 ml)				
	10,600	26.4	6.6	1.6	0.4
Periembryonic stab	—	100	67	67	67
Allantoic membrane	—	100	100	83	33
Yolk sac	—	83	33	20	0
Allantoic cavity	—	33	0	0	
Albumen	50	0			
Air cell	0				

and into the allantoic cavity. The greatest susceptibility is observed when the virus is introduced onto the allantoic membranes or by means of a periembryonic stab (*Fox and Laemmert, 1947*).

Malformations. No clearly characteristic morphological changes were observed in infected embryos (*Fox and Laemmert, 1947*). Retarded development was often associated with generalized edema and grossly mottled liver. There was irregular occurrence of focal necrosis in the muscles and an increase in foci of myelopoiesis.

BACTERIA

Botulinus

Growth. Nineteen-day-old embryos inoculated with *Clostridium botulinum* toxin (30 μg/egg) intravenously in the allantois on the seventh and twelfth days of incubation weighed 45 per cent less than controls (*Drachman, 1964*).

Weight of Organs. Botulinus infection causes considerable reduction in the weight of lower limb muscle and of other organs (*Drachman, 1964*). Relative weights on the nineteenth day in embryos injected with toxin on the seventh day of development are shown below.

Organ	Reduction in Weight (% of control)
Limb muscle	81
Heart	27
Liver	17

Mortality. Systemic administration of botulinus toxin in minute quantities produces death by respiratory paralysis in most animals; the chicken embryo readily survives doses which would be lethal to many other adult birds (*Drachman, 1964*). This is due presumably to the special mechanism of respiratory gas exchange, which in the avian embryo is accomplished by passive diffusion across the chorioallantoic membrane (*Romanoff, 1960, p. 566*).

Malformations. Embryos injected with botulinus toxin show severe ankylosis of multiple joints (*arthrogryposis multiplex congenita*), which results from skeletal immobilization during embryonic development. Apart from a slightly shortened upper beak, the abnormalities are limited to atrophy and degeneration of the skeletomuscular system (*Drachman, 1964*).

On gross observation, all skeletal muscles of the body and limbs are strikingly shrunken and fatty.

Cholera

Mortality. Embryonic death occurs very uniformly and rapidly following intravenous injection with *Vibrio cholerae* (Fig. 165*A*). The time of

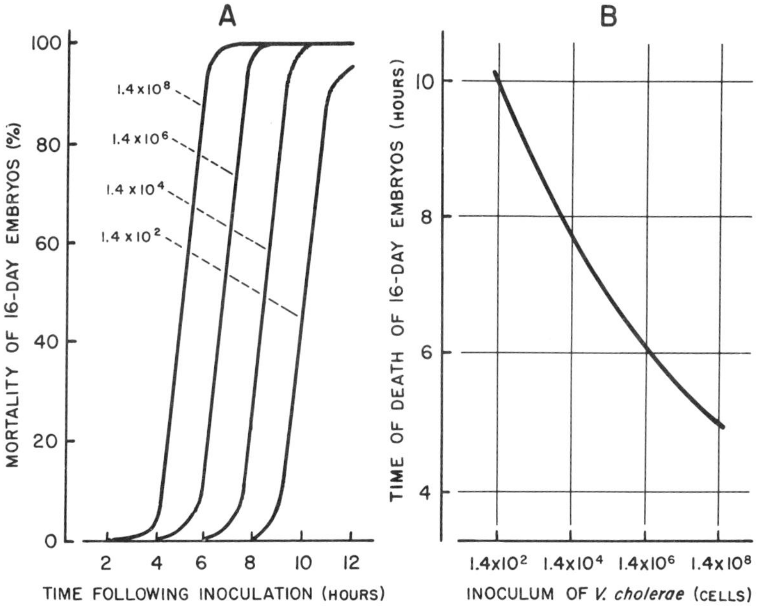

Fig. 165. Effect of intravenous inoculation with cholera bacteria (*Vibrio cholerae*) on the mortality of the chicken embryo. (With some modifications after Finkelstein and Ramm, 1962.) *A*, pattern of embryonic death following administration of various doses of the inoculum; *B*, the time of death (plotted as the time of 50 percent lethality) in relation to inoculum of viable cells.

death (plotted as the time of 50 per cent mortality; see Fig. 165*B*) is inversely proportional to the size of the inoculum (viable cells).

Sixteen-day-old embryos, fully susceptible to intravenous injection of cholera organisms, are virtually resistant to administration by the allantoic route (*Finkelstein and Ramm, 1962*).

Malformations. Embryos subjected to intravenous injection of *Vibrio cholerae* are almost covered with hemorrhagic areas. The most striking histologic feature is hemorrhagic necrosis of the mesonephros (intermediate kidney) of embryos up to 15 days of age (*Finkelstein and Ramm, 1962*). Dead embryos show degenerative changes.

Meningococcal Endotoxin

The chicken embryo is susceptible to injected meningococcal endotoxin.

Mortality. The number of deaths occurring within 18 hours after inoculation (Fig. 166*A*) indicates a striking susceptibility to the cocci.

Ten-day-old embryos are highly susceptible (Fig. 166*B*), while

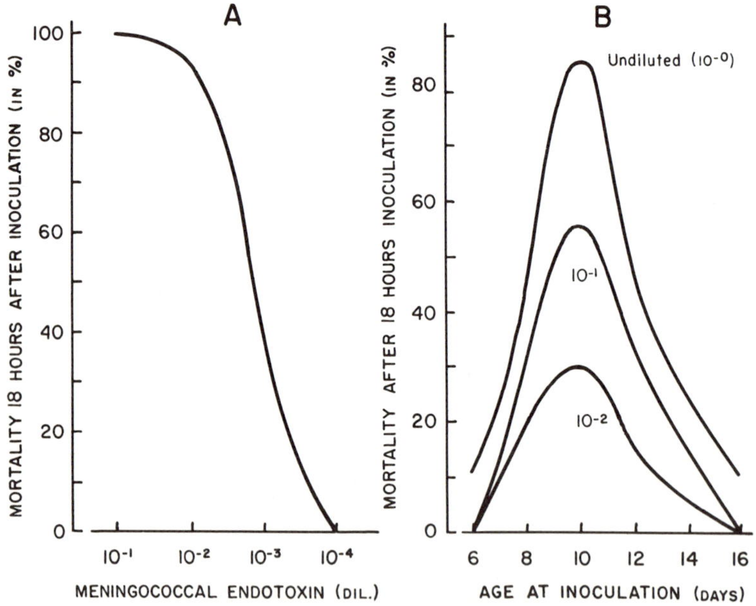

Fig. 166. The lethal effect of meningococcal endotoxin on the developing embryo. (With some modifications after Smith and Thomas, 1956.) *A*, effect of various dilutions of the endotoxin, in 10-day embryos; *B*, effect of age of the embryo at the time of inoculation with different dilutions of endotoxin on the allantoic membrane.

those younger or older than 10 days show much less reactivity to any of the doses of endotoxin. At 6 and 16 days, the embryos appear to be completely unsusceptible (*Smith and Thomas, 1956*).

Malformations. The susceptibility of the embryo to infection with meningococci depends upon its stage of development. The embryo reacts to the infection with more or less specific lesions (*Buddingh and Polk, 1939*).

In the 12-day-old embryo, a meningococcal septicemia develops, with production of generalized lesions.

In older embryos, 14 to 15 days, a more selective localization takes place in the cranial sinuses, lungs, and meninges, with development of an acute inflammatory reaction.

Pleuropneumonia

Growth. Many pleuropneumonia-treated embryos are dwarfed (*Chute and Cole, 1954*), as is shown by measurement of the middle toe.

Age of Embryo (days)	Measurement of Toe (Infected/Control) (cm)
13	0.9/0.98
18	1.5/2.04
20	1.6/2.14

Mortality. The pattern of embryonic mortality following inoculation (*Moulton and Adler, 1957*) is shown in the accompanying table.

Time after Inoculation (days)	Embryonic Mortality (%)
2	0.0
3	2.1
4	16.7
5	10.4
6	6.3
7	6.3
8	4.2

The specific pattern of embryonic mortality did not appear to be associated with the species of birds used.

Malformations. Infected embryos dying between 8 and 14 days of incubation show extensive hemorrhage and edema (*Chute and Cole, 1954*).

Ecchymotic hemorrhages are noticed frequently in the skin of the head, neck, and ventral thoracic region.

Embryos which die between 14 and 21 days of incubation have hemorrhage and edema. Edema is prevalent in the head, dorsal neck, sublingual, and lumbar and thoracic vertebral regions.

The incidence of lesions in the tibiofemoral joints (*Moulton and Adler, 1957*) following inoculation with pleuropneumonia organisms is shown in the accompanying table.

Time after Inoculation (days)	Incidence of Gross Joint Lesions (%)
4	0
5	20
6	60
7	100

In the development of these arthritic lesions from the earliest stage of hyperemia, necrosis precedes cellular invasion of the joint cavity.

The relative numbers of lesions observed in individual organs of chicken embryos infected with pleuropneumonia organisms (*Chute and Cole, 1954*) are tabulated below.

Organ	Total Number of Lesions
Yolk sac	120
Liver	105
Joint	105
Brain	101
Lung	91
Heart	88
Allantois	79
Kidney	67

Hyperemia, hemorrhage, periarteritis, arteritis, and periphlebitis are vascular changes of frequent occurrence in all organs. In the lungs, periarteritis occurs in more than 51 per cent. In the liver, lymphocytic foci are most frequent (53 per cent); next is periphlebitis (28 per cent). Present in the heart are pericaritis, myocarditis, and endocarditis. See also Summary Table 15, under Pleuropneumonia.

Staphylococcus

The chicken embryo is highly susceptible to some strains of staphylococci and relatively resistant to others.

Mortality. With endocarditis-producing *Staphylococcus epidermis*, the 50 per cent lethal dose (LD_{50}) for 8-day chicken embryos, observed 48 hours after inoculation, may vary for different bacterial strains (*Lyles, Sharp, Nabors, Larsen, and Potter, 1965*), as is shown by a few examples in the table below.

Strain	LD_{50}
A	317
B	4,790
C	12,350

An overall decrease in embryonic mortality with increase in age at inoculation is illustrated in Fig. 167. This trend had been confirmed by some previous observations (*Smith and Thomas, 1956; Buddingh and Henigst, 1960; Finkelstein and Ramm, 1962; Gardner, Lyles, and Lankford, 1964; McCabe, 1964*).

Malformations. Virulent strains of staphylococci are found in large concentrations in organs (*insert* of Fig. 167).

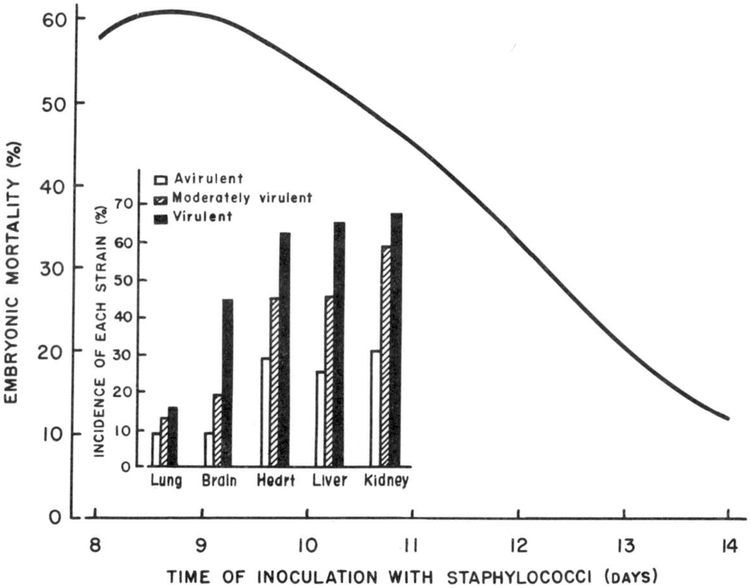

Fig. 167. Virulence of *Staphylococcus aureus* inoculated at various stages of development. (With some modifications after Lyles, Sharp, Nabors, Larsen, and Potter, 1965.) *Insert:* relative frequency of invasion by staphylococci in the tissues of different organs within 48 hours. (From the same authors.)

High counts of bacteria in the kidney, liver, and heart are usually associated with extensive hemorrhage and tissue damage.

For further details, see Summary Table 15, under Staphylococcus.

Streptobacillus

Mortality. The effect of inoculation of S. *moniliformis* into the allantois of 12- and 14-day-old embryos on subsequent embryonic mortality (*Buddingh, 1944*) is shown in the table below.

Time after Inoculation (days)	Embryonic Mortality (%)	
	Inoculated at 12 Days	Inoculated at 14 Days
1	10	4
2	85	27
3	100	66
4	—	100

Malformations. In the inoculated embryos, all lesions observed are confined to the joints (*Buddingh, 1944*).

Embryos sacrificed 24 hours following inoculation reveal localized lesions exclusively in the synovial lining of the joints.

At the 48-hour stage, the affected joints show involvement of the entire synovial membrane.

At 72 hours after inoculation, there is an increase in the destruction of the synovial lining.

Sections through the liver on the second and third days also reveal marked dilatation of the sinusoids.

Tetanus Toxin

Malformations. When 24-hour chicken embryos are injected with 0.05 ml of tetanus toxin (under the vitelline membrane) and examined after 17 hours, some morphological changes are apparent (*Corliss, Fedinec, and Robertson, 1966*). Serial sections of the embryo disclose concentrations of lesions almost exclusively in neural tissue. These tetanus-induced lesions include encephalosis, myeloschisis, and platyneuria at various levels of the neural tube. (See also *Gray and Worthing, 1941.*)

For the incidence of various anomalies see Summary Table 15, under Tetanus.

PROTOZOA

Coccidium

Mortality. The mortality rate in embryonated eggs inoculated with coccidia depends more upon the age of the embryo than on the route of inoculation (*Burke, Salvin, and Gerloff, 1952*). This is demonstrated by the data in the accompanying table.

Age at Inoculation (days)	Embryonic Mortality (% dead on the 5th day)
3	100
5	91
7	80
8	73
9	24
10	19
11	15
12	7

It is obvious that 3- to 8-day-old embryos are more susceptible to infection of 0.1 ml of this spore suspension than are 9- to 12-day-old embryos.

Plasmodium

Infection of chicken embryos with *Plasmodium gallinaceum* has been accomplished by allowing mosquitoes to feed on the allantoic blood vessels, by intravenously inoculating parasite-containing blood, or by injecting into the yolk sac an emulsion of brain containing exo-erythrocytic forms (*Haas, Wilcox, and Ewing, 1945*).

Mortality. The response of developing chicken embryos inoculated intravenously with these parasites at the age of 11 to 12 days, was measured by the rate of mortality (*Haas, Feldman, and Ewing, 1945*). The majority of deaths from malaria infection occurred on the ninth day after inoculation.

Malformation. The presence of a green color (of the pigmented forms) in infected embryonic membranes and fluids and in the liver and spleen of the embryo itself indicated the extent of malaria infection (*Haas, Feldman, and Ewing, 1945*).

In duck embryos (as in chicken embryos, *McGhee, 1949b*), natural immunity plays the most important role in resisting experimentally induced blood infection with *Plasmodium gallinaceum* (*McGhee, 1949a*).

FUNGI

Mycoplasma

Mortality. Observations on the pathogenicity of this fungus (*Yamamoto and Ortmayer, 1966*) after inoculation into the air sac of the chicken embryo on the seventh day indicate that, in general, embryonic mortality increases with successive passages, as is shown below.

Number of Passages	Embryonic Mortality (%)
1–5	3.9
6–9	24.4
10–14	78.5
15–19	58.1

Malformations. In developing turkey eggs, inoculated with various quantities of mycoplasma (as a colony-forming unit) on the thirteenth day of incubation, incidence of air sac lesions ranges from 37 to 100 per cent (*Yamamoto and Bigland, 1966*).

The abdominal air sacs are affected less frequently than the clavicular and thoracic air sacs.

The minimum infectious dose of mycoplasma is about 0.685 colony-forming units per embryo, which produces air sac lesions in 50 per cent of embryos; thus very few organisms are necessary to produce airsacculitis.

NEOPLASMS

Neoplasms have been shown to occur in association with congenital malformations.

Tumor

Growth. On the fourth day, embryos from mammary tumor-inoculated eggs are usually smaller than those from noninoculated eggs incubated for the same time (*Armstrong and Ham, 1947*).

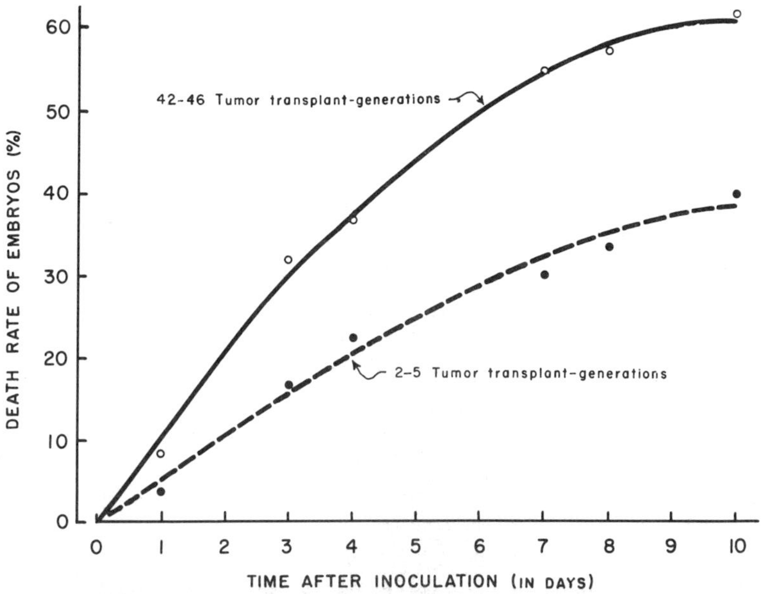

Fig. 168. The effect of repeated tumor passage through the yolk sac on the mortality rate of the chicken embryo after tumor inoculation. (After Taylor and Carmichael, 1949; see also Armstrong and Ham, 1947.)

Weight of Organs. When tumor-bearing embryos are opened, their livers and hearts are found to be larger than those of the controls (*Armstrong and Ham, 1947*).

Mortality. It has been established (*Armstrong and Ham, 1947; Taylor and Carmichael, 1949*) that repeated tumor passages through the yolk sac increase the tumor's lethality (Fig. 168).

The lethal properties gained in passages are enhanced by increasing the amount of tumor tissue given by yolk sac inoculation at 4 days (*Armstrong and Ham, 1947*). This is illustrated in the accompanying tabulation.

Time after Inoculation (days)	Embryonic Mortality at 17 Days (%)	
	With 0.1 cm^3 Tumor	With 0.01 cm^3 Tumor
9	22	22
11	29	23
13	65	45
15	95	55
17	—	71

Malformations. In about 5 per cent of the embryos of the group receiving the highest transplant (Fig. 168) various morphological changes, with pathological manifestations, are present.

Among the chief pathological symptoms in the affected embryos are hypochromic anemia, enlarged flabby hearts, and enlarged livers showing peripheral necrosis (*Armstrong and Ham, 1947*).

It has been remarked (*Karnofsky, Ridgway, and Patterson, 1952*) that the developing chicken embryo has no defense against the growth of heterologous tissue, and that it will supply blood vessels and connective tissue for the growth of transplanted tumors.

DISEASES OF UNKNOWN ORIGIN

Leukemia

Growth. The weight of the embryo is not affected by injection of leukemia cells until the time of hatching, when it is appreciably lower than normal.

Weight of Organs. Shortly after injection of leukemia cells, the liver and spleen become enlarged (Fig. 169*A* and *B*). Splenic enlargement is always proportionally greater than that of the liver.

Mortality. With intravenous injection on the tenth day, massive tumor may develop, and the embryos then die shortly before hatching time, or a few days thereafter (*Clarkson, Katz, Dann, and Karnofsky, 1964*). A toxic effect is associated with the leukemia-induced splenomegaly.

It has been observed (*Brandly, Thorp, and Prickett, 1949*) that when leukemic blood of an adult chicken is injected into eggs at 10 to 14 days of development, the route of administration has some effect on the mortality of the embryos at hatching. This is illustrated by the data in the table below.

Route of Injection	Incidence of Mortality (%)
Chorioallantoic	70.5
Intravenous	66.0
Allantoic sac	47.0
Amniotic sac	44.1
Yolk sac	31.0

Malformations. After intravenous injection of leukemia cells into the 10-day-old chicken embryo, some of the cells proliferate to form easily

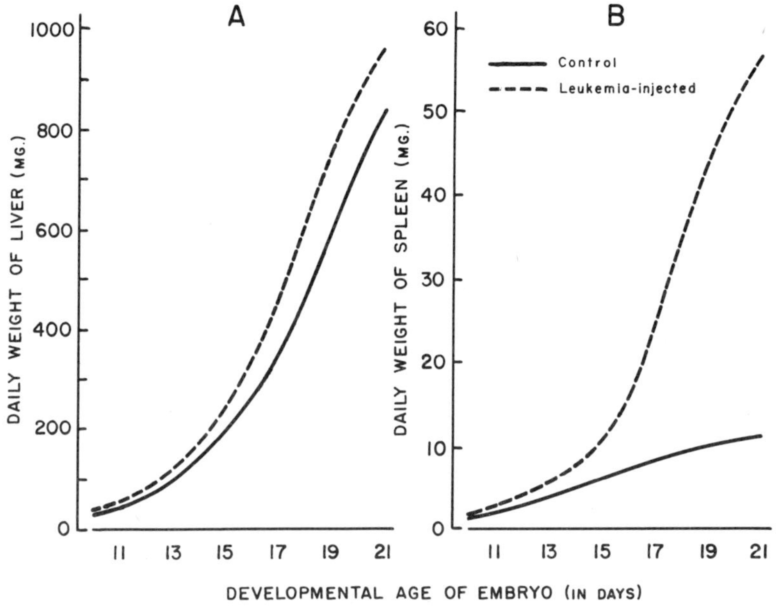

Fig. 169. Effect of injection of leukemia cells on the growth of embryonic liver (*A*) and spleen (*B*). (Plotted, with some modification, from the data of Clarkson, Katz, Dann, and Karnofsky, 1964.) *Note:* approximately 3 million cells were injected intravenously into the 10-day-old chicken embryos.

identifiable nodules. This occurs as early as the fifteenth day of incubation and is clearly observed on the twentieth day. In a few instances nodules proliferate to form a massive tumor (*Clarkson, Katz, Dann, and Karnofsky, 1964*).

In association with the splenic enlargement, there is marked granulocytosis and development of mild anemia.

The route of injection of leukemic blood from adult chickens into 10- to 14-day-old developing eggs influences the incidence of embryonic disease (*Brandly, Thorp, and Prickett, 1949*), as shown in the accompanying table.

Route of Injection	Incidence of Gross Lesions (%)
Intravenous	46.0
Chorioallantoic	31.8
Amniotic sac	11.2
Allantoic sac	2.0
Yolk sac	0.0

Summary Table 15 Effect of Pathogenic Afflictions on Toxicity and Teratogeny in the Avian Embryo

Age (days) Treatment	Age (days) Examination	Malformations (%)	Mortality (%)	Incidence of Typical Syndromes	Reference
Bronchitis virus					
10	19	50		(Doses: 0.2 cm³ inoculum/egg, in allantoic sac) Dwarfing (to one-half size); jaundice (33%); distended cloaca (25%); frequently, incomplete closure of the abdomen. Abnormalities of organs: extensive hepatitis with hemorrhages, marked discoloration (33%); lung one-half normal size; swelling and edema of kidneys; heart reduced in size; spleen enlarged (twice normal size); deformed leg and tarsal joints.	Loomis, Cunningham, Gray, and Thorp (1950)
Cancerous serum					
5–12				(Doses: from 0.25 to 0.5 ml/egg, intravenously) Enhanced growth; edema: generalized or confined to the occipital region; malformations of beak; bilateral anophthalmia.	Lustig (1954)
9–11	14–16	52		(Doses: 0.01 ml in 0.1 ml solution/egg, intravenously) Local malformations affecting: the osseous system (brachymelia, hemimelia, absence of parietal, occipital, and maxillary bones); the central nervous system (exencephalia, excerebration, microcephaly, anencephaly); the eyes (anophthalmia); the beak (crossed parrot beak and brachygnathy). Malformations of general character: edema, occasional inhibition of growth, general hemorrhages, and anemia.	Lustig and Fiszer (1956)

				(Doses: 0.1 ml/egg, in allantoic vein)	Fiszer-Szafarz (1967)
10		25	76	Most often: general edema. Other abnormalities: submaxillary edema, delay in development; malformations of hind limbs (atrophy, fragility of articulations); abnormal curvature of the cervical vertebrae; cerebral hemorrhage; all these accompanied by large zones of necrosis, and sometimes by enlarged liver and large flabby heart.	
Influenza A virus				(Doses: virus-infected allantoic fluid, from undiluted-10^{-7}/egg)	Hamburger and Habel (1947)
2	4	86	50	Disproportionately small head, in particular, and brain (microcephaly, micrencephaly).	
				(Doses: 0.03 ml, 10^{-4} to 10^{-6} solution/egg, close to embryo)	Shear, Heath, Imagawa, Jones, and Adams (1955)
2	4	29–100	70–100	Extreme flexion of body axis (97%+); anomalies of head region (95%); near-complete collapse of mesencephalon and telencephalic hemispheres; frequently, flattened and misshapen eyes.	
				(Doses: 0.05 ml of 10^{-7} ID_{50}/egg, beneath the blastoderm)	Cruz, Campillo-Sainz, Munoz-Armas, Pascual-Chavez, and Zamudio-Bandera (1963)
¾	Every 2 days	45	99	Retarded development: defective heart (10 to 36%); other defects (15 to 35%); profound alterations of the nervous system (microcephaly), sensory organs and general body shape. (See Fig. 161, on mortality.)	
Mumps virus				(Doses: 0.03 ml, diluted 1.01 to 1.0/egg, close to embryo)	Hamburger and Habel (1947)
2	3–7	16	57	Chief malformation: twist of the axis.	

Summary Table 15 ***(Continued)***

Age (days) Treatment	Age (days) Examination	Malformations (%)	Mortality (%)	Incidence of Typical Syndromes	Reference
Mumps virus				(Doses: 0.05 ml, diluted $10^{1.8}$ to $10^{5.8}$/egg, under vitelline membrane)	Williamson, Blattner, and Simonsen (1957)
2	7		18–58	Growth inhibition; cataracts (5 to 85%); mild twist of axis. (See Fig. 162.)	
				(Doses: 0.05 ml, diluted $10^{5.8}$ to $10^{6.5}$/egg, under vitelline membrane)	Robertson, Williamson, and Blattner (1964)
1½	3–14			High incidence of cataracts at early stages of development (up to 88%). (See Fig. 162 on effects of dose and time of inoculation.)	
Newcastle disease virus				(Doses: 0.01 ml/egg, over blastoderm)	Blattner and Williamson (1951)
2	3	45–90		Retardation in growth; abnormalities in lens and auditory sac (otocyst), ranging from reduction in size to complete absence.	
				(Doses: 0.05 ml of 10^7/egg, near embryo)	Robertson, Williamson, and Blattner (1955)
1½–3½		50–100	37–65	Defects of neural tube, lens, auditory vesicles, visceral arches, limb buds, and olfactory pits—consisting of cytoplasmic degeneration followed by nuclear degeneration (see Fig. 163).	
Pleuropneumonia organisms				(Doses: nearest to LD_{50}, in yolk sac)	Chute and Cole (1954)
7	8–21			Retardation in growth, hemorrhage, and edema. Gross lesions, in descending order of incidence: hyperemia, generalized edema, hemorrhage and necrotic foci in the liver; edema of the head, neck and joints; hemorrhage of kidney.	

				(Doses: inoculated into yolk sac)	Moulton and Adler (1957)
8	8–16	0–100	25	Pathology of arthritis: gross joint involvement (20%) with joint lesions and cutaneous hemorrhages over the pectoral region; hyperemia and necrosis of synovial-cell layer, affecting the articular cartilage and the joint cavity.	
Staphylococcus				(Doses: 0.1 ml/egg, in allantoic membrane)	Lyles, Sharp, Nabors,
8	10–17		15–62	Effects on organs: kidney—frequent hemorrhage; liver—hemorrhage with distortion of structure; heart—possible cardiac involvement; brain—hemorrhagic encephalosis, cerebral lesions, hemorrhage and necrosis of nervous tissue; lung—always spotty. (See Fig. 167, on mortality.)	Larsen, and Potter, (1965)
Tetanus Toxin				(Doses: 0.05 ml/egg, into vitelline membrane)	Corliss, Fedinec, and
1½	2¼	98		Necrosis (17%); neural anomalies of cephalic region (55%), somite region (66%), caudal region (84%).	Robertson (1966)
Toxoplasma				(Doses: from 0.005 to 0.15 ml/egg)	Lutz-Ostertag and Senaud
1–2	3–7	93	69	Microcephaly (64%); anomalies of central nervous system (60%); microphthalmia (53%); somatic anomalies (52%); atrophy of lower part of trunk (47%); anomalies of limb buds (46%); lordosis (34%).	(1966)

X

Traumatic Disturbance

Disturbance, injury or shock
In embryonic early life
Could be a lasting stumbling-block
To health and strength to weather strife.

This chapter deals primarily with the causes of some embryonic abnormalities and deaths of traumatic origin.

In early developmental stages the embryo is highly susceptible to mechanical injury. The consequent developmental alterations may often be so severe that they can hinder further development and result in permanent malformations and in death.

Many experimental interventions in embryonic development, such as mechanical interference by positioning, agitation, pressure, and various surgical manipulations of the egg, may result in elimination or disfiguration of parts, or modification of some complex developmental processes of the embryo, with lethal and teratological effects.

The effects of experimentally produced localized damage cannot be confined completely to the damaged area, because of functional interaction of various body parts, particularly on the cellular level, through the blood and other circulating body fluids.

MECHANICAL INTERFERENCE

Mechanical injury, in the widest sense, accounts for many structural changes and prenatal deaths. Such injury could be a direct assault on the embryo itself, or it could be indirect, through handling of eggs during incubation, damage of egg covering, or interference with development by some exterior or interior physical means.

Position of the Egg

Mortality. It is generally known among students of avian embryology that eggs incubated with the small (sharp) end up yield a higher incidence of embryonic mortality.

In 1197 eggs positioned small end up during the first 15 days of incubation (*El-Ibiary, Shaffner, and Godfrey, 1966*), mortality was significantly higher, as is shown in the tabulation below.

	Embryonic Mortality at Hatching (%)
Control	15.3
Experimental (sharp end up)	25.5

Earlier studies (*Byerly and Olsen, 1931*) indicated that eggs incubated small end up have a large incidence of various malpositions. Among these, the largest percentage are positions with the head in the small end of the egg (about 60 per cent), which are the main cause of increased embryonic mortality (see Fig. 15, various types of malpositions, in Chapter II).

Mechanical Agitation of Eggs

Turning. Turning of eggs, as a preventive measure to reduce embryonic mortality, has been observed in nature (*Réaumur, 1749; Dareste, 1891*) and was practiced in artificial incubation by Egyptians for many centuries.

Experimentally, it has been determined (*Robertson, 1961a*) that in unturned eggs there is a very high incidence of embryonic mortality. It could be reduced immediately by turning and further improved by increase in turning, as is shown below.

Turnings per Day	Embryonic Mortality at Hatching (%)
0.0	86.2
0.5	37.2
2	22.9
6	22.3
12	18.1
96	16.5
144	14.6

Shaking. Pivotal displacement or mechanical turning (tilting back and forth through a 90-degree arc) at various frequencies of oscillation (*Randles and Romanoff, 1954*) produced the following incidences of embryonic mortality:

Treatment	Embryonic Mortality at Hatching (%)
Control (turned every 2 hours)	9.5
216 oscillations/min	30.5
345 oscillations/min	80.0
431 oscillations/min	100.0

Similar shaking before incubation at 120 to 216 oscillations/min (*Landauer and Baumann, 1943*), produced an embryonic mortality of 72.8 per cent in the experimental group, as against 18.9 per cent in the controls.

Much of the mortality which occurred after shaking was due to such factors as loosening of the outer and inner shell membranes, causing the air cell to wander (cf. movable air cell), and rupture of the yolk sac membrane, that is, destroying the developing egg as a physicochemical entity (*Romanoff, 1967, Chapter V, pp. 178–232*).

When eggs at various stages of embryonic development were shaken for 7 min at the rate of 229 oscillations/min through a distance of 3 in. (*Olsen and Byerly, 1938*), total embryonic mortality was affected as shown in Fig. 170. The effect of shaking was highly noticeable only when the direction of force was parallel to the short axis of the egg.

The mortality attributed to treatment increased rapidly to a level of over 40 per cent at the age of 4 to 11 days, then rapidly decreased to zero.

The treated embryos died within 72 hours as a result of rupture of the yolk sac or hemorrhage caused by disruption of the allantois from the overlying shell membrane.

It has been observed (*Landauer and Baumann, 1943*) that eggs

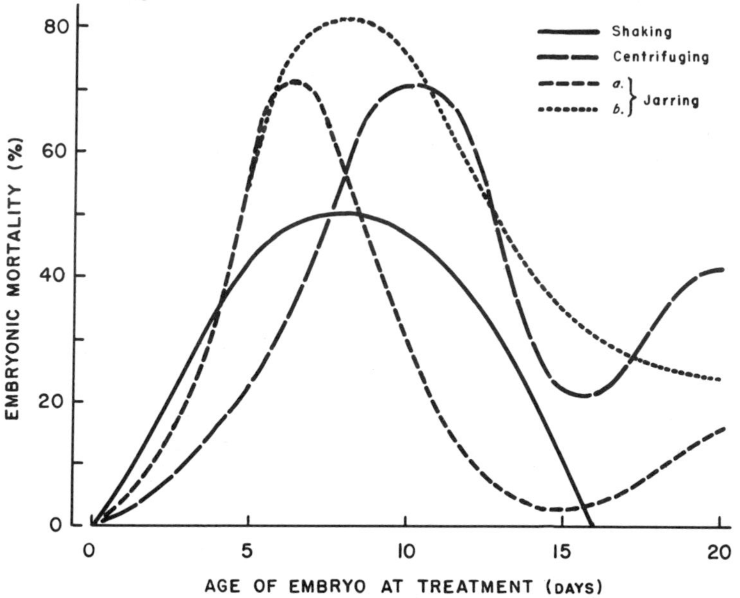

Fig. 170. Effect of mechanical disturbance of incubating eggs on the embryonic mortality, as a function of time of treatment. (After Olsen and Byerly, 1938.)

which are shaken, especially before incubation, produce a greatly increased number of rumpless (anourous) embryos. The average occurrence of this malformation in experimental eggs was 6.9 per cent, and in controls, 1.3 per cent.

Malformations of the embryo produced by shaking (*Landauer and Baumann, 1943*) are not specific, but incidences are considerably increased above the control level, as can be seen in the accompanying table.

Type of Deformity	Shaken (%)	Control (%)
Anencephaly	0.12	0.00
Anophthalmia	1.27	0.48
Microphthalmia	0.58	0.28
Malformation of beak	1.97	1.11
Cerebral hernia	0.35	0.07
Rudimentary legs	0.69	0.00
Ectopia viscerum	0.12	0.00
Taillessness (anouria)	6.9	1.3

Jarring. A 490-gram weight was dropped from a distance of 6 cm onto the tabletop where the incubator rested (*Stiles and Watterson, 1937*). The shocks were delivered at the rate of 15 blows per minute at 15-min intervals, from the fourth to the twelfth hours of incubation. The jarring reduced the hatchability to approximately zero (from 155 fertile eggs, only one chick hatched).

Examination of jarred embryos revealed that the vitelline circulation and the central nervous system were most seriously affected.

In another experiment (*Olsen and Byerly, 1938*) striking the blunt or the sharp end of the egg three times against the palm of the operator's hand (in many cases, the force was sufficient to produce a tremulous air cell in the egg) increased the embryonic mortality up to 80 per cent (see Fig. 170). It can be noted that the curve of mortality is higher for group *b*, in which force was applied to the large end of the egg, than for group *a* (Fig. 170), in which it was applied to the small end.

It has been demonstrated (*Knox and Olsen, 1936*) that fertile eggs with movable air cells (caused by separation of the two egg membranes, outer and inner, between which the air cell is formed) do not hatch as well as normal eggs.

Condition of Air Cell	Embryonic Mortality at Hatching (%)
Movable	43.5
Normal	22.7

This condition occurs when the eggs are jarred extensively (as in rough handling or transportation).

Centrifugation. The incidence of embryonic mortality attributable to centrifuging of fertile eggs for 1 min at an average force of 710 dynes is shown in Fig. 170. The curve is based on results in eggs with the blunt ends placed in the centrifuge cups so that the direction of force was exerted against the large end of the eggs (*Olsen and Byerly, 1938*).

The mortality curve shows that the heaviest mortality occurred in eggs treated at the tenth day, reaching a maximum of about 75 per cent.

The same general trend is noticeable, but to a much lesser degree, in the group centrifuged with the force exerted on the small end of the egg.

Chicken embryos of various ages that were centrifuged for 3 to 5 min (at 1000 revolutions per minute) suffered all kinds of malformations (*Papilian and Nana, 1932*).

At the age of 8 to 10 days they included chiefly deformities of heart and skull, such as ectopia, large openings, and blebs.

In further studies (*Papilian and Nana, 1933*) there were observed numerous cysts on the extraembryonic membranes, which also can occur spontaneously (see *Romanoff, 1960, p. 1118*).

Dynamite Blasts. Groups of eggs containing embryos of various ages were placed close to charges of dynamite (*Olsen and Byerly, 1938*).

The effect of the shock on mortality was not so noticeable during the 72 hours after treatment as at the end of incubation. The average total mortality was as high as 30 to 40 per cent only in groups treated between 0 and 14 days.

Removal of Eggshell

During the development of a method for continuous observation of the embryo from the first day to hatching (*Romanoff, 1931; 1933b*), a portion of the shell (over 1 in. in diameter), with adjacent shell membrane and egg membrane (*Romanoff and Romanoff, 1949, p. 112*), was permanently removed. The opened egg was then artificially provided with the normal atmospheric conditions of an intact egg—a supply of fresh air and control of evaporation by a special arrangement—as is shown in Fig. 171.

Mortality. As a result of these disturbances, embryos in eggs opened prior to incubation cannot survive beyond the ninth day (Fig. 172).

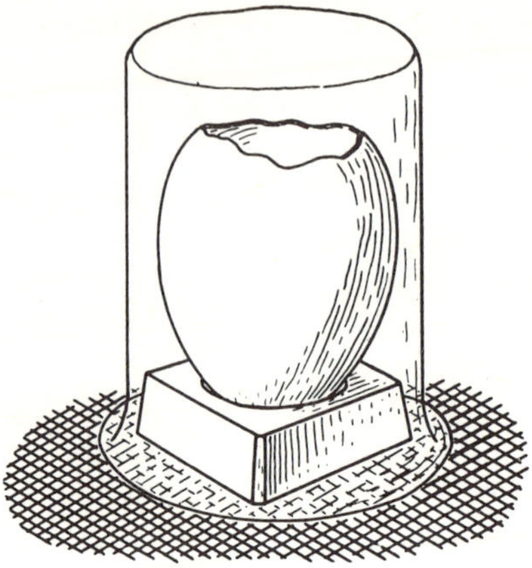

Fig. 171. A diagram of the opened chicken egg set on an embryological square watchglass and covered with a high-form 150 cm³ beaker. The ensemble is set on a wire-screened platform in the incubator. (After Romanoff, 1931, 1933b.) *Note:* this method of cultivation of the embryo in an opened egg, with precise control of temperature and humidity, and with the egg membrane removed (not made transparent by use of water or oil), permits continuous gross observation of the same individual under normal and experimental conditions.

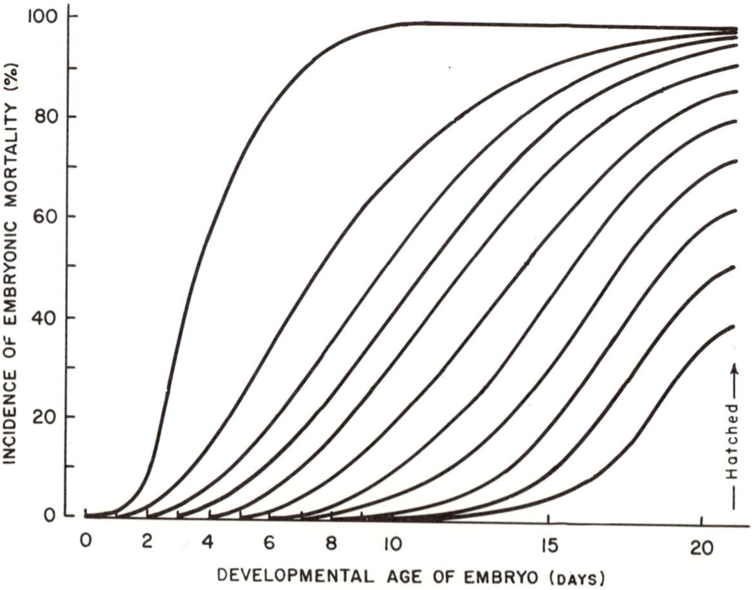

Fig. 172. Effect of a window in the eggshell, with removal of adjacent shell membranes (Fig. 171), on the incidence of embryonic mortality during the postoperative period of incubation. Eggs were opened at different stages, from 0 to 10 days, and kept in well-balanced environmental conditions. (Reconstructed from the experimental data of Romanoff, 1931, 1933b, and Romanoff, unpublished.)

They die from causes similar to those in unturned eggs (cf. turning of eggs).

Embryos in eggs opened on the second day and later usually survive until hatching time, and some hatch and show vitality like normal chicks. The incidence of embryonic mortality due to this disturbance is noticeably reduced with successive stages of embryonic development.

Formation of a window only 1 cm^2 at 24 hours resulted in 59.2 per cent mortality during the following 24 hours of incubation (*McCallion and Clarke, 1959*).

Again, eggs with holes of about 1.5 cm^2 have a heavy embryonic mortality (*Ancel, 1956*). The values for the incidence of embryonic mortality before and after 72 hours of incubation in eggs opened at various stages of development are shown below.

Age at Opening (hours)	Incidence of Mortality (%)	
	Before 72 Hours	After 72 Hours
0	21.0	65.2
13	21.4	56.2
20	17.1	32.0
26	5.8	27.7
34	6.6	3.7

Malformations. Without special control of environmental conditions, opening a window in the shell and shell membrane of the chicken egg at about 24 hours of incubation produces abnormalities of the central nervous system (*McCallion and Clarke, 1959*), which result from failure of the neural plate to close (acleiencephaly).

In a somewhat similar experimental procedure (*Ancel, 1956*), detailed teratogenic effects were determined for the embryos still alive after 72 hours of incubation, as shown in the accompanying table.

Malformation	Incidence of Malformation (%)	
	Opened at 0 Hours	Opened at 24 Hours
Omphalocephaly	6.3	0.0
Acleiencephaly	2.6	2.5
Cleiencephaly	1.1	0.7
Bilateral	7.4	1.6
Microphthalmia		
Left	4.7	3.3
Right	2.1	0.0

Laminar Materials

It has been observed (*Mitolo, 1963*) that the superimposition of small pieces of various laminar materials upon the 24-hour avian blastoderm caused the onset of severe degenerative processes in the posterior part of the embryo. The materials used in this treatment are filter paper, shell membrane (either outer or inner), and tissue paper (3×4 mm).

Mortality. A high percentage of embryos (33 to 50%) stop developing during the first hours after the treatment, and die. Later, the death rate decreases and no longer exceeds that in the control group.

Malformations. The caudal half of the embryo undergoes striking degenerative processes which diminish the development of some parts of the body. Among the most commonly observed anomalies are ectrosomy, posterior brachymelia, taillessness (anouria), and brachyury.

Besides constant deficiencies of the caudal region, other malformations have occurred, mainly spina bifida and micromelia. Sometimes the wing bud is reduced to a rudiment of varying size and shape; the hind limbs are always reduced in size and more seriously malformed.

The teratological effect of laminar materials is probably due to (1) pressure on the blastoderm, (2) physical contact with the vitelline membrane and its adhesion, and (3) hindrance of the cellular respiration (*Mitolo, 1963*).

Suspensions of Minute Particles

Colloidal suspensions (in water or saline) of minute particles of sand, alumina, attapulgus clay, or ground glass, injected into the amniotic cavity of the 5-day-old embryo, resulted in death of about two thirds of embryos by the twelfth day, with incidence of various types of malformations ranging from 46 to 79 per cent (*Williamson, Blattner, and Lutz, 1963*).

Among the abnormalities observed were encephalocele and eyelid defects. Short upper beak and crossed beak were also seen with significant frequency. In addition, alumina and attapulgus clay, particularly in the higher doses (0.1 and 1.0 mg/egg of each, respectively), produced severe twisting and distortion of the body axis and appendages.

Similar studies have been extended to the effects of carbon suspensions and colloidal silica (*Williamson, Blattner, and Robertson, 1967*). The incidence of mortality and of abnormal development in embryos inoculated with various doses of carbon suspension is shown in the tabulation below.

Dose of Carbon (mg/egg)	Incidence (%)	
	Mortality	Malformations
0.0	62	0
0.001	52	3
0.01	75	17
0.1	86	50

The incidence of mortality and abnormalities among the embryos inoculated with various doses of colloidal silica is shown below.

Dose of Silica (mg/egg)	Incidence (%)	
	Mortality	Malformations
0.0	29	0
0.008	55	17
0.08	61	52
0.8	68	40

The defects observed with carbon suspension and colloidal silica are exophthalmia, encephalocele or exencephaly, external edematous blebs, ectopic viscera, and crossed or short upper beak.

Frequently, the silica-treated embryos also show fusion of the appendages to the body wall, with concomitant malformations of the affected limbs.

Pressure on the Blastoderm

Simple mechanical pressure on the blastoderm is capable of producing deviations and aberrant development (*Gallera, 1934*). A small bubble introduced into the subgerminal cavity underneath the blastoderm creates a pressure which provokes a series of characteristic deviations in embryonic development. These become manifest in abnormal development of the nervous system, the dorsal cord of the cephalic region, and the brain region.

SURGICAL PRODUCTION OF ANOMALIES

Direct surgical manipulation of the developing embryo is one of several physical methods known to produce congenital anomalies in experimental animals. Among these are disfiguration of parts and abnormal growth of tissues displaced by trauma.

Procedures in Avian Embryology

Among the surgical techniques frequently used in experimental avian embryology are the following:

Castration—(a) by X-rays (*Wolff and Wolff, 1951*); (b) by electrodiathermic coagulation (*Groenendijk-Huijbers, 1965*)—sinistral castration. (*Note:* it is not possible to castrate the avian embryo by a surgical technique—*Wolff and Wolff, 1951.*)

Destruction, of brain, of left ovary (*Wolff, 1938*).

Excision, of notochord, brain (*Schowing, 1968a, b, c*).

Explantation, of a whole embryo (*Klein, 1965*).

Extirpation, of wing bud (*Warren, 1934*).

Grafting, of mesonephros and metanephros (*Zwilling, 1955; 1956a and b*).

Implantation, of thyroid (*Willier, 1924*).

Inversion, of wings (*Saunders and Gasseling, 1959*).

Isolation, of part of the wing (*Warren, 1934*); of head components (*Silver, 1962*); of presumptive liver region (*Le Douarin, 1964b*).

Separation (partitioning) of limbs (*Murillo-Ferrol, 1963a*).

Transection, of blastoderm (*Rogulska, 1968*).

Transplantation, of renal region (*Miclea, 1962*).

Experimentally, several specific malformations have been obtained using surgical methods, as listed below.

Acleiencephaly (30%)—by opening the brain from the anterior rhombencephalon forward, at 40 hours (*Nelsen, 1960*).

Agenesia of limbs—by separating the prospective limb bud from the corresponding somites and inserting a sheet of mica (*Murillo-Ferrol, 1963a*).

Agenesia of sternum—by placing an obstacle between local area and prospective sternum (*Murillo-Ferrol, 1963b*).

Anouria (rumplessness)—by a transverse cut through the body of the 2-day-old embryo (*Zwilling, 1945*).

Aplasia of desmocranium—by local operative opening of the rhombencephalic roof (*Klika and Jelinek, 1961a*).

Cyclopia—by removal of the telencephalon and reversal of bilateralization in somite stage (*Rogers, 1964*).

Ectopia of heart—by a tearing of the pericardium and of the thin body wall (*Rychter, 1963*).

Exencephaly—by covering with a layer of mesenchymal cells (*Klika and Jelinek, 1961b*).

Hypermorphosis of brain—by extirpation of neuromere (*Bergquist, 1960*).

Spina bifida—by splitting of the spinal cord (*Fowler, 1953*).

In addition, several other deformities have been obtained surgically.

Mirror-twin duplication of the wing bud (75%)—by reversal of the apical zone in a 3.5-day-old embryo (*Saunders and Gasseling, 1963*).

Shrinkage of the brain vesicles—by damage of internal organization (*Klika and Jelinek, 1961a*).

Interference with cardiogenesis—by separating the cardiac tube of the 3-day-old embryo for 24 to 60 hours (*Gessner, 1966*).

Local damage of the heart loop and aortic arches—by elimination or suppression of aortic arches (*Rychter, 1963*).

SPECIFIC SURGICAL EFFECTS

Transection of Blastoderm. When an unincubated blastoderm (of the Pekin duck) is transected either perpendicularly or parallel to the long axis of the egg (*Rogulska, 1968*) and incubated for 2 to 4 days, the resulting two developing embryos are always smaller than control embryos, retarded in development, and often abnormal.

(See also Chapter I, on Potentialities for Maldevelopment.)

Lesions in the Nervous System. Lesions in the nervous system (operation at 75 hours of incubation) may cause shortening of the long bones of the limbs, dislocation of joints, and emaciation of muscle (*Schowing and Celers, 1965*).

Destruction of Brain. It has been observed (*Wolff, 1938*) that after destruction of the brain there is no retardation in the development of the embryo or in the differentiation of the organs.

(cf. effects of removal of brain and its parts in Table 16.)

Table 16 Effect of Some Surgical Operations of the Head on Teratogeny in the Avian Embryo

Age (days) Operation	Age (days) Examination	Nature of Operation and Results	Reference
Surgery on brain		(Complete removal of the brain)	Schowing (1968a)
2	14	Elimination of frontal and parietal bones; squamosals reduced in size.	
		(Removal of the notochord only)	Schowing (1961)
2	14	Reduction of the base plate and basioccipital.	
		(Removal of the prosencephalon only)	Schowing (1968b)
2	14	Reduction of the frontal bones. (See Romanoff, 1960, p. 238, Fig. 93.)	
		(Removal of the mesencephalon only)	Schowing (1968b)
2	14	Weaker effect than removal of other parts of the brain.	
		(Removal of the rhombencephalon only)	Schowing (1968b)
2	14	Reduction of parietals and the basal plate.	
		(Removal of telencephalon)	Rogers (1964)
1–2		Production of cyclopia (4%).	
		(Removal of fore- and midbrain)	Schowing (1965)
1–3		Production of perfect cyclopia (33%).	
Surgery of neural tube		(Splitting of roof plate of spinal cord)	Fowler (1953)
1–3		Increase in volume (75%) of the gray matter of the spinal cord, due to mechanically produced spina bifida.	
Surgery on eye		(Removal and additional grafting)	Silver (1962)
1		Following removal of one primary optic vesicle—reduction in size of orbit. Following removal of both primary optic vesicles—both orbits much reduced in size, including the interorbital septum. After grafting of an additional optic vesicle on one side—formation of a "composed eye."	

Decapitation

Growth. Decapitation of the embryo by ablation of the free-head region during the second day of development has an effect upon the general body weight, especially during the last half of incubation (*Vogel, 1965*).

Incubation Age (days)	Weight of Embryo (grams)	
	Decapitated	Control
14	8.7	11.9
15	10.0	14.3
16	11.1	17.0
17	12.6	19.9
18	14.8	22.0

Mortality. Of 235 decapitated embryos (*Vogel, 1965*), 94, or approximately 40 per cent, survived to 14, 16, and 18 days, the dates of sacrifice.

Removal of Parts of Brain

The removal of different parts of the brain leads to a reduction or disappearance of skeletal elements of the neurocranium. The effect is determined by the importance of the excised region (*Schowing, 1968c*). The cranial vault depends on the brain; the palatines and parasphenoidal rostrum depend upon the presence of the face—each region of the brain acts on neighboring bone anlage.

(For further details see Table 16, on the effects of surgery on the head.)

Removal of Lenses (Lentectomy)

Removal of lenses of 4-day-old embryos (*Zinn, 1970*) inhibits the enlargement of the vitreous body, as is shown by the data on globe diameter in the table below.

Incubation Time (days)	Globe Diameter (mm)	
	Normal	Experimental
4	1.8	1.8
5	2.8	—
8	6.1	3.2
11	8.3	4.2
14	9.4	4.5
17	9.9	4.9

After lentectomy, the following changes occurred in the corneal or perilimbic zones:

Postoperative Time	Postoperative Changes
0–2 hours	No change
2–4 hours	Appearance of a few scattered fibroblasts
4–8 hours	Formation of a sheet of tissue (5 to 8 cells thick)
5–6 days	Formation of retrocorneal mesenchyme (20 cells thick)
7–14 days	Association with collagen fibrils

At 17 days of incubation, the collagen fibrils within this corneal stroma have a cross-sectional diameter up to three times greater than that of fibrils in a normal embryo.

Removal of Pituitary (Hypophysectomy)

Embryos which have been hypophysectomized (at 1½ days, by removal of the forehead—see *insert* of Fig. 173), undergo noticeable changes, especially during the latter half of incubation period (*Fugo, 1940*).

Growth. In general, the operated embryos are smaller than controls incubated for the same length of time. By the twelfth day of development, weight of experimental embryos is approximately 50 per cent that of controls (Fig. 173). By the end of the normal incubation period, the experimental embryos weigh, on an average, less than one third as much as controls (*Fugo, 1940; Vogel, 1965*).

The hypophysectomized embryos are incapable of indrawing the yolk sac at the termination of incubation. Yet embryos remain alive within the egg for as long as 8 days beyond the normal incubation time (up to 29 days from the beginning of incubation).

Size of Organs. The thyroids of hypophysectomized embryos are smaller than the glands of controls at the same age. They remain permanently at the size attained by the fourteenth day (*Fugo, 1940*).

The testes are smaller than normal at about the thirteenth day of incubation.

The left ovary is distinctly deficient in cortical development.

Mortality. About 3 per cent of the operated embryos (with very severe edema) died on about the tenth day of development. Among embryos

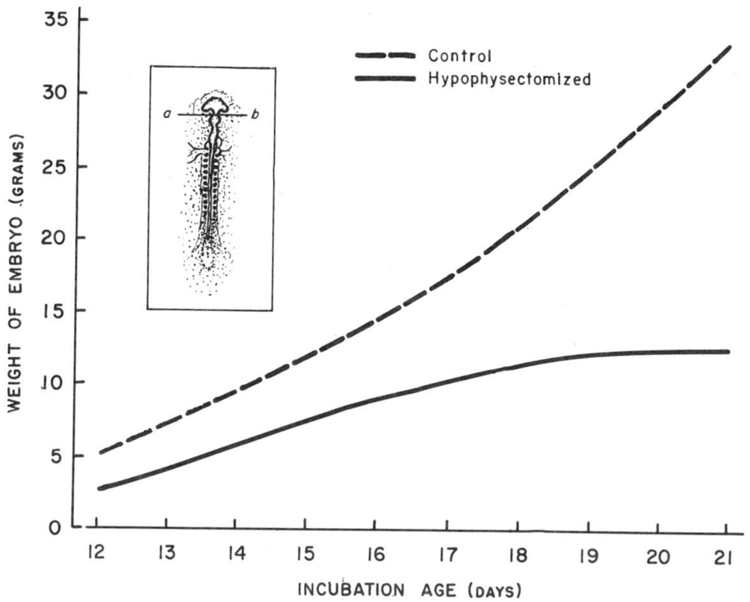

Fig. 173. Effect of removal of the pituitary gland of the 1½-day embryo on the subsequent growth during the incubation period. (Plotted after the data of Fugo, 1940.) *Insert:* the embryo at the stage of operation, with a line (*a* to *b*) indicating the approximate site of operation—the removal of the forehead in hypophysectomy.

living beyond the twelfth day, the death rate was highest on the fourteenth to sixteenth days of incubation (*Fugo, 1940*).

The total mortality of the operated embryos was about 77 per cent.

Malformations. In general, the limb bones of hypophysectomized embryos are shorter than those of controls, although when correlated with the body size, the proportional difference is insignificant.

Most of the experimental embryos are abnormal as a result of operation. Eyes, nostrils, and combs are entirely missing. In a few cases even the lower beak is absent.

The site of the excision, in the majority of cases, is completely covered by epidermis (*Fugo, 1940*).

Removal of Left Ovary (Sinistral Ovariectomy)

In 4-day-old embryos, the left gonadal primordium was destroyed by local electrodiathermic coagulation (*Groenendijk-Huijbers, 1965*).

Growth of Right Ovaries. The growth of the remaining right ovaries was studied from 12 to 20 days. The increase in surface area of right ovaries, of sinistrally castrated embryos and of controls, is shown in the accompanying table.

Age of Incubation (days)	Weight of Right Ovaries (in planimetric units)	
	Controls	Left Castrated
12	14.3	18.5
14	15.0	21.2
16	16.3	25.0
18	18.0	
20	20.0	

It is apparent that there is compensatory growth of the right ovary after sinistral ovariectomy, since normally only the left ovary is functional. Microscopically, the compensatory growth is accompanied by active mitosis with an increase in number of cells in the medullary cord.

Extirpation of Limb Bud

Complete extirpation of the posterior limb bud of the chicken embryo at the age of 65 hours, accompanied by excision of some of the pelvic girdle anlagen, results in a deficiency of some elements of the girdle (*Rogulska, 1965*).

The embryo with excised limb buds are usually smaller than the controls.

XI

Aberrant Twinning

Abilities to multiply
At times give rise to monstrous twins . . .
For some aberrances deny
The equal chance for normal kins.

Prenatal duplicity in birds is not less common than in other animals. Complete duplication, with perfect twins, is rare, although numerous cases of separate and apparently normal twin embryos in the early stages have frequently been observed. Some twin embryos survive the incubation period, but only a very few are able to hatch.

Twinning is accompanied by various developmental hazards. Double and deformed embryos are not uncommon under natural conditions, but most die early during development.

Double monsters are largely formed either of incompletely separated and regenerated bilateral halves of an original single embryo, or by a partial fusion of separate embryonic axes.

This chapter deals with the analysis of various forms of spontaneously occurring and experimentally induced prenatal duplicity in birds.

SPONTANEOUS TWINNING

The literature on sporadic twinning in Aves is voluminous. Many embryologists who have handled chick embryos in large numbers have encountered sporadic cases of duplicity. The distribution of various forms of spontaneous duplicity in 81 embryos is shown in Table 17. Among them appeared only one pair of perfect twins.

In general, published papers describe a range from normal twins to a great variety of double monsters. Most of these observations were made during the early stages of embryonic development, extending from the primitive streak to the fourth day of incubation. Prior to 1900, there had been recorded (*Bryce, 1900*) 95 such cases of multiple formation. Since that time, many observations have been added on normal and monstrous twin embryos which were alive or dying at hatching.

Abnormalities in Early Stages

Double and deformed embryos, although not uncommon under natural conditions of incubation, usually die during the early days of develop-

Table 17 Distribution of Various Forms of Spontaneous Duplicity in Birds (Chickens and Ducks)*

Duplicate Abnormalities	Distribution (%)
Pygomelus (rump)**	33.75
Pelvamelus (pelvis)	22.50
Deradelphus (skin)	10.00
Opodymus (head)	10.00
Omphalomelus (limb-thorax)	5.00
Melomelus (rudimentary limbs)	3.75
Psoadelphus (loin muscle)	3.75
Thoracomelus (limb)	2.50
Thoracomelus (chest)	2.50
Cephalothoracopagus monosymmetros (head-thorax)	1.25
Duplicate anus (double anus)	1.25
Gastromelus (stomach)	1.25
Ischiopagus disymmetros parasiticus (heap)	1.25
Ischiopagus monosymmetros (heap)	1.25
Perfect twins	1.25

* Rearranged from Komai and Nakamura (1938).
** The point of contact between twin embryos.

ment. A few typical samples of such maldevelopment are shown in Fig. 174.

It has been generally observed that such twinning is often associated with delayed development, especially at the time of gastrulation. At that time, the process of formation of the axis of symmetry, or the bilateral axis, may be so interfered with that the primordia of the two halves may become partially or unevenly separated (*Newman, 1923*).

TYPES OF COMPLETE DUPLICITY

In birds, at least three types of duplicity can be recognized. Complete twin embryos may arise:

I. From a large double-yolked egg (*binovular twins*).
II. From a normal-sized egg (*monovular twins*).
  a. With two blastoderms on one yolk (*bizygotic twins*).
  b. With one blastoderm on one yolk (*monozygotic twins*).

These three types of duplicity are semidiagrammatically illustrated in Fig. 175*A*, *B*, and *C*, respectively. Ordinarily, twins from double-

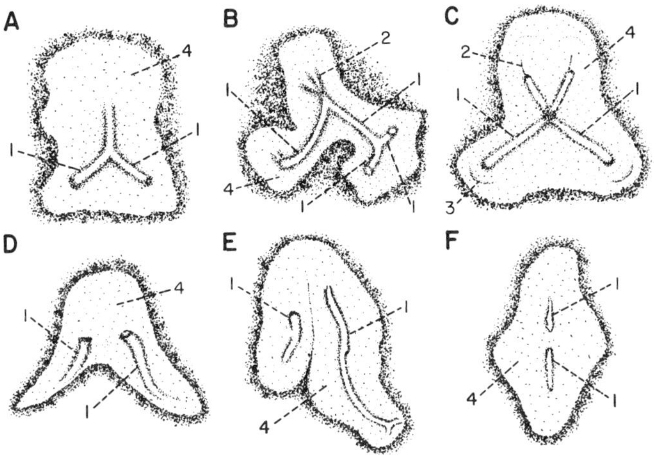

Fig. 174. A few cases of spontaneously occurring abnormal duplicities in the early stages of development in Aves. All these monsters correspond to the primitive streak stage of about 16 hours, and show a delay in development from 8 to 12 hours. *A*, a double monster with posterior forking—*duplicitas posterior* (redrawn after Kaestner, 1899); *B*, a triple monster—*triplicitas* (after Fortuyn, 1927); *C*, four primitive streaks fused at a common point, which represents the posterior end of the four structures (after Tannreuther, 1919); *D*, a pair of twins—in the goose (after Grundmann, 1900); *E*, a pair of twins derived from the fusion of two unequal blastoderms (after Kaestner, 1899); and *F*, a double arrangement of the primitive streak with the embryos arising directly opposite each other, as the result of a primary and a secondary gastrulation (after Burckhardt, 1888). *1*, primitive streak; *2*, head process; *3*, head fold; *4*, area pellucida.

yolked eggs have individual amnions, chorioallantoic membranes, and umbilical cords leading to their respective yolk sacs; they can be of either sex and are true fraternal twins.

Twins from a single-yolked egg with double blastoderms, under normal conditions, also have individual amnions and chorioallantoic membranes; their umbilical cords, either individual or joint, lead to the same yolk sac. Normally, under double fertilization, they may be of different sexes.

Twins from a single-yolked egg with a single blastoderm have individual amnions (or occasionally joint amnions), and joint chorioallantoic membranes and umbilical cord leading to the same yolk; these twins are of the same sex and are true identical twins. (For deviations in sex see the topic Freemartins in Aves, p. 388.)

Incidence of Multiple-Yolked Eggs. The occurrence of double-yolked eggs is very common among various domesticated birds. Féré (*1899*) estimated that there is a double-yolked egg in every 2000 normal eggs; Curtis (*1915*) put the incidence at 1 in every 531 eggs; and Jeffrey,

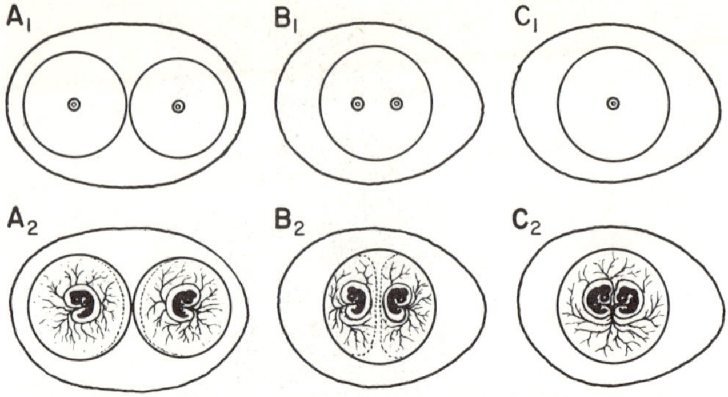

Fig. 175. Semidiagrammatic illustrations showing patterns of normal twinning in Aves. A_1 and A_2, binovular (from a double-yolked egg), showing the fresh egg (Romanoff and Romanoff, 1949, pp. 272–283) and developing embryo (Panum, 1860; Dahlgren, 1941); B_1 and B_2, monovular (dizygotic), showing a fresh egg (Romanoff and Romanoff, 1949, p. 285) and developing embryos (Jackson, 1965); C_1 and C_2, monovular (monozygotic), showing a fresh egg (Dareste, 1891, pl. xiv), and developing embryos (Hess, Juhn, and Jull, 1947). Komai and Nakamura (1938) observed a case of female twins united by a single umbilical cord (at hatching time). See also Freemartins in Aves, p. 388.

Fox, and Smyth (*1953*), from the examination of 7552 pullet eggs, found that the ratio of double-yolked to normal is 1 to 363. (For a detailed study of double-yolked and other multiple-yolked eggs, the reader is referred to *The Avian Egg, Romanoff and Romanoff, 1949, pp. 276–284.*)

The occurrence of triple-yolked eggs is less than 0.1 per cent of all multiple-yolked eggs (*Curtis, 1915*). (For further details see *Romanoff and Romanoff, 1949, pp. 283–284.*)

Twins from Double-Yolked Eggs

When double-yolked eggs are fertilized, they give rise to twin embryos which develop separately. Some reach full size, but only a few can hatch, often needing outside help.

From the study of 78 pairs (see Table 18) of such twins from double-yolked eggs which weighed on an average over 75 grams (about 15 grams heavier than normal), it was found that only three pairs (about 3.8 per cent) could hatch (*Romanoff and Klevay,* unpublished). The weight of such twins of various species, in relation to the weight of their eggs, is shown in Table 18.

Weight and Mortality of the Twin Embryos. It is obvious that the size of double-yolked eggs determines the size of their twin embryos.

Table 18 The Weight of Twin Embryos from Double-Yolked Eggs

Species	Weight of Eggs (grams)	Weight of Twins* (grams)		Reference
		Small	Large	
Chicken	—	21.0♀	29.0♂	Jeffrey, Fox, and Smyth (1953)
Chicken	77.7	22.1	23.5	Romanoff and Klevay (unpublished)
	80.7	23.0	26.1	
	83.6	23.2	31.6	
Duck	99.8	35.5	41.0	Sumulong (1925)
Goose	221.0	65.3♂	78.9♀	Krizenecky (1926)

* With the reserve yolk sac drawn into body cavity.

The weights of the pairs of twin embryos at the time of death are shown in Fig. 176. Among embryos dying between 18 and 23 days, about 56 per cent were fully developed. It appears that many of these embryos ruptured the yolk sac and were drowned in the struggle to pip the shell, or were suffocated because their position in relation to the air cell was inverted.

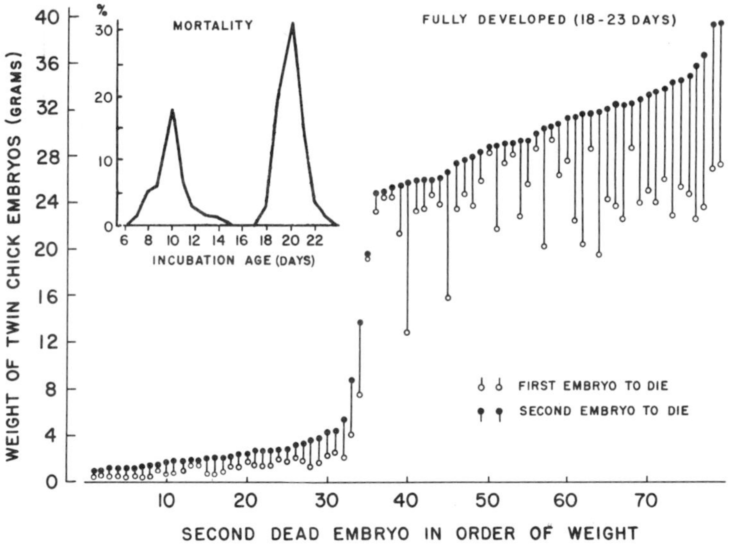

Fig. 176. Weights of twin embryos (of double-yolked eggs) arranged in order of increasing weight of the second embryo to die. (From the data of Romanoff and Klevay, unpublished.) *Insert:* distribution of mortality of the second twin embryos during the developmental period. (The same authors.)

Table 19 Incidence of Hatchability of Double-Yolked Eggs

Species	Double-Yolked Eggs Observed (number)	Live Twin Chicks Hatched (pairs)	Reference
Chicken	10	9	Bernard (1850)
Pigeon	36	5	Hollander and Levi (1940)
Chicken	208	1	Jeffrey, Fox, and Smyth (1953)
Chicken	500	16	Lutz and Lutz-Ostertag (1959)
Chicken	220	7	Monkman (1963)
Chicken	164	40	Fechheimer and Jaffe (1966)
Chicken	78	3	Romanoff and Klevay (unpublished)

The incidence of hatchability of double-yolked eggs is shown in Table 19.

The difference in weights between the twin embryos indicates that the first embryo dies 1 to 5 days earlier than the second. However, Levi (*1957*) cited a case in which one twin pigeon died at 7 days of age, whereas the other survived to hatching time, the seventeenth day.

The distribution of mortality among the second twin embryos (see *insert* of Fig. 176) shows that between the two peaks of incidence of embryonic mortality—that is, between 14 and 18 days of incubation—there are no deaths. A similar distribution of mortality among twin embryos of 164 double-yolked eggs was observed by Fechheimer and Jaffe, (*1966*), as is shown below.

Stage of Development (days)	Mortality of Twin Embryos (%)
1–6	28.0
4–13	3.8
14–20	68.1

It seems that the curve of mortality, especially of the embryos dying first (see *insert* of Fig. 176), closely follows the normal distribution curve for mortality in the avian embryo (see Fig. 13, in Chapter II).

Twins from Single-Yolked Eggs

Monovular twins, from a single-yolked egg, are less frequent than binovular ones, from a double-yolked egg (Table 20). However, such twins have been observed in several species of birds, including doves (*Riddle, 1923*); ducks (*Forsheim, 1908; Ruth, 1918; Komai and Nakamura, 1938*); geese (*Grundmann, 1900; Krizenecky, 1926*); guinea fowls (*Daniel, 1959*); ostriches (*Duerden, 1910*); pigeons (*Riddle, 1923; Hollander and Levi, 1940*); quail (*Lutz and Lutz-Ostertag, 1964; Homma and Jinno, 1968*); turkeys (*Olsen, 1962*); and chickens (numerous observations).

Incidence of Twinning in Single-Yolked Eggs. Complete monovular twins (Fig. 175*C*) are very rare, accounting for perhaps not over 2 per cent of the total twins observed.

In twins which reach hatching time, it cannot be accurately determined whether they are from two blastoderms (fraternal twins), or from one blastoderm (identical twins). From observing several sets of single-yolked eggs (Table 21), it is obvious that in each set the twin embryos vary in weight. One embryo is always heavier than the other. Furthermore, when sex has been determined all embryos appear to be of one sex, female.

Twins from Two Blastoderms in One Yolk. The occurrence of single-yolked eggs with two blastoderms (Fig. 175*B*) has been a well-known phenomenon (see *Romanoff and Romanoff, 1949, p. 285*). Some of these eggs presumably produce twin embryos, as has happened in pigeons

Table 20 Incidence of Monovular Twins in Early Development (Including Complete and Incomplete Twinning)

Species	Fertile Eggs (number)	Type of Twinning (number): Complete	Incomplete: Posterior	Incomplete: Anterior	Reference
Duck	500	1	2	1	Ruth (1916)
Dove / Pigeon	20,000	8	?	?	Riddle (1923)
Chicken R.I.R.	64,716	2	63	27	Byerly and Olsen (1934)
Chicken Other	57,616	1	31	12	Byerly and Olsen (1934)
Chicken	670		12	1	Kondo and Ikoto (1953)
Turkey	5,504	3	?	?	Olsen (1962)
Quail	516	4	? 2	?	Homma and Jinno (1968)

Table 21 The Weight of Twin Embryos from Single-Yolked Eggs (At Hatching Time)*

Species	Weight of Twin Embryos (grams)			Reference
Chicken	29.0		25.0	Komai and Nakamura (1938)
	16.0		11.0	Nalbandov (1942)
	17.0		15.2	Hess, Juhn, and Jull (1947)
	13.2		12.2	Hess, Juhn, and Jull (1947)
		31.0		Kushner and Panfilova (1955)
	15.0		13.7	Munro (1965)
Quail	2.3		2.1	Homma and Jinno (1968)
	3.0		2.6	Homma and Jinno (1968)
	4.2		3.4	Homma and Jinno (1968)

* Without yolk sac.
Note: All twins were found to be females (see also the topic Freemartins in Aves, p. 388).

(*Hollander and Levi, 1940*) when two breeds were crossed, resulting in a difference in the down of the twins. In another study, two quail embryos were hatched from an egg with two blastoderms (*Lutz, 1965*).

Twins from Single-Yolked Eggs with One Blastoderm. The formation of two complete embryos from a single blastoderm (Fig. 175*C*) is a rare occurrence. Monozygotic twins are of the same sex, male or female. But it is not easy to discriminate them from twins from two blastoderms, which are under the influence of vascular anastomoses (see Freemartins).

Twins from a single blastoderm could result either from a double gastrulation or from longitudinal fission of an early blastoderm.

Freemartins in Aves. It is very difficult for twin embryos arising from one blastoderm, or even from two adjacent blastoderms, to develop without fusion of some vitelline vessels of the yolk sac. The suggestion has been made (*Ruch, 1962*) that such fusion could also occur in some double-yolked eggs.

With the establishment of such anastomotic communication, through arteries and veins, between twin embryos of opposite sexes, there would be hormonal interchange and, inevitably, masculinization of genetic females and feminization of genetic males, as has been shown experimentally (see *Romanoff, 1960, pp. 842–853; Wolff and Lutz-Ostertag, 1961*).

From the study of 14 cases of spontaneous freemartins in birds (chicken and ducks), Lutz and Lutz-Ostertag (*1959*) were able to ascertain reciprocal action of hormones in twins of both sexes. The Müllerian duct of the female embryo partially regressed and lost its connection with the cloaca. The male embryo showed transformation of its gonads (Table 21).

Triplets in Aves

Triplets in Aves are many times rarer than twins. Apparently, there is no record of triplets hatching from any type of duplicity. Triplets of an early age have been described by several investigators: two sets by Dareste (*1891, pl. XIV*), and one set each by Fortuyn (*1927*), Hsu (*1930*), and Newman (*1940*). The observed triplets' ages varied from about 1 to 2 days.

TYPES OF INCOMPLETE DUPLICITY

The formation of various types of monsters has often been observed in monozygotic twins, that is, those from a single-yolked egg with one blastoderm. When bilateral halves of an originally single embryo are incompletely divided into two embryos, the result is formation of a double monster (Fig. 177*A* to *J*). This separation could be either posterior (katadidymus) or anterior (anadidymus).

Incomplete twinning (Table 20) has a much higher incidence than complete twinning, and posterior forking is more frequent than anterior.

It has been suggested (*Waddington, 1941*) that posteriorly forked twins originate from double invagination, whereas anterior forking is probably due to splitting of the invaginating endoderm or of the elongating primitive streak.

Many of these incomplete monstrous embryos die in very early stages of development, and some degenerate and thus escape notice.

Conjoined Twinning

The vast majority of monstrous twins are of the conjoined rather than the separate type. The former are, as a rule, the product of a relatively late developmental interruption affecting the final steps in formation of the embryonic axis.

A retardation in the early stages of development may cause fusion of the head process (*Bollinger, 1928*), and retardation in the later stages

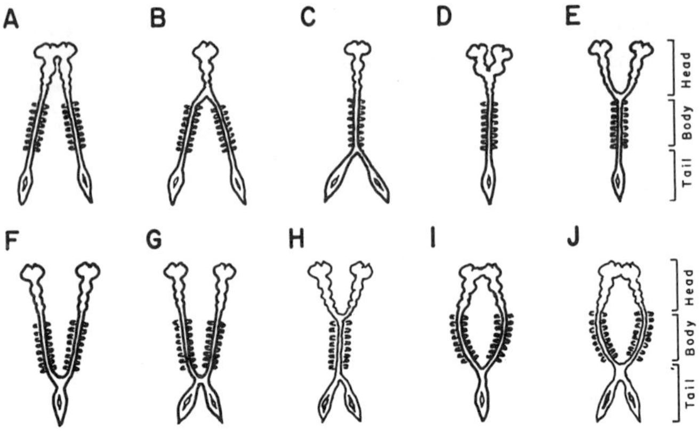

Fig. 177. A semidiagrammatic reconstruction of various types of symmetrical incomplete twinning (*diplopagus*) in Aves, exemplified by 24-hour embryos (with 8 somites). *A, B,* and *C,* posterior twinning (*katadidymus*); *D, E,* and *F,* anterior twinning (*anadidymus*); *G* and *H,* anteroposterior twinning; *I* and *J,* medial twinning. This diagram is reconstructed from the observations of Gerlach (1882); Lucas (1890); Dareste (1891); Kaestner (1898); Ruth (1918); Tannreuther (1919); Bollinger (1928); Newman (1940); Frost (1947); Lutz (1948); and Lutz and Lutz-Ostertag (1963).

of development, at the primitive streak stage, may cause fusion of posterior parts (*Tannreuther, 1919*). (See Fig. 177*A* and *F*.)

Autosite–Parasite Condition

When two embryos arise from two separate points of gastrulation, often one becomes primary and the other secondary (*Newman, 1923, p. 87*). The smaller, secondary, embryo evidently becomes partially inhibited by the larger, primary, embryo; this destines it to be a mere parasite on the body of the larger twin.

Conjoined twins can also arise asymmetrically; then the smaller embryo may become a parasite of the larger embryo.

DUPLICITY OF EMBRYONIC ORGANS

Heart. In birds, partial duplication of parts is frequent, but the formation of double hearts is not common. Figure 178 shows three embryos, each with two distinct hearts. Judging from the single well-developed vitelline vein of each heart, the two hearts began as two parallel endocardial tubes (*Romanoff, 1960, pp. 683–687*), which failed to unite into a common tube, and thus each independent tube resulted in a distinct, normal heart.

Tails and Limbs. Duplicity of the tail is merely a phase of bilateral twinning involving a posterior region, the tail bud (*Newman, 1923, p. 9*).

Twinning of limbs, however, involves certain complexities that are not present in bilateral twinning. Examples of duplicity in wings and legs are illustrated in Fig. 179. In the specimens presented, there is considerable variation in the attachment of limbs.

Together with the duplication of limbs, there can be a duplication of some portions of the digestive tube (*Chidester, 1927*), shown in Fig. 180.

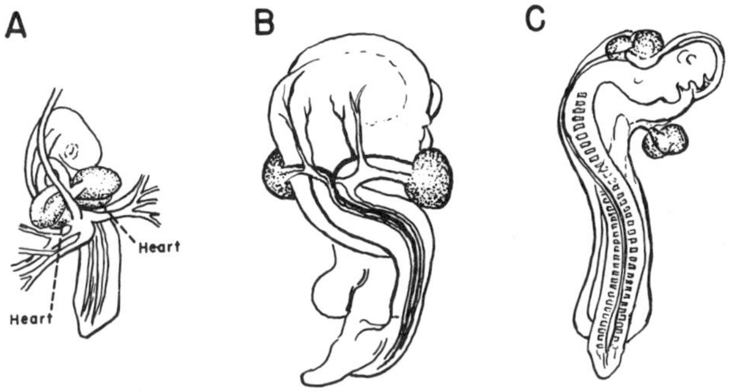

Fig. 178. Three cases of a double heart in the early embryo. *A,* after Warynski (1886); *B,* after Wücher (1900); *C,* after Tannreuther (1919).

Fig. 179. Some typical specimens of posterior duplicity in birds. (Rearranged after Chidester, 1927.) *1, 2, 3,* and *4,* with two wings and four legs; *5,* duckling with two wings and four legs; *6,* chick with four wings and four legs.

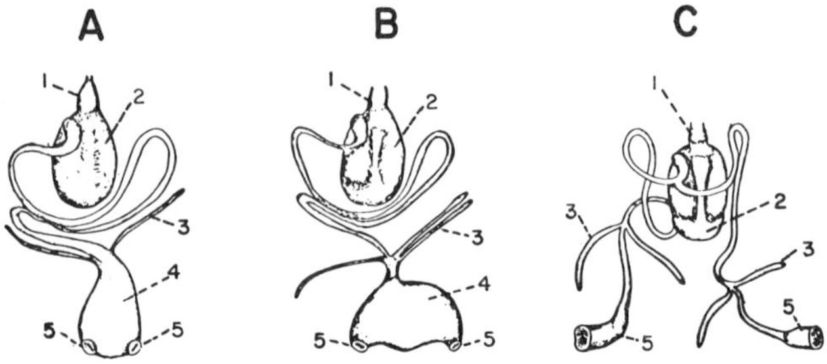

Fig. 180. Duplicity of digestive system of the newly hatched chick. (Rearranged after Chidester, 1927.) *A,* double cloacae; *B,* double cloacae with an enormous rectum; *C,* digestive tubes double behind the gizzard—the tubes are mirror images of each other.

PARTHENOGENESIS AND ABORTIVE DEVELOPMENT

Abortive parthenogenesis in the eggs of birds has been studied frequently for over 70 years (see *Romanoff, 1960, pp. 126–127*). It usually occurs in eggs of virgin hens.

In general, parthenogenetic development is delayed and abnormal; in most instances, it proceeds to formation of an anidean embryo (*Kosin and Sato, 1960*), or results in various types of irregular duplicity (*Olsen, 1962*). The latter phenomenon was observed in 1071 infertile turkey eggs incubated, as is shown in the accompanying table.

Parthenogenetic Duplicity	Incidence (number)	Incidence (%)
Twins	30	2.80
Triplets	8	0.75
Quadruplets	1	0.09
Octuplets	1	0.09
Incomplete (posteriorly)	3	0.28

One set of twins attained the size of a 7-day embryo; others, about that of 3-day embryos. All duplications occurred in the eggs of a group of 47 hens.

Parthenogenetic development is less frequent in chickens, especially in white Leghorns (*Olsen, 1966*).

Examination of 35,939 infertile chicken eggs, incubated for 8 to 10 days, showed that the incidence of parthenogenetic development varies among different breeds and strains from 0 to 6.38 per cent, with an average for all chickens of about 3.2 per cent.

Bartelmez and Riddle (*1924*) found some evidence of parthenogenesis in pigeon eggs.

Growth of Parthenogenetic Embryos. The comparative body weights of parthenogenetic and normal turkey embryos from the tenth to the twenty-fifth day of incubation (*Olsen, 1969*) are shown in the table below.

Incubation Age (days)	Weight of Embryo (grams)	
	Parthenogenetic	Normal
10	0.2	0.5
15	2.0	5.0
20	10.0	18.0
25	21.0	37.0

The difference in development is attributed to the characteristic delay in initial development of the blastoderms in unfertilized eggs when they are placed in the incubator.

EXPERIMENTAL PRODUCTION OF POLYEMBRYONY

A number of experimental procedures have been used in the production of polyembryony in chickens, turkeys, ducks, and quail. The most popular method has been fissuration, or division of the unincubated fertilized blastoderm into several equal parts (*Twiesselmann, 1938; Wolff, 1948; Lutz, 1949:* Fig. 181*A* to *C*). This technique has successfully produced normal twins, triplets, and quadruplets, as well as incomplete twinning (*duplicitas symmetros*), both posterior and anterior (see Fig. 177).

It has been observed (*Lepori, 1967*) that transverse sections of unincubated duck blastoderm, besides having embryogenetic potentialities, produce twins and triplets with great variability of orientation (Fig. 182). The majority of embryos are in the normal position, with their heads directed to the left (*Romanoff, 1960, pp. 141–143*), but some have inverted posture with an inverted heart (about 3 per cent), or inverted posture, but not an inverted heart.

Similar results were observed in fractional parts cultivated *in vitro* (*Spratt and Haas, 1960*). By subdividing the unincubated blastoderm, two, three, or four essentially complete embryos have been obtained.

Doubling in the chicken embryo has been produced by mere mechanical injury of the blastoderm (*Twiesselmann, 1935*). A higher incidence of duplication (about 8.2 per cent) was induced in the chicken

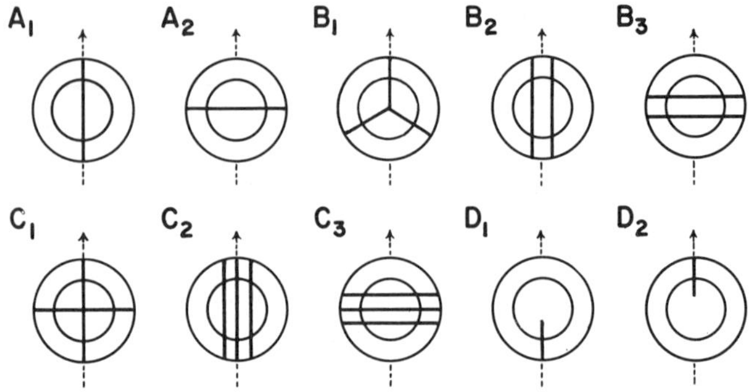

Fig. 181. A scheme of experimental production of twinning in birds by fissuration of the unincubated blastoderm. (Recombined chiefly after Wolff, 1948, and Lutz, 1948, 1949, 1964.) A_1 and A_2, production of twins by dividing the blastoderm into two equal parts, either along the presumptive axis of the future embryo (shown by an arrow), or perpendicular to it, respectively; B_1, B_2, and B_3, production of triplets by dividing the blastoderm into three equal parts, as indicated on the diagram; C_1, C_2, and C_3, production of quadruplets; D_1 and D_2, production of incomplete twin embryos by incomplete separation of the posterior part of the blastoderm (forming a double monster, *katadidymus*) and by partial separation of the anterior part (forming a double monster, *anadidymus.*)

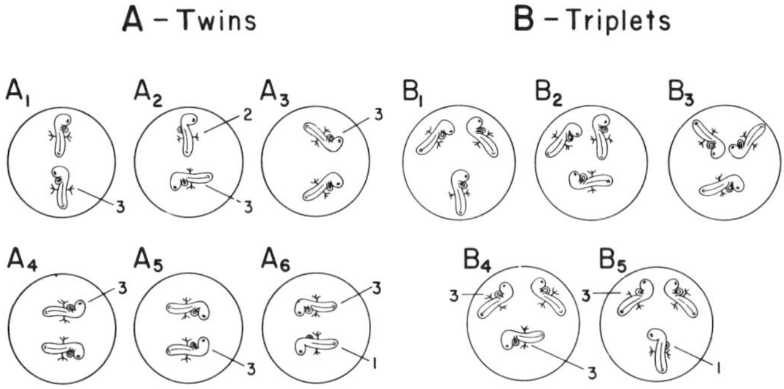

Fig. 182. Orientation of twins and triplets produced experimentally by dividing the blastoderm into two or three equal sections. (After Lepori, 1967.) Series *A*, orientations of twin embryos; *B*, orientations of triplets. *1*, inverted posture; *2*, inverted heart (*situs cordis invertus*); *3*, inverted posture and heart.

embryo by subjecting the hens to hypothermia before and during ovulation, and during the early cleavage stages (*Sturkie, 1946*). Chilling of eggs at —1 to 4.5° C. for 2 to 8 weeks before incubation resulted in a threefold increase in the incidence of twinning (*Olsen and Haynes, 1948*), in comparison with the level of about 0.1 per cent spontaneous duplication given by Byerly and Olsen (*1934*).

Appendix

GROWTH OF BONES *IN VITRO* VS. *IN VIVO*

It has been demonstrated (*Fell and Mellanby, 1955*) that all long bones develop according to the same general plan *in vitro* as *in vivo*, but histogenesis is not identical. There are also differences among the normal growth rates of the various rudiments (Fig. 183). The tibia is the most rapidly growing bone and the radius the slowest.

Increase in length and weight of the embryonic chicken tibiotarsus *in vivo* and *in vitro* is shown in *inserts a* and *b* of Fig. 183.

These observations suggest that the matrix of skeletal elements may be deficient in one or more components.

COMBINED EFFECTS—SYNERGISM

Although harmless alone, small amounts of potentially teratogenic substances, such as (a) 300 μg/egg of sulfanilamide and (b) 1.5 μg/egg of 6-aminonicotinamide (*Landauer and Clark, 1964a*) may in combination produce powerful synergism, as is shown below.

	Incidence (%)		
	With *a*	With *a* and *b*	With *b*
Mortality	15.8	38.7	12.8
Malformations			
Micromelia	0.0	79.3	0.0
Beak defects	0.0	61.9	0.5
Ataxia	0.0	17.2	0.0

(See also Fig. 150 for combined effect of cortisone and insulin.)

PREINCUBATION AGE OF FERTILE EGGS

Prior to incubation, fertilized eggs undergo various physical and chemical changes (*Kaufman, 1939; Romanoff, 1940a; Romanoff and Romanoff, 1949, pp. 654–689*) which progressively impair the development of the

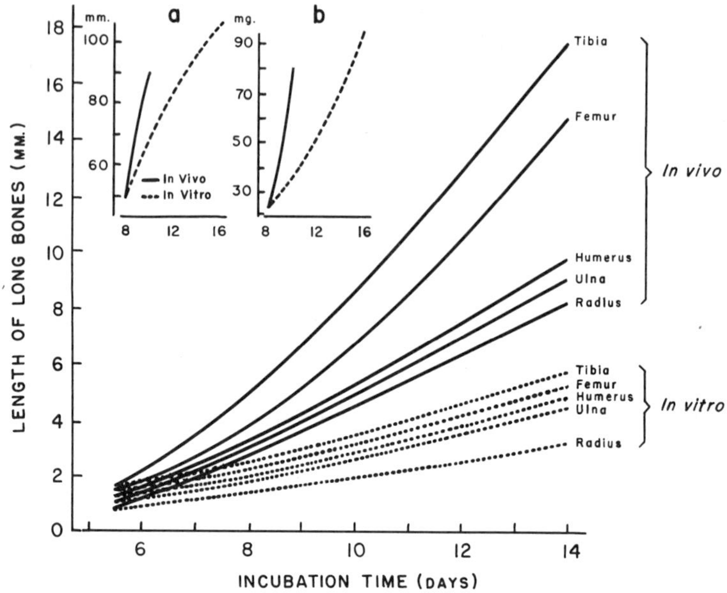

Fig. 183. Average growth curves of the long bones of the chicken embryo *in vivo* and *in vitro*. (After Fell and Mellanby, 1955.) *Insert: a,* comparative length of 8-day tibiotarsus *in vivo* and *in vitro; b,* comparative weight of tibiotarsus *in vivo* and *in vitro* (after Schryver, 1966.)

embryo, even in eggs held under optimal environmental conditions. (See Chapter IV, on the effects of atmospheric environment.)

Delay in Initial Development. Fertile eggs stored for a length of time before incubation show delayed embryonic development (*Kaufman, 1938; Arora and Kosin, 1966*). Consequently, embryos which survive the effect of aging usually hatch 1 day later than those from fresh eggs. The effect on the early growth of the embryo is well illustrated by the development of somites in Japanese quail, *Coturnix coturnix japonica* (*Manner and Granik, 1967*) as is shown below.

Preincubation Age of Eggs (hours)	Number of Somites at 33 Hours
0–24	21
25–48	19
49–72	17
73–96	16

Cytological studies of blastoderms from eggs held for extended periods and then incubated for ten hours (*Arora and Kosin, 1967*) reveal changes in mitotic index and the presence of various types of nuclear irregularities, as is shown in the tabulation below.

Length of Storage (days)	Types of Nuclei (%)		
	Mitotic	Fragmented	Necrotic
1	5.3	1.8	3.6
7	4.5	2.7	3.9
14	2.8	5.7	4.9
21	2.6	5.2	8.2

The frequency of fragmented nuclei increases as the period of holding is extended to 14 days or beyond, whereas the opposite is true for mitotic figures.

Embryonic Mortality

It has frequently been demonstrated (since *Waite, 1919*) that with increase in preincubation age of fertile eggs there is an accelerating increase in embryonic mortality (Fig. 184).

Under identical conditions, with the most favorable temperature of about 14° C, eggs of different species of birds also show a wide variation in susceptibility to aging. Turkey eggs, for example, are highly resistant, while duck eggs are more susceptible, as is shown in the *insert* of Fig. 184.

Incidence of Abnormalities

Prolonged preincubation storage causes a marked increase in the frequency of abnormal embryos, especially those with defects in the head and tail regions (*Kosin and Mun, 1965*). Histological studies indicate that the embryonic axis is shortened. Neural and mesodermal differentiation and heart formation are inhibited (*Mookerjee, 1953*). After 15 to 20 days of aging, the axial structures of many embryos are entirely suppressed.

Delay in early differentiation of turkey blastoderms has been observed, as well as a 16 to 35 per cent incidence of malformations (*Arora and Kosin, 1966*). Numerous types of abnormalities were encountered during early embryonic development (Fig. 185).

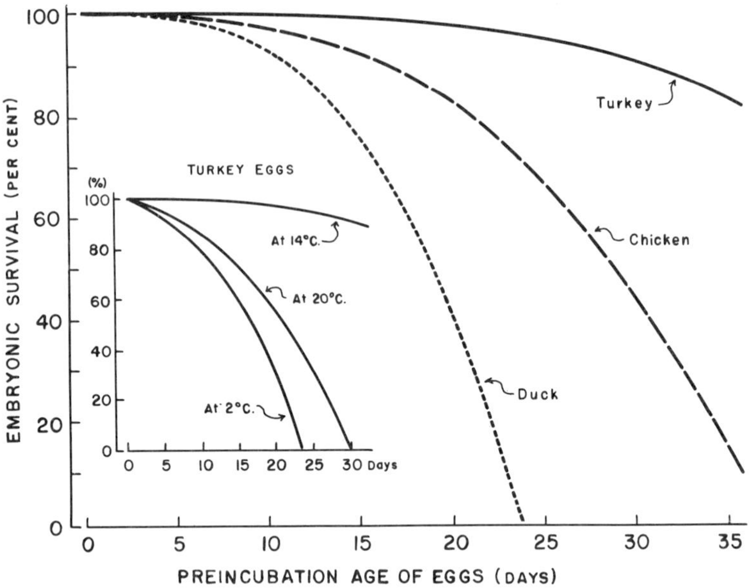

Fig. 184. Incidence of survival of the avian embryo with the increase in preincubation age of eggs held under the optimum temperature of 14° C. (On chicken, after Scott, 1933; Warren, 1934; Olsen and Haynes, 1948; Arora and Kosin, 1966; on turkey, after Scott, 1933; on duck, after Penionzhkevich, 1934.) *Insert:* effect of temperature on incidence of embryonic survival in turkey eggs (after Scott, 1933). For a detailed assessment of the ability of the avian blastoderm to withstand various conditions of preincubation storage, including temperature, the reader is referred to Romanoff, 1960 (*The Avian Embryo,* pp. 196–270, and particularly Fig. 77).

VARIABILITY IN EGG CHARACTERISTICS

The physical characteristics of the egg are determined genetically and modified by some extraneous elements of the environment during its formation (especially by the nutrition of the hen). Consequently, there is considerable variation in physical characteristics among the eggs within a flock of hens and, in lesser degree, among different eggs of the same hen. Some of the characteristics which deserve special consideration are analyzed numerically, specifically by frequency distribution.

Frequency Distribution

Egg Weight. There is obvious variation in the size of eggs laid in 1 day by a flock of hens (Fig. 186). Among the smallest is one egg weigh-

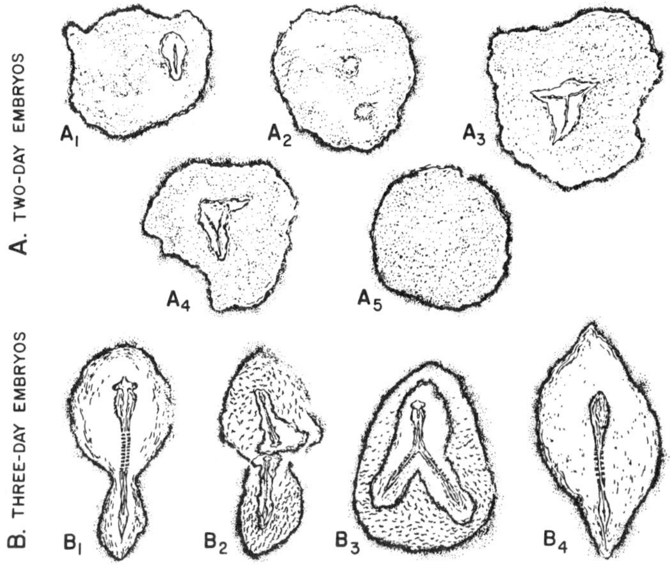

Fig. 185. Representative cases of some early developmental abnormalities in blastoderms of turkey eggs held in storage for over 14 days before incubation. (After Arora and Kosin, 1966.) At 2 days of incubation: A_1, with acentrally differentiated embryonal area; A_2, with two distinct sites of differentiating areas; A_3, three embryos at an early notochord stage; A_4, twin embryos with a common primitive streak; A_5, a blastoderm without any recognizable embryonic structure (anidean). At 3 days of incubation: B_1, incomplete differentiation of the area pellucida; B_2, head-to-head twinning; B_3, posterior twinning; B_4, excessive lateral differentiation of the area pellucida.

ing 35 grams, and among the largest, two eggs which each weigh about 90 grams and which presumably contain multiple yolks.

Egg Weight, Volume, Specific Gravity, Length, Breadth, and Shape. Frequency distributions of these egg characteristics are shown in Fig. 187.

Number of Pores and Permeability of Eggshell to Air. The frequency distributions of the average number of pores (Fig. 188) and the permeability of the eggshell (Fig. 189) show that these factors are subject to great variation.

Breaking Strength of Eggshell. The experimental data show that the breaking strength of the chicken eggshell varies greatly among individual eggs. The frequency distribution curve (Fig. 190) indicates that the average strength is about 4.5 kg.

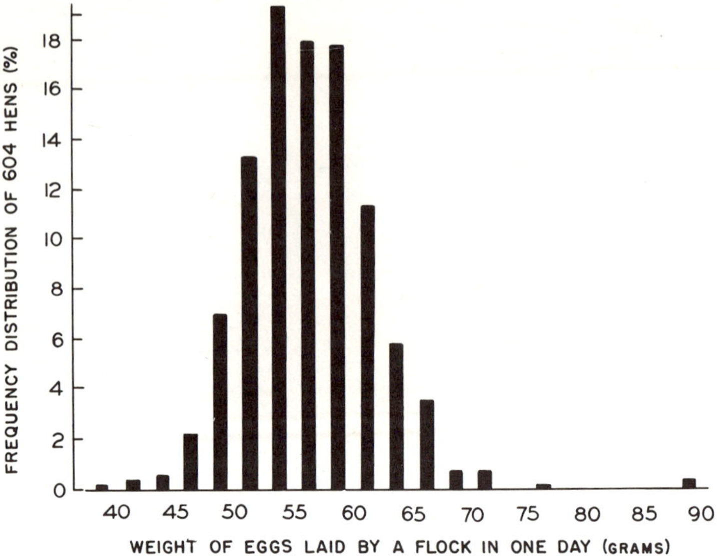

Fig. 186. **Frequency distribution of a flock of 604 hens according to the size of their eggs laid in one day.** (Romanoff, unpublished.)

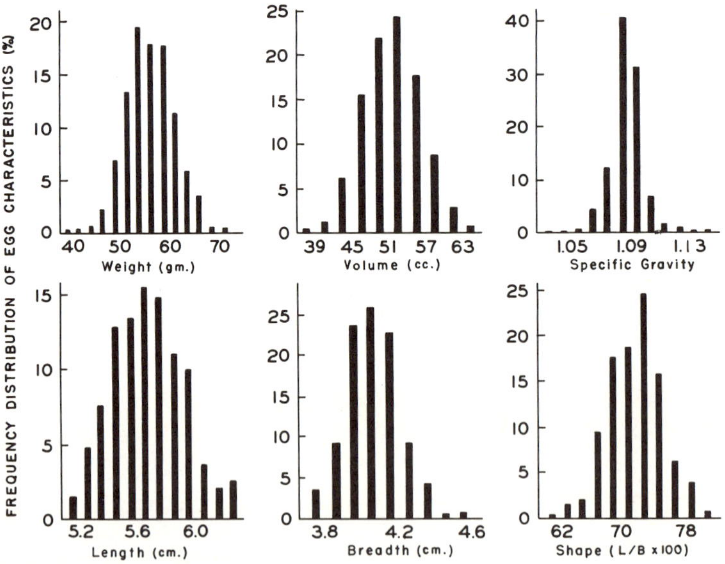

Fig. 187. **Frequency distributions of various egg characteristics, determined on 605 chicken eggs.** (Romanoff, unpublished.)

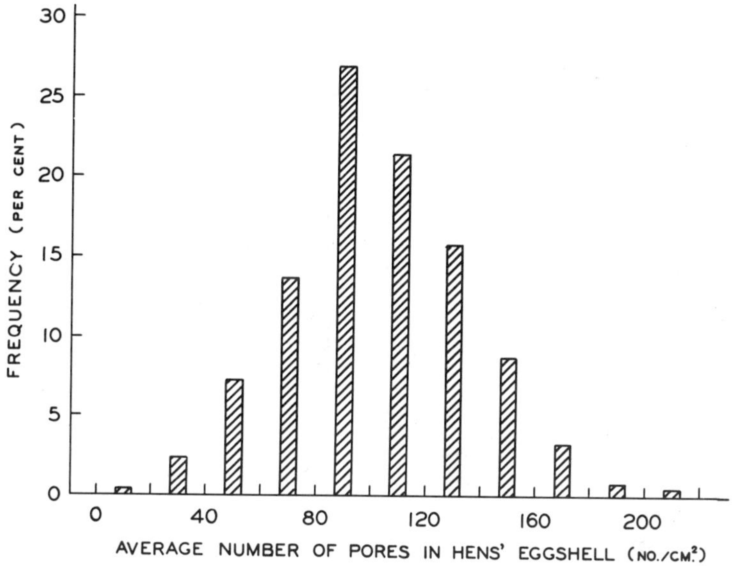

Fig. 188. Frequency distribution of 793 eggs from 260 hens according to number of pores. (Romanoff and Romanoff, 1949, pp. 166–168; and Romanoff, unpublished.)

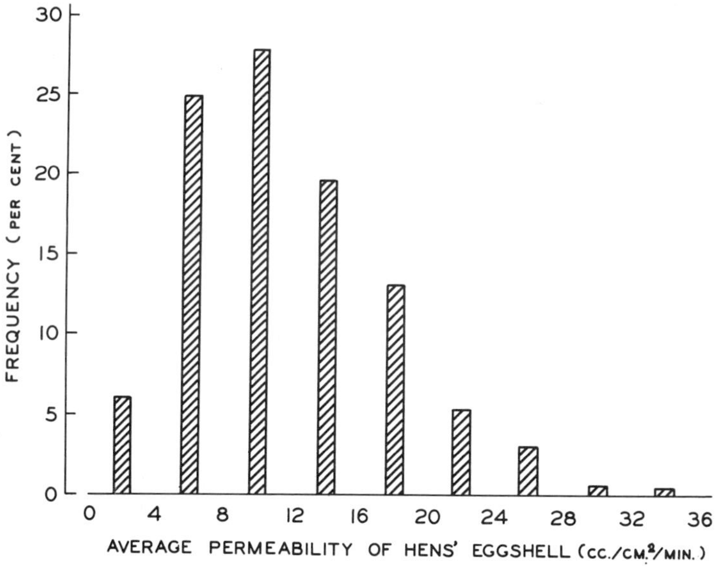

Fig. 189. Frequency distribution of 259 hens according to the average permeability of their eggshells. (After Romanoff and Romanoff, 1949, p. 384.)

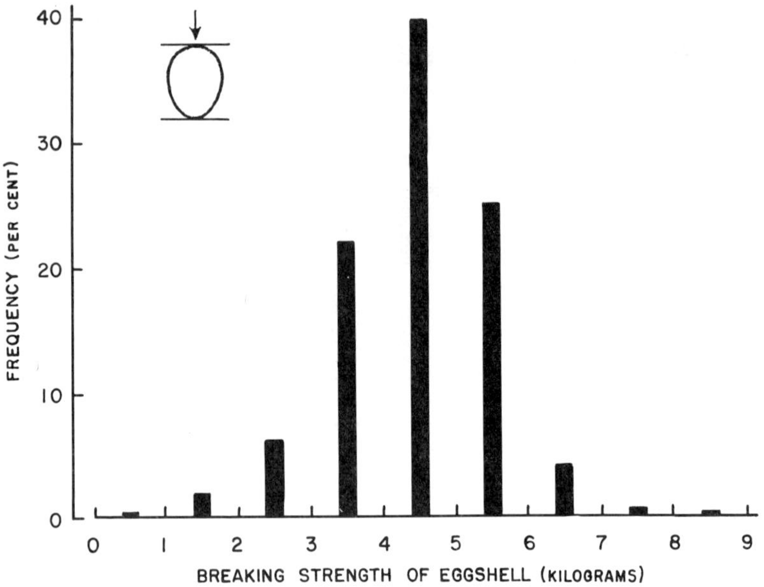

Fig. 190. Frequency distribution of breaking strength of 3998 chicken eggshells. (Plotted from the data of Romanoff, 1929a.)

The average strength of eggshell is characteristic for each individual hen, and normally rises with increase in rate of egg laying (*Romanoff, 1929a*).

EFFECTS OF EGG CHARACTERISTICS

Numerous observations show that, in general, eggs either structurally abnormal or having structural characteristics that deviate from the average values for the species do not hatch as well as normal.

Egg Weight

Mortality. It has been observed that the incidence of embryonic mortality is higher in eggs whose weight deviates from the average for the species. The summaries of the experimental data on the egg weight of mature birds, including chickens (*Dunn, 1922; Jull and Haynes, 1925; Axelsson, 1932; Godfrey, 1936; Hébert and Laugier, 1943*), turkeys (*Marble and Morgolf, 1936; Byerly and Marsden, 1938; Insko, MacLaury, and Baute, 1943; Parker, 1946; Funk, 1950*), and ducks

(*Rendel, 1943*), given in the accompanying table, demonstrate the relationship between egg weight and embryonic mortality.

Egg Weight (grams)	Incidence of Embryonic Mortality (%)		
	Chicken	Turkey	Duck
40–50	25.4		
50–60	5.4		
60–70	17.4	15.4	38.4
70–80	65.3	8.3	33.1
80–90		3.3	43.0
90–100		14.9	

The average weight of the eggs of each species is as follows: of the chicken, 50 to 60 grams; of the turkey, 80 to 90 grams; and of the duck, 70 to 80 grams (see *Romanoff and Romanoff, 1949, p. 62*).

Thickness of Eggshell

Observations on 1434 chicken eggs (*Rauch and Steinke, 1953*) show that extremely thin or thick eggshells have a negative effect in terms of embryonic mortality, in comparison to eggshells of normal (average) thickness, as is demonstrated in the accompanying tabulation.

Thickness of Eggshell (mm)	Incidence of Mortality (%)
0.18–0.22	39
0.22–0.26	21
0.26–0.30	12
0.30–0.34	20
0.34–0.38	42
0.38–0.42	50

Porosity of Eggshell

The eggshell plays an important role, as a medium for gaseous exchange between the developing embryo and the atmosphere, and in the control of evaporation from the egg. A relationship between the rate of weight loss of the egg and the survival of the embryo has been shown experimentally. The accompanying table gives composite values of three sets of observations on a total of over 7900 eggs (*Axelsson, 1932, 1954*).

Porosity of Eggshell (mg loss/24 hour/at 22° C)	Embryonic Mortality (%)
60	47.7
100	33.2
140	25.7
180	28.6
200	40.2
240	44.2

There is an optimum condition for embryonic survival. The peak may differ slightly among various investigations primarily because eggshell porosity can change according to nutrition of the hen and humidity control during incubation.

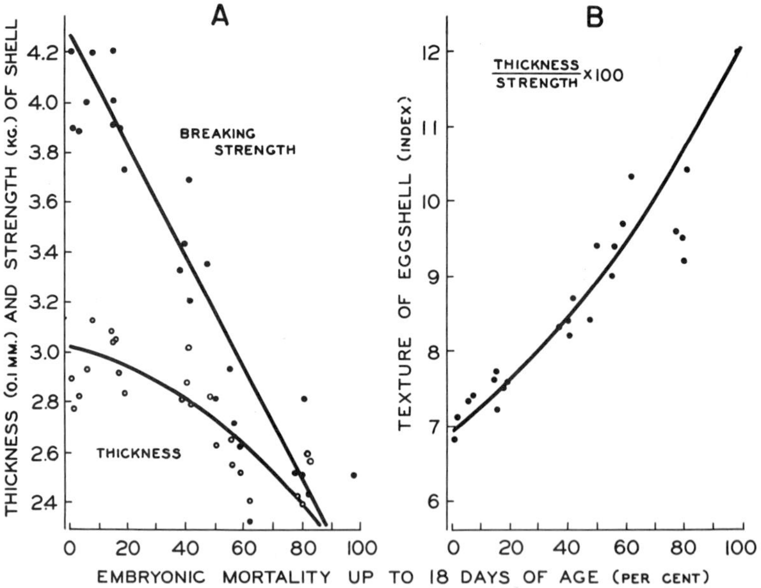

Fig. 191. Apparent effect of the strength and thickness of the eggshell on the incidence of embryonic mortality at 18 days of age. (Romanoff, 1929a, and Romanoff, unpublished.) *A*, relationship between the breaking strength and thickness of eggshells and embryonic mortality; *B*, relationship between the index of the texture of eggshells and the incidence of embryonic mortality.

Other Relationships

Strength and Thickness of Eggshell. There is a definite relationship between the breaking strength and thickness of the eggshell and the rate of mortality (Fig. 191*A* and *B*).

Weight of Embryo and Yolk Sac. The embryonic weight and the weight of the reserve yolk sac at 18 days are closely related to the rate of embryonic mortality (Fig. 192*A* and *B*).

MATERNAL AGE AND EMBRYONIC MORTALITY

It has been demonstrated that embryonic mortality correspondingly increases with the age of the mother hen. It also has been shown (*Insko, Steele, and Wightman, 1947*) that there is a proportional increase in

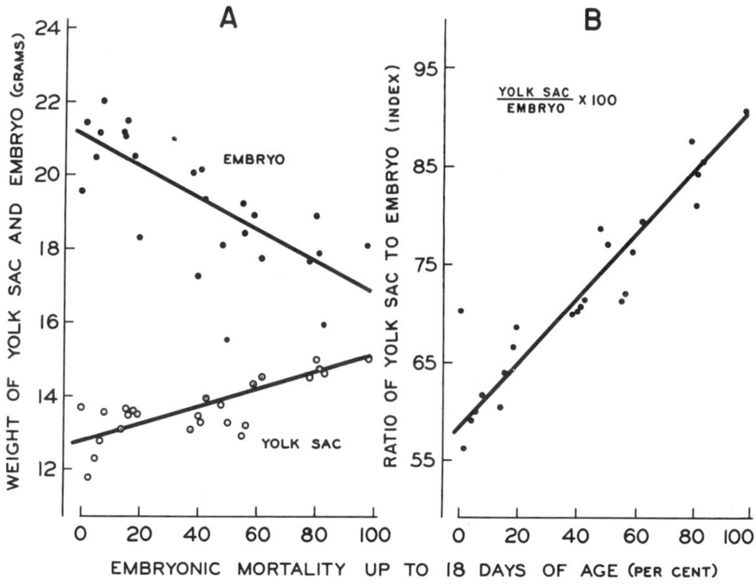

Fig. 192. Observed trends of variations in the weights of 18-day-old embryos and their yolk sacs in relation to the incidence of embryonic mortality among 1099 eggs laid by 27 hens. (Romanoff, unpublished.) *A*, relationship of the weights of the embryos and their yolk sacs, respectively, to the mortality at 18 days of age; *B*, relationship between the ratio of the yolk sac weight to the weight of the embryo and the rate of mortality at 18 days of age.

weak chicks accompanying the increase in embryonic mortality due to aging of the hen, as is shown in the table below.

Age of Hen (years)	Embryonic Mortality (%)	Weak Chicks (%)
1	20.5	1.4
2	24.5	3.4
3	28.3	4.1
4	31.4	5.0
5	41.8	5.9
6	42.5	8.4
7	53.3	12.9

There is some evidence (*Meirelles de Miranda, 1956*) that there is also a rise in the rate of embryonic malformations with maternal age in Aves.

MATERNAL WEIGHT AND EMBRYONIC MORTALITY

Observations on two breeds of turkeys, Bronze and White Holland (Funk, 1950), show a definite relationship between the body weight of the mother and the incidence of embryonic mortality in her fertile eggs, as is shown below.

Weight of Dams (kg)		Total Mortality at Hatching (%)	
Bronze	White Holland	Bronze	White Holland
	5.5		23.7
	6.5		30.2
7.5	7.5	15.7	33.1
8.5		29.2	
9.5		41.5	

Glossary

GREEK MEDICAL TERMS OF CONGENITAL MALFORMATIONS*

Ablepharia—reduction or absence of the eyelids.
Acardia—absence of the heart.
Acephalia—absence of the head.
Achondroplasia—inadequate endocondral bone formation.
Acrania—partial or complete absence of the cranium, or skull.
Agenesia (*see also* Aplasia)—faulty or incomplete development.
Amelia—absence of limbs.
Anadidymus—in twin embryo, a fusion of the lower limbs into one.
Anencephalia—absence of the brain.
Anideus—a blastoderm without apparent embryo.
Ankylosis—complete immobility of a joint.
Anophthalmia—absence of eyes.
Anouria—reduction of tail end of the skeleton.
Aplasia (*see also* Agenesia)—lack of development of an organ.

* These terms are based chiefly on descriptive classifications by Dareste (1891), Wolff (1936a), Ancel (1950), and other French avian teratologists. In the text, the suffix "ia" as in "Hypodactylia" is often anglicized to "y."

Brachycephalia—a disproportionately short head.
Brachydactylia—abnormally short digits and toes.
Brachygnathia—abnormally short mandible.
Brachyuria—shortened abdomen.
Buphthalmia—enlargement of the eye.

Celosomia—hernia of the viscera.
Cephalodymus—a twin embryo with a single or united head.
Cephalothoracopagus—a twin embryo united from the crown of the head to the middle of the abdomen.
Chondrodystrophia—abnormally developed cartilage.
Coelosoma—eventration of the viscera through a body cleft.
Coloboma—a fissure of any part of the eye.
Cranioschisis—a fissure of the cranium.
Cyclopia—a fusion of both eyes medially into one.

Diastematomyelia—splitting of a part of the spinal cord.
Diplocardia—a separation, by a fissure, of the two lateral halves of the heart.
Diplopagus—twin embryos with equal component parts, the symmetrical equivalents of one another.
Diprosopus—an embryo with two faces and heads.
Dystopia—malposition or displacement of any organ.

Ectopia—a morbid malposition or traumatic displacement of an organ or part.
Ectrodactylia—absence of all or only part of a digit.
Ectromelia—absence of one or more limbs.
Ectrosomia—miscarriage of the body.
Ectrosyndactylia—absence of some digits and cohesion of those remaining.
Encephalocele—hernial protrusion of the brain through a fissure in the skull.
Exencephalia—exposure of the whole or a part of an imperfect brain.
Exophthalmos—abnormal protrusion of the eyeball.

Gastromelus—an embryo with a supernumerary leg(s) attached to the abdomen.
Gastroschisis—an open fissure of the abdominal cavity.

Hemimelia—absence or imperfection of the distal half of a limb.
Hemisomus—an embryo with a lateral half of the body absent or maldeveloped.
Heteradelphus—unequally developed twin embryos, when the parasite is joined to the abdomen of the autosite.
Heterauxesis—growth of one organ disproportionately to another.
Heterotaxia—displacement of parts and organs of the body.
Hyperdactylia—presence of more than normal number of fingers and toes.

Hyperencephalus—an embryo without the cranial vault, displaying the exposed brain.
Hypertrophia—an excessive enlargement or overgrowth of an organ or part.
Hypodactylia—a deficiency in the number of digits, fingers or toes.
Hypophalangism—less than the normal number of phalanges of a finger or a toe.
Hypoplasia—defective or incomplete development.

Ischiopagus—twin embryos fused end-to-end at the hips, with two bodies extending in opposite directions.

Katadidymus—twin embryos in which the upper parts are fused.
Keratoconus—a conical protrusion of the cornea.
Kyphoscoliosis—a forward and lateral curvature of the spinal column.
Kyphosis—a pronounced forward curvature of the spine.

Lordosis—curvature of the spine in lumbar region.

Macrocephalia—extremely large size of the head.
Melodidymus—an embryo with two sets of limbs.
Melomelus—an embryo with normal limbs and imperfectly developed supernumerary limbs.
Micrencephalia—abnormally small brain.
Microcephalia—abnormally small head.
Microcrania—abnormally small skull.
Micrognathia—an undersized jaw, particularly the mandible.
Micromelia—abnormally small limbs.
Micromyelia—abnormal smallness of the spinal cord.
Microphthalmia—abnormally small eyes.
Microsomia—abnormally small size of the embryo.
Monomicrophthalmia—smallness of only one eye.
Monophthalmia—presence of only one eye.
Myeloschisis—exposure of the neural canal on the surface (associated with spina bifida).
Myodystrophia—(associated with muscular dystrophy).

Nanomelia—undersized limbs.

Omphalopagus—twin embryos united at the umbilicus.
Opocephalus—twin embryos without mouth and nose and united by the ears.
Opodidymus—twin embryos with fused heads, but with two distinct faces.
Otocephalus—twin embryos joined below the face.

Perocephalus—an embryo with imperfectly formed head.
Perocormus—an embryo with a grossly deformed body.

Perodactylia—imperfect fingers or toes.
Peromelia—defective limbs.
Phocomelia—absent or rudimentary limbs.
Platymeria—lateral compression of the femur.
Polydactylia—presence of supernumerary digits on a wing or a foot.
Polyembryony—the occurrence of two or more embryos from the same ovum.
Prognathism—abnormal projection of one or both jaws.
Prosophthalmia—forward projection of eyes.
Pseudarthrosis—a false joint.
Pseudencephalus—a partial development of the brain.
Pseudosymmelia—an apparent fusion of the lower limbs.
Pygodidymus—an embryo with double pelvis and hips.
Pygomelus—an embryo with a supernumerary limb attached at the region of the buttock.
Pygopagus—twin embryos joined together back to back at the sacrum and coccyx.

Rachschisis—a fissure of the spinal column (spina bifida).

Scoliosis—a lateral curvature of the spine.
Sirenomelus—an embryo with fused legs.
Strophosomia—twisting of the body.
Symmelia—a fusion of the lower limbs.
Syndactylia—a fusion of two or more fingers or toes (a webbing between adjacent digits).
Synophthalmia—a complete or partial fusion of the two eyes into one (see Cyclopia).

Thoracodelphus—an embryo possessing four legs.
Thoracodidymus—twin embryos joined at the thorax.
Thoracomelus—an embryo with a supernumerary wing or leg attached to its chest (or thorax).

Bibliography*

Abbott, U. K. and Goetinck, P. F. (1960), *J. Heredity, 51:*161–166 (*52*).

Abbott, U. K. and MacCabe, J. A. (1966), *J. Heredity, 57:*207–211 (*51*).

Abbott, U. K., Taylor, L. W., and Abplanalp, H. (1960), *J. Heredity, 51:*195–202 (*46, 53*).

Ackermann, W. W. and Francis, T. (1950), *Proc. Soc. Exptl. Biol. Med., 74:*123–126 (*333*).

Ackermann, W. W. and Taylor, A. (1948), *Proc. Soc. Exptl. Biol. Med., 67:*449–452 (*42, 265*).

Adams, A. E. (1958), *Anat. Record, 131:*445–463 (*301, 323*).

Adams, A. E. and Bennett, M. (1953), *J. Exptl. Zool., 123:*315–333 (*311, 322*).

Adams, A. E. and Bull, A. L. (1949), *Anat. Record, 104:*421–443 (*307, 309, 310, 311, 323, 324*).

Adams, A. E. and Buss, J. M. (1952), *Endocrinology, 50:*234–253 (*309, 311, 323, 325*).

Adamstone, F. B. (1931), *J. Morphol. Physiol., 52:*47–90 (*186*).

Adhikari, S. (1967), *Folia Biol., 15:*231–235 (*279*).

Agarwal, I. P. (1956), *J. Animal Morphol. Physiol., 3:*63–74 (*270*).

Agarwal, I. P. and Monga, J. N. (1958a), *Indian J. Med. Res., 46:*647–652 (*288*).

Agarwal, I. P. and Monga, J. N. (1958b), *J. Anat. Soc. India 7:*15–18 (*271*).

* The italicized numbers in parentheses refer to pages of the text.

Agarwal, I. P., Monga, J. N., and Agarwal, L. D. (1959), *J. Anat. Soc. India, 8*:56–60 (*238, 271*).

Ajitomi, S. (1959), *Nippon Yakurigaku Zasshi, 55*:1021–1050 (*274*).

Aleksandrov, P. N., Bogdanova, V. A., and Chernukh, A. M. (1966), *Patol. Fiziol. Eksperim. Terapiya, 10*(6):53–56 (*179, 247, 252*).

Aleksandrov, V. A. and Stanzhevskaya, T. I. (1969), *Arkh. Anat. Gistol. Embriol., 57*(12):42–46 (*256*).

Allen, S. C. (1963), *Aerospace Med., 34*:897–899 (*99*).

Allenspach, A. L. (1970), *J. Morphol., 13*:89–101 (*51*).

Alsop, F. M. (1919), *Anat. Record, 15*:307–331 (*30, 78*).

Ancel, P. (1945), *Ann. Endocrinol., 6*:1–24 (*42, 273*).

Ancel, P. (1950), *La Chimioteratogenese chez les Vertebres. Paris,* 397 pp. (*4, 230, 319, 409*).

Ancel, P. (1955), *J. Embryol. Exptl. Morphol., 3*:335–354 (*250, 257, 272, 273*).

Ancel, P. (1956), *Arch. Anat. Microscop. Morphol. Exptl., 45*:203–217 (*216, 369*).

Ancel, P. (1958), *J. Embryol. Exptl. Morphol., 6*:335–344 (*42, 80, 82*).

Ancel, P. (1959), *J. Embryol. Exptl. Morphol., 7*:330–334 (*74, 82*).

Ancel, P. and Lallemand, S. (1941), *Compt. Rend. Soc. Biol., 135*:221–223 (*78*).

Ancel, P. and Lallemand, S. (1943a), *Arch. Physiol. Biol., 16*:43–45 (*273*).

Ancel, P. and Lallemand, S. (1943b), *Arch. Physiol. Biol., 17*:3–4 (*270*).

Ancel, P. and Vintemberger, P. (1925), *Compt. Rend. Soc. Biol., 92*:172–174 (*119*).

Ancel, P. and Wolff, Et. (1934), *Compt. Rend. Acad. Sci., 199*:1071 (*122*).

Andersen, K. T. (1932), *Arch. Zool. Ital., 16*:888–899 (*86*).

Anderson, K. (1940), *Am. J. Pathol., 16*:137–155 (*333*).

Andrews, F. N. and Schnetzler, E. E. (1945), *Endocrinology, 37*:382–384 (*305, 311*).

Armstrong, M. and Ham, A. (1947), *Cancer Res., 7*:481–490 (*352, 353, 354*).

Arora, K. L. and Kosin, I. L. (1966), *Poultry Sci., 45*:958–970 (*398, 399, 400, 401*).

Arora, K. L. and Kosin, I. L. (1967), *Biol. Bull., 133*:303–309 (*399*).

Arrington, L. R., Cruz, R. A. S., Harms, R. H., and Wilson, H. R. (1967), *J. Nutr., 92*:325–330 (*172, 174, 245*).

Asayama, S. (1965), *J. Biol. Osaka Univ., 16*:59–68 (*317*).

Asayama, S. and Maruoka, C. (1967). *Zool. Mag.* (*Osaka Univ.*), *76*:265–270 (*317*).

Ash, W. J. (1966), *J. Heredity, 57*:137–141 (*52*).

Asmundson, V. S. (1936), *J. Heredity, 27*:401–404 (*52*).

Asmundson, V. S. (1942a), *Proc. Soc. Exptl. Biol. Med., 50*:120–123 (*53*).

Asmundson, V. S. (1942b), *J. Heredity, 33*:328–330 (*52*).

Asmundson, V. S. (1944), *J. Heredity, 35*:295–299 (*52*).

Asmundson, V. S. (1945), *J. Heredity, 36*:173–176 (*23, 51*).

Asmundson, V. S. and Julian, L. M. (1956), *J. Heredity, 47*:248–252 (*54*).

Asmundson, V. S. and Pun, C. F. (1956), *J. Exptl. Zool.* *131*:225–237 (*45, 51*).

Atkinson, R. L., Ferguson, T. M., Quisenberry, J. H., and Couch, J. R. (1955), *J. Nutr.*, *55*:387–398 (*187*).

Axelsson, J. (1932), *Lunds Univ. Arsskr. Avd. 2, Part I*:1–196 (*404, 406*).

Axelsson, J. (1954), *Ann. Roy. Agr. Coll. (Sweden)*, *21*:81–103 (*406*).

Bang, F. B. (1942), *J. Exptl. Med.*, *76*:263–270 (*341*).

Bang, F. B. (1943), *J. Exptl. Med.*, *77*:337–344 (*331*).

Barnett, B. D., Richey, D. J., and Morgan, C. L. (1957), *Proc. Soc. Exptl. Biol. Med.*, *95*:101–104 (*224, 225, 252*).

Barott, H. G. (1937), *U.S. Dept. Agr. Tech. Bull.*, *553*:1–46 (*90, 91, 97, 102, 103*).

Bartelmez, G. W. and Riddle, O. (1924), *Am. J. Anat.*, *43*:45–76 (*393*).

Bearse, G. E. and Miller, M. W. (1937), *Poultry Sci.*, *16*:39–43 (*163*).

Beaudette, F. R. and Hudson, C. B. (1937), *J. Am. Vet. Med. Assoc.*, *90*(N. S. 43):51–58 (*329*).

Beaudoin, A. R. (1961), *J. Embryol. Exptl. Morphol.*, *9*:14–21 (*209, 210, 211*).

Beaudoin, A. R. and Wilson, J. G. (1958), *Proc Soc Exptl. Biol. Med.*, *97*:85–90 (*278*).

Becher, H. (1939), *Verhandl. Anat. Ges.*, *47*:144–154 (*105*).

Beer, A. E., Scott, M. L, and Nesheim, M. C. (1963), *Brit. Poultry Sci.*, *4*:243–253 (*158, 159, 184*).

Beohar, P. C., Goldhaber, P., and Hertig, A. T. (1964), *Indian J. Med. Res.*, *52*:263–272 (*237, 265*).

Bergquist, H. (1960), *J. Embryol. Exptl. Morphol.*, *8*:69–72 (*373*).

Bergström, R. M., Erilä, T., and Pirskanen, R. (1967), *Experientia*, *23*:767–768 (*263*).

Bernard, M. (1850), *Compt. Rend. Soc. Biol.*, *1*:9–10 (*386*).

Bernier, P. E. (1951), *Records Genet. Soc. Am.*, *20*:89 (*52*).

Bernier, P. E., Taylor, L. W., and Gunns, C. A. (1951), *Hilgardia*, *20/21*:529–628 (*18, 19, 20*).

Beskrovni, M. A. (1941), *Exptl. Med. (Ukraine)*, *8*(2):45–53 (*265*).

Bethke, R. M., Record, P. R., Kick, C. H., and Kennard, D. C. (1936), *Poultry Sci.*, *15*:326–335 (*168, 186*).

Bevelander, G., Nakahara, H., and Rolle, G. (1960), *Develop. Biol.*, *2*:298–312 (*274*).

Beyer, R. E. (1952), *Endocrinology*, *50*:497–503 (*301, 302, 322*).

Biagi, G. L. (1965), *Boll. Soc. Ital. Biol. Sper.*, *41*:1313–1314 (*240*).

Biagi, G. L. and Graziani, G. (1966), *Boll. Soc. Ital. Biol. Sper.*, *42*:834–838 (*314*).

Biener, J., Balaceanu, M., and Albu, N. (1960), *Acad. Rep. Populare Romine*, *11*:485–489 (*314*).

Biggers, I. D., Webb, M., Parker, C., and George, M. (1957), *Nature*, *180*:825–828 (*179*).

Biocca, E. (1942), *Rev. Brasil Biol., 2:*143–146 (*207*).

Bissonnette, T. H. (1941), *J. Heredity, 32:*139–144 (*54*).

Blackwood, U. B. (1960), *Proc. Soc. Exptl. Biol. Med., 104:*373–378 (*30*).

Blackwood, U. B. (1961), *Poultry Sci., 40:*3–9 (*228, 256*).

Blackwood, U. B. (1962), *J. Embryol. Exptl. Morphol., 10:*315–336 (*229, 256, 257*).

Blamberg, D. L., Blackwood, U. B., Supplee, W. C., and Combs, G. F. (1960), *Proc. Soc. Exptl. Biol. Med., 104:*217–220 (*176, 178, 188*).

Blattner, R. J. and Williamson, A. P. (1951), *Proc. Soc. Exptl. Biol. Med., 77:*619–621 (*339, 358*).

Blattner, R. J., Williamson, A. P., and Simonsen, L. (1958), *Proc. Soc. Exptl. Biol. Med., 97:*560–564 (*226, 255*).

Blattner, R. J., Williamson, A. P., Simonsen, L., and Robertson, G. G. (1960), *J. Pediatrics, 56:*285–293 (*199, 235, 236, 253, 263, 277*).

Bless, A. A. and Romanoff, A. L. (1943), *J. Cellular Comp. Physiol., 21:*117–121 (*110, 112, 149*).

Bloom, S. E. (1969), *Chromosoma, 28:*357–369 (*40*).

Bodit, F., Stoll, R., and Maraud, R. (1966), *Compt. Rend. Soc. Biol., 160:*960–963 (*262, 273, 276*).

Boland, J. (1954), *Brit. J. Radiol., 27:*680–687 (*110, 112, 114, 115, 117, 127*).

Boland, J. and Tranter, F. W. (1957), *Brit. J. Radiol., 30:*350–351 (*3, 112, 143*).

Bollinger, O. P. (1928), *Proc. Penn. Acad. Sci., 2:*106–107 (*389, 390*).

Booker, E. E. and Sturkie, P. D. (1949), *Poultry Sci., 28:*147–148 (*301, 321*).

Bordier, P. and Galimard, J. (1905), *Lyon Med. 37:*1368–1371 (*119*).

Boriskin, V. V., Oblapenko, P. V., Rol'nik, V. V., and Savin, B. M. (1962), *Dokl. Akad. Nauk SSSR.*, 143–475; (AIBS) Translation Nos. 1–6 (*99*).

Borsuk, I. (1964), *Folia Morphol., 15:*297–305 (*217, 274*).

Bower, R. K., Haberman, S., and Minton, P. D. (1970), *J. Pharmacol. Exptl. Therap., 171:*314–324 (*270*).

Bowman, P. (1967), *J. Embryol. Exptl. Morphol., 17:*425–431 (*260*).

Brandly, C. A., Thorp, F., and Prickett, C. O. (1949), *Poultry Sci., 28:*486–498 (*202, 354, 355*).

Braun, S. (1954), *Acta Morphol. Acad. Sci. (Hungary), 4:*61–83 (*263*).

Brettschneider, (1952), *Strahlentherapie, 87:*517–527 (*148*).

Broekhuizen, E. G. and Albada, Ir. M. van (1953), *Genetica, 26:*415–429 (*40, 50*).

Bryce, T. H. (1900), *Proc. Roy. Soc. Edinburgh, 22:*622–630 (*381*).

Bucciante, L. (1931), *Arch. Exptl. Zellforsch., 11:*397–423 (*87*).

Bucciante, L. (1933), *Arch. Exptl. Zellforsch., 14:*56–138 (*87*).

Buckner, G. D., Martin, J. H., and Peter, A. M. (1929), *Kentucky Agr. Expt. Sta. Bull., 291:*1–23 (*170*).

Buddingh, G. J. (1944), *J. Exptl. Med., 80:*59–64 (*350*).

Buddingh, G. J. and Henigst, W. (1960), *Am. J. Pathol., 37:*477–495 (*349*).

Buddingh, G. J. and Polk, A. D. (1939), *J. Exptl. Med., 70:*485–498 (*347*).

Bueker, E. D. and Platner, W. S. (1956), *Proc. Soc. Exptl. Biol. Med., 91*:539–543 (*250, 254, 266, 270*).

Buno, W. and Goyena, H. (1955), *Proc Soc. Exptl. Biol. Med., 89*:622–625 (*287*).

Burckhardt, R. (1888), *Arch. Anat. Physiol.* (*Anat. Abt.*):431–432 (*383*).

Burke, R. C., Salvin, S. B., and Gerloff, R. K. (1952), *Proc. Soc. Exptl. Biol. Med., 81*:91–94 (*351*).

Bursian, A. V. (1960), *Mater. Evol. Physiol., 4*:265–273 (*149*).

Butt, E. M., Pearson, H. E., and Simonsen, D. G. (1952), *Proc. Soc. Exptl. Biol. Med., 79*:247–249 (*245*).

Byerly, T. C. (1926), *Anat. Record, 33*:319–325 (*13*).

Byerly, T. C. (1930a), *J. Morphol. Physiol., 50*:341–359 (*37*).

Byerly, T. C. (1930b), *Proc. World Poultry Congr.* (*London*), *4*:178–186 (*28, 29, 30, 33*).

Byerly, T. C. and Jull, M. A. (1932), *J. Exptl. Zool., 62*:489–498 (*53*).

Byerly, T. C., Knox, C. W., and Jull, M. A. (1934), *Poultry Sci., 13*:230–238 (*39*).

Byerly, T. C. and Marsden, S. J. (1938), *Poultry Sci., 17*:298–300 (*404*).

Byerly, T. C. and Olsen, M. W. (1931), *Poultry Sci., 10*:281–287 (*364*).

Byerly, T. C. and Olsen, M. W. (1934), *Science, 80*:247–248 (*387, 394*).

Byerly, T. C., Titus, H. W., and Ellis, N. R. (1933), *J. Agr. Res., 46*:1–22 (*161, 179*).

Calame, S. (1961), *Arch. Anat. Microscop. Morphol. Exptl., 50*:299–308 (*11*).

Calvert, C. C. (1967), *Poultry Sci., 46*:967–973 (*181, 182*).

Cameron, J. M. (1962), *J. Pathol. Bacteriol., 84*:438–439 (*273*).

Caskey, C. D. and Norris, L. C. (1939), *Proc. Soc. Exptl. Biol. Med., 40*:590–593 (*187*).

Caskey, C. D. and Norris, L. C. (1940), *Proc. Soc. Exptl. Biol. Med., 44*:332–335 (*42*).

Caskey, C. D., Norris, L. C., and Heuser, G. F. (1944), *Poultry Sci., 23*:516–520 (*187*).

Catizone, O. and Gray, P. (1941), *J. Exptl. Zool. 87*:71–82 (*245*).

Caujolle, F. M. E., Caujolle, D. H., Cros, S. B., and Calvet, M. M. J. (1967), *Ann. N.Y. Acad. Sci., 141*:110–126 (*259*).

Caujolle, F., Caujolle, D., Cros, S., Dao-Huy-Giao, M., Moulas, F., Tollon, Y., and Caylus, J. (1965), *Bull. Trav. Soc. Pharm. Lyon, 9*:221–235 (*219, 250*).

Cervenansky, J., Kalman, E., Micek, J., Maar, D., Makovsky, V., and Skrovina, S., (1964), *Act. Chi. Ort. Traum. Ceh., 31*:304–312 (*194, 249*).

Champion, L. R. and Schnabel, M. L. (1966), *Mich. State Univ. Agr. Expt. Sta. Quart. Bull., 48*:384–396 (*253*).

Chang, C. Y., Witschi, E., and Ponseti, I. V. (1955), *Proc. Soc. Exptl. Biol. Med., 90*:45–50 (*224, 252*).

Chapman, A. O. (1954), *J. Morphol., 95*:451–470 (*137, 138*).

Chen, J. M. (1954), *J. Physiol., 125*:148–162 (*294*).

Chernukh, A. M. and Aleksandrov, P. N. (1965), *Patol., Fiziol. Eksperim. Terapiya, 9:*41–44 (*196*).

Chidester, F. E. (1927), *Am. Naturalist, 61:*272–279 (*391, 392*).

Chung, S. W. (1965), *New Med. J.* (*Seoul, Korea*), *8:*59–77 (*137, 138*).

Chute, H. L. and Cole, C. R. (1954), *Am. J. Vet. Res., 15:*108–118 (*347, 348, 358*).

Clapp, N. K. (1964), *Radiation Res., 22:*457–462 (*3, 116, 128*).

Clark, T. B., Runnels, T. D., and Livesay, E. A. (1944), *U.S. Egg & Poultry Mag., 50:*418–420, 430–431 (*38*).

Clarkson, B., Katz, A., Dann, M. A., and Karnofsky, D. A. (1964), *J. Natl. Cancer Inst., 32:*471–495 (*354, 355*).

Clearwaters, K. P. (1954), *J. Comp. Neurol., 101:*317–330 (*12*).

Clegg, D. J. (1964), *Food Cosmet. Toxicol., 2:*717–729 (*214, 220*).

Colasanti, G. (1875), *Atti. Reale Accad. Lincei, 2:*543–544 (*89*).

Cole, R. K. (1942), *J. Heredity, 33:*82–86 (*46, 53*).

Cole, R. K. (1957), *J. Heredity, 48:*173–175 (*54*).

Cole, R. K. (1967), *J. Heredity, 58:*141–146 (*40, 43, 50*).

Collado, J. J. (1965), *Anales Desar., 13:*196–206 (*337*).

Conner, M. H. and Shaffner, C. S. (1953), *J. Heredity, 44:*223–224 (*54*).

Conti, G. and Aita, M. (1966), *Arch. Biol., 77:*613–644 (*233, 234, 261*).

Conti, G. and Milio, G. (1964a), *Experientia, 20:*110–117 (*216*).

Conti, G. and Milio, G. (1964b), *Experientia, 20:*282–284 (*214*).

Conti, G. and Milio, G. (1964c), *Experientia, 20:*331–335 (*215*).

Conti, G. and Milio, G. (1965), *J. Embryol. Exptl. Morphol., 14*(Pt. 1):37–51 (*259*).

Corliss, C. E., Fedinec, A. A., and Robertson, G. G. (1966), *Anat. Record, 154:*221–223 (*350, 359*).

Couch, J. R., Cravens, W. W., Elvehjem, C. A., and Halpin, J. G. (1948a), *J. Nutr., 35:*57–72 (*153, 183*).

Couch, J. R., Cravens, W. W., Elvehjem, C. A., and Halpin, J. G. (1948b), *Anat. Record, 100:*29–48 (*42, 155, 183*).

Couch, J. R., Cravens, W. W., Elvehjem, C. A., and Halpin, J. G. (1949), *Arch. Biochem., 21:*77–86 (*153, 154, 183*).

Cravens, W. W., McGibbon, W. H., and Sebesta, E. E. (1944), *Anat. Record, 90:*55–64 (*42, 154, 183*).

Cravens, W. W., Sebesta, E. E., Halpin, J. G., and Hart, E. B. (1942), *Proc. Soc. Exptl. Biol. Med., 51:*106–108 (*157*).

Cravens, W. W. and Snell, E. E. (1949), *Proc. Soc. Exptl. Biol. Med., 71:*73–76 (*160, 200*).

Crawley, J. F. (1948), *J. Immunol., 59:*83–94 (*331*).

Creger, C. R. and Colvin, L. B. (1967), *Radiation Res., 32:*131–137 (*140*).

Cros, S., Tollon, Y., and Oreglia, J. (1965), *Compt. Rend. Soc., Biol., 159:*2055–2060 (*279*).

Cruz, M. V., Campillo-Sainz, C., and Munoz-Armas, S. (1966), *Circulation Res., 18:*257–262 (*70, 85*).

Cruz, M. V., Campillo-Sainz, C., Munoz-Armas, S., Pascual-Chavez, A., and Zamudio-Bandera, R. (1963), *Circulation Res.*, *13*:572–579 (*214, 337, 357*).

Cruz, S. R., and Romanoff, A. L. (1944), *Physiol. Zool.*, *17*:184–187 (*90, 91, 95, 96*).

Cunningham, B. (1936), *J. Elisha Mitchell Sci. Soc.*, *52*:269–273 (*105, 106*).

Cunningham, B. (1938), *Science*, *87*:90–91 (*105*).

Cunningham, B., and Stanfield, W. (1928), *J. Elisha Mitchell Sci. Soc.*, *44*:63–65 (*320*).

Curtis, M. R. (1915), *J. Agr. Res.*, *3*:375–385 (*383, 384*).

Cury, M. and Montenegro, M. A. (1969), *Rev. Med. Chile*, *97*:233–237 (*198, 258*).

Cutler, I. E. (1925), *J. Heredity*, *16*:352–356 (*50*).

Czaja, M. (1939), *Proc. World Poultry Congr.* (*Cleveland*), *7*:55 (*39*).

Dagg, C. P. and Karnofsky, D. A. (1955), *J. Exptl. Zool.*, *130*:555–572 (*226, 227, 255*).

Dagg, C. P., Karnofsky, D. A., Stock, C. C., Lacon, C. R., and Roddy, J. (1955), *Proc. Soc. Exptl. Biol. Med.*, *90*:489–494 (*42, 276*).

Dahlgren, K. V. O. (1941), *Fauna och Flora*, *5*:203–215 (*384*).

Dalton, A. J., Sergeyeva, M. A., and Selye, H. (1940), *Can. Med. Assoc. J.*, *42*:188–190 (*317*).

Danchakoff, V. (1933), *Arch. Entwicklungsmech. Organ.*, *127*:542–568 (*119*).

Danforth, C. H. (1932), *Proc. Soc. Exptl. Biol. Med.*, *30*:143–145 (*42*).

Daniel, J. C. (1959), *J. Heredity*, *50*:176 (*387*).

Dareste, C. (1891), *Recherches sur la Production Artificielle des Monstrosités*. Paris, 590 pp. + 16 pl. (*3, 4, 76, 364, 384, 389, 390, 409*).

Davis, H. J., Norris, L. C., and Heuser, G. F. (1938), *Poultry Sci.*, *17*:87–93 (*48, 161, 162*).

Dawson, E., Ferguson, T. M., Deyoe, C. W., and Couch, J. R. (1962), *Poultry Sci.*, *41*:1639 (Abstr.) (*184*).

Delaplane, J. P. and Stuart, H. O. (1941), *Rhode Island Univ. Agr. Expt. Sta. Bull.*, *284*:1–20 (*330*).

Delphia, J. M. and Eveleth, D. F. (1961), *J. Avian Diseases*, *5*:177–182 (*78*).

Dessi, P., Rizzoli, C., and Cestari, A. (1959), *Ric. Sci.*, *29*:1446–1451 (*260*).

Deuchar, E. M. (1952), *J. Anat.*, *86*:443–458 (*78*).

DeWitt, J. B. (1956), *J. Agr. Food Chem.*, *4*:863–866 (*212*).

Diwan, B. A. (1967), *Current Sci.* (*Univ. Poona, India*), *36*:312–313 (*251*).

Diwan, B. A. and Batra, B. K. (1969), *Acta Haematol.*, *41*:55–61 (*110*).

Dmitrenko, E. V. (1954), *Byul. Eksperim. Biol. Med.*, *37*:59–61 (*2*).

Doina, I., Onea, O., and Marcu, T. (1969), *Romania Inst. Biol. Agr., Ser. Med. Vet. Zool.*, *23/24*:27–39 (*40, 132*).

Dollander, A., Marconnet, R., and Autelin, R. (1962), *Compt. Rend. Soc. Biol.*, *156*:890–892 (*279*).

Donner, L., Chyle, P., and Sainerova, H. (1969), *J. Heredity*, *60*:113–116 (*40*).

Dougherty, J. E. (1926), *Am. J. Physiol.*, *79*:39–43 (*89*).

Drachman, D. B. (1964), *Science*, *145*:719–721 (*344, 345*).

Duerden, J. E. (1910), *Agr. J. South Africa*, *37*:713 (*387*).

Dugnat, J. M. (1965), *Compt. Rend. Soc. Biol.*, *159*:1545–1547 (*100, 101*).

Dunachie, J. F. and Fletcher, W. W. (1969), *Ann. Appl. Biol.*, *64*:409–423 (*212*).

Duncker, H. (1924), *Z. Ind. Abstr. Vererb.*, *32*:363–376 (*53*).

Dunn, L. C. (1922), *Conn (Storrs) Agr. Expt. Sta. Bull.*, *109*:89–114 (*404*).

Dunn, L. C. (1923a), *Am. Naturalist*, *57*:345–349 (*53*).

Dunn, L. C. (1923b), *Poultry Sci.*, *2/6*:199–204 (*30*).

Dunn, L. C. (1925), *J. Heredity*, *16*:127–134 (*41*).

Dunn, L. C. (1927a), *Arch. Entwicklungsmech. Organ.*, *110*:341–365 (*179*).

Dunn, L. C. (1927b), *Proc. World Poultry Congr. (Berlin)*, *5*:609–617 (*38*).

Dunn, L. C. and Landauer, W. (1934), *J. Genetics*, *29*:217–243 (*54*).

Duraiswami, P. K. (1952), *J. Bone Joint Surg.*, *34-B*:646–698 (*296, 297, 318*).

Duraiswami, P. K. (1954), *Bull. Johns Hopkins Hosp.*, *95*:57–67 (*256*).

Duraiswami, P. K. (1955), *J. Bone Joint Surg.*, *37-A*:277–294 (*245, 249, 265, 268, 315, 318*).

Duraiswami, P. K. (1961), *Indian J. Med. Res.*, *49*:631–634 (*33, 263*).

Elford, F. C. (1921), *Can. Dept. Agr. Rept.*, 3–29 (*89*).

Elias, S. (1963), *Studii Cercetari Stiinte Med.*, *10*:249–271 (*210*).

El-Ibiary, H. M., Shaffner, C. S., and Godfrey, E. F. (1966), *Poultry Sci.*, *45*:419–420 (*364*).

Engel, R. W., Phillips, P. H., and Halpin, J. G. (1940), *Poultry Sci.*, *19*:135–142 (*184*).

Essenberg, J. M. (1932), *Anat. Record*, *52*:12 (Abstr.) (*119*).

Essenberg, J. M. (1935), *Radiology*, *25*:739–744 (*112, 114, 120*).

Essenberg, J. M. and Zikmund, A. (1938), *Radiology*, *31*:94–103 (*109, 129*).

Evans, H. J. (1953), *Proc. Soc. Exptl. Biol. Med.*, *83*:31–34 (*286, 293, 294*).

Everdingen, Van C. and Noyons, W. A. G. (1940), *Ned. Tijdschr. Geneesk.*, *84*:4370–4380 (*149*).

Fabre, R. and Girault, M. (1957), *Compt. Rend. Acad. Sci.*, *244*:535–538 (*244*).

Fargeix, N. (1969), *Arch. Anat. Microscop. Morphol. Exptl.*, *58*:355–373 (*10*).

Fargeix, N. (1970), *Ann. Embryol. Morphogen.*, *3*:107–131 (*11*).

Faucounau, N., Stoll, R., and Maraud, R. (1963a), *Compt. Rend. Soc. Biol.*, *157*:780–782 (*275*).

Faucounau, N., Stoll, R., and Maraud, R. (1963b), *Compt. Rend. Soc., Biol.*, *157*:1564–1566 (*277*).

Fechheimer, N. S. and Jaffe, W. P. (1966), *J. Reprod. Fertility*, *12*:363–364 (*386*).

Feldman, G. L., Ferguson, T. M., and Couch, J. R. (1959), *J. Exptl. Zool., 140:*191–206 (*260*).

Feldman, G. L., Ferguson, T. M., Rigdon, R. H., Reid, B. L., Cross, M. S., and Couch, J. R. (1958), *Proc. Soc. Exptl. Biol. Med., 98:*646–648 (*259*).

Fell, H. B. (1960), *Proc. Nutr. Soc., 19:*50–54 (*164*).

Fell, H. B. and Mellanby, E. (1952), *J. Physiol. (London), 116:*320–349 (*165, 185*).

Fell, H. B. and Mellanby, E. (1955), *J. Physiol., 127:*427–447 (*397, 398*).

Féré, C. (1894), *Compt. Rend. Soc. Biol., 46:*646 (*3, 220*).

Féré, C. (1899), *Compt. Rend. Soc. Biol., 51:*921–922 (*3, 383*).

Ferguson, T. M., Atkinson, R. L., and Couch, J. R. (1954), *Proc. Soc. Exptl. Biol. Med., 86:*868–871 (*168, 186*).

Ferguson, T. M. and Couch, J. R. (1954), *J. Nutr., 54:*361–370 (*185*).

Ferguson, T. M., Rigdon, R. H., and Couch, J. R. (1955), *A.M.A. Arch. Pathol., 60:*393–400 (*165, 167, 186*).

Ferguson, T. M., Rigdon, R. H., and Couch, J. R. (1956), *Arch. Ophthalmol., 55:*346–355 (*187*).

Ferguson, T. M., Rigdon, R. H., and Couch, J. R. (1957), *Endocrinology, 60:*13–21 (*165*).

Ferguson, T. M., Swanson, A. A., Couch, J. R., Feldman, G. L., and Rigdon, R. H. (1960), *Am. J. Ophthalmol., 49*(Pt. 2):1165–1167 (*187*).

Ferguson, T. M., Whiteside, C. H., Creger, C. R., Jones, M. L., Atkinson, R. L., and Couch, J. R. (1961), *Poultry Sci., 40:*1151–1159 (*169*).

Finkelstein, R. A. and Ramm, G. M. (1962), *J. Infect. Diseases, 111:*239–249 (*4, 345, 346, 349*).

Fischer, J. L. (1969), *Ann. Embryol. Morphogen., 2:*227–244 (*103*).

Fiszer-Szafarz, B. (1967), *Cancer Res., 27A:*191–197 (*357*).

Fleming, A. and Allison, V. D. (1924), *Lancet, 206*(I):1303–1306 (*2*).

Flemister, L. J. and Cunningham, B. (1940), *Growth, 4:*63–71 (*106*).

Forsheim, A. (1908), *Anat. Hefte, (Abt. 1), 37:*117–142 (*387*).

Fortuyn, van, C. E. D. (1927), *Anat. Record, 34:*233–236 (*383, 389*).

Fowler, I. (1953), *J. Exptl. Zool., 123:*115–151 (*373, 374*).

Fox, J. P. and Laemmert, H. W. (1947), *Am. J. Hyg., 46:*20–40 (*342, 343, 344*).

Franciscis, P., de, and Boccalatte, F. (1962), *Nature, 193:*989–990 (*244*).

Franciscis, P., de, Finelli, R., Gargiulo, F., and Renda, G. (1960), *Boll. Soc. Ital. Biol. Sper., 36:*1454–1455 (*314*).

Franciscis, P., de, Gargiulo, F., Manzo, E., and Renda, G. (1960), *Boll. Soc. Ital. Biol. Sper., 36:*1451–1453 (*317*).

Franciscis, P., de, and Landauer, W. (1959), *Nature, 184:*101–103 (*304, 305*).

Franke, K. W., Moxon, A. L., Poley, W. E., and Tully, W. C. (1936), *Anat. Record, 65:*15–22 (*175, 244, 247, 248*).

Fraps, R. M. (1945), *Science, 101:*339 (*105*).

Frost, D. (1947), *Biol. Review (N.Y.), 9:*34–37 (*390*).

Fugo, N. W. (1940), *J. Exptl. Zool., 85:*271–298 (*376, 377*).

Funk, F. M. (1950), *Poultry Sci., 29:*64–66 (*404, 408*).

Funk, F. M. and Biellier, H. V. (1944), *Poultry Sci., 23:*538–540 (*58*).

Gallera, J. (1934), *Bull. Intern. Acad. Polonaise, Ser. B., Sci. Nat., 1934:*73–83 (*371*).

Gallera, J. (1951), *Acta Anat. (Basel), 11:*549–585 (*92, 97*).

Garcia, A. A. (1965), *Anales Desar., 13:*75–95 (*276*).

Gardiner, E. E., Rogler, J. C., and Parker, H. E. (1960), *Poultry Sci., 39:*1111–1115 (*173*).

Gardner, E. W., Lyles, S. T., and Lankford, C. E. (1964), *J. Infect. Diseases, 114:*412–416 (*349*).

Gebhardt, D. O. E. (1968), *Teratology, 1:*153–162 (*233, 234, 278*).

Genazzani, E., Pelagalli, G., Pagnini, G., and Di Carlo, R. (1965), *Atti. Sic. Ital. Sci. Vet., 19:*207–211 (*199, 264, 268, 275*).

Gentry, R. F. (1958), *Avian Diseases, 2:*76–82 (*196, 198, 269*).

Gerlach, J. (1882), *Die Entstehungsweise der Doppelmissbildungen bei den höheren Wirbelthieren.* F. Enke Publishing Co., 233 pp. (*390*).

Gerlinger, P., Ruch, J. V., and Clavert, J. (1962), *Compt. Rend. Acad. Sci., 255:*3229–3231 (*257*).

Gessner, I. H. (1966), *Circulation Res., 18:*625–633 (*373*).

Giacomini, C. (1894), *Arch. Ital. Biol., 22:*169–183 (*3*).

Gillis, M. B., Heuser, G. F., and Norris, L. C. (1948), *J. Nutr., 35:*351–363 (*157*).

Gilman, P. K. and Baetjer, F. H. (1904), *Am. J. Physiol., 10:*222–224 (*111, 119*).

Godfrey, A. B. (1936), *Poultry Sci., 15:*294–297 (*404*).

Goetinck, P. F., Abbott, U. K., and Kratzer, F. H. (1957), *Poultry Sci., 36:*455–457 (*201, 268*).

Goff, O. E. and Tumlin, J. T. (1950), *Assoc. Southern Agr. Workers, 47th Am. Conv., Biloxi, Miss. Feb.,* p. 151 (*39*).

Goff, R. A. (1959), *J. Exptl. Zool., 141:*477–497 (*33, 114, 115, 116*).

Goff, R. A. (1962), *J. Exptl. Zool., 151:*177–200 (*123, 124*).

Goldring, I. P. (1956), *Radiation Res., 5:*390–403 (*128*).

Grabowski, C. T. (1957), *Am. J. Anat., 101:*101–133 (*11, 12*).

Grabowski, C. T. (1961), *Am. J. Anat., 109:*25–36 (*90, 94, 95*).

Grabowski, C. T. (1964), *J. Exptl. Zool., 157:*307–325 (*94*).

Grabowski, C. T. (1966), *J. Embryol. Exptl. Morphol., 15:*113–118 (*243, 244, 247*).

Grabowski, C. T. and Paar, J. A. (1958), *Am. J. Anat., 103:*313–347 (*33, 90, 91, 92, 93, 94, 99*).

Grafnetterova, J., Grosi, E., Fumagalli, R., Morganti, P., and Grafnetter, D. (1966), *Neoplasma, 13:*251–258 (*255*).

Grau, C. R., Austic, R. E., and Matteson, G. C. (1965), *Science, 148:*1743–1745 (*180*).

Gray, P. and Worthing, H. (1941), *J. Exptl. Zool., 86:*423–438 (*350*).

Graziani, G. and Biagi, G. L. (1966), *Boll. Soc. Ital. Biol. Sper.*, *42*:838–840 (*283, 314*).

Greenberg, J. and LeHam, Q. N. (1969), *Can. J. Zool.*, *47*:539–542 (*269*).

Greiff, D. (1952), *Ann. N.Y. Acad. Sci.*, *55*(Art. 2):254–266 (*65*).

Grobstein, C. (1955), *J. Exptl. Zool.*, *130*:319–340 (*11*).

Grodzinski, Z. (1931), *Compt. Rend. Ass. Anat.* (*Warsaw*), 3–7 (*59*).

Grodzinski, Z. (1934), *Arch Entwicklungsmech. Organ.*, *131*:653–671 (*89*).

Groenendijk-Huijbers, M. M. (1965), *Anat. Record*, *153*:93–105 (*372, 377*).

Grossowicz, N. (1946), *Proc. Soc. Exptl. Biol. Med.*, *63*:151–152 (*307, 324*).

Gruenwald, P. (1944), *Anat. Record*, *88*:67–81 (*52*).

Gruenwald, P. (1958), *Amer. J. Pathol.*, *34*:77–103 (*246*).

Grundmann, E. (1900), *Anat. Hefte*, *14*:257–300 (*383, 387*).

Gruninger, R. P. and Spink, W. W. (1967), *Proc. Soc. Exptl. Biol. Med.*, *124*:1277–1279 (*204*).

Grüter, M. (1916), *Arch. Exptl. Pathol. Pharmakol.*, *79*:337–360 (*226*).

Gutierrez, L. S. (1965), *Anales Desar.*, *13*:13–32 (*12*).

Guyer, M. F. (1909), *Biol. Bull.* *16*:193–198 (see Romanoff, 1960, pp. 109–111) (*38*).

Guzman-Garcia, J. and Laguna, J. (1965), *Ciencia* (*Mexico*), *24*:50–60 (*322*).

Haas, V. H., Feldman, H. A., and Ewing, F. M. (1945), *Public Health Rept.*, *60*:577–582 (*351*).

Haas, V. H., Wilcox, A., and Ewing, F. M. (1945), *J. Natl. Malaria Soc.*, *4*:279–284 (*351*).

Hadjinsky, P. (1962), *Bull. Inst. Morphol.* (*Sofia*), *6*:55–66 (*120*).

Halick, J. V., Reid, B. L., Brown, C. L., and Couch, J. R. (1953), *J. Nutr.*, *50*:331–340 (*165*).

Halverson, A. W., Jerde, L. G., and Hills, C. L. (1965), *Toxicol. Appl. Pharmacol.*, *7*:675–679 (*246*).

Hamburger, V. (1938), *J. Exptl. Zool.*, *77*:379–398 (*9*).

Hamburger, V. and Habel, K. (1947), *Proc. Soc. Exptl. Biol. Med.*, *66*:608–617 (*335, 336, 339, 357*).

Hammett, F. S. and Wallace, V. L. (1928), *J. Exptl. Med.*, *48*:659–665 (*245*).

Hanan, E. B. (1928), *Proc. Soc. Exptl. Biol. Med.*, *25*:422–425 (*301*).

Hansborough, L. A. (1947a), *Growth*, *11*:77–88 (*185, 186*).

Hansborough, L. A. (1947b), *Growth*, *11*:177–184 (*184*).

Harrison, J. R. (1957), *Physiol. Zool.*, *30*:187–197 (*80*).

Harrison, J. R. and Klein, I. (1954), *Biol. Bull.*, *106*:48–59 (*79, 80*).

Hays, F. A. (1944), *Am. Naturalist*, *78*:54–58 (*52*).

Heath, H. D., Shear, H. H., Imagawa, D. T., Jones, M. H., and Adams, J. M. (1956), *Proc. Soc. Exptl. Biol. Med.*, *92*:675–682 (*331, 333, 334, 335, 336, 337, 338*).

Hébert, J. and Laugier, H. (1943), *Rev. Can. Biol.*, *2*:113–119 (*404*).

Herrmann, H., Clark, E. M., and Landauer, W. (1963), *Acta Embryol. Morphol. Exptl.*, *6*:169–174 (*51*).

Hess, C. W., Juhn, M., and Jull, M. A. (1947), *Poultry Sci., 26:*648–650 (*384, 388*).

Hill, W. G., Lloyd, G. L., and Abplanalp, H. (1963), *J. Heredity, 54:*188–190 (*51*).

Hinchliffe, J. R. and Ede, D. A. (1968), *J. Embryol. Exptl. Morphol., 19:*327–339 (*53*).

Hinrichs, M. A. (1927), *J. Exptl. Zool., 47:*309–342 (*145, 146, 149*).

Hirayama, T. (1960), *Kyosai Iho, 9:*95–106 (*315*).

Hoffstadt, R. E. and Tripi, H. B. (1946), *J. Bacteriol. 52:*617–627 (*342*).

Hollander, W. F. (1945), *J. Heredity, 36:*297–300 (*40, 47, 50*).

Hollander, W. F. (1948), *J. Heredity, 39:*289–292 (*52*).

Hollander, W. F. and Levi, W. M. (1940), *Auk, 57:*326–329 (*386, 387, 388*).

Hollander, W. F. and Levi, W. M. (1942), *J. Heredity, 33:*385–391 (*43, 46, 53*).

Homma, K. and Jinno, M. (1968), *Poultry Sci.,* 130–133 (*387, 388*).

Honjo, S. and Kuth, T. (1959), *Japan, J. Vet. Sci., 21:*9–18 (*303, 310, 321, 324*).

Horowitz, A. (1903), *Baumgarten's Jahresber., 19:*984–985 (*2*).

Hsu, H. F. (1930), *Biol. Lab. Ser. Zool.* (*Sci. Soc. China*), *6:*75–93 (*389*).

Humphrey, R. M. and Sinclair, W. K. (1962), *Radiation Res., 16:*384–393 (*143*).

Hunt, E. A. (1932), *J. Exptl. Zool., 62:*57–92 (*9*).

Hunt, E. L. (1951), *Science, 113:*386–387 (*273*).

Hutt, F. B. (1929), *Proc. Roy. Soc. Edinburgh, 49:*118–130 (*28, 29*).

Hutt, F. B. (1930), *Proc. World Poultry Congr.* (*London*), *4:*195–202 (*33*).

Hutt, F. B. and Child, G. P. (1935), *J. Heredity, 25:*341–350 (*53*).

Hutt, F. B. and Greenwood, A. W. (1929), *Proc. Roy. Soc. Edinburgh, 49:*145–154 (*30, 33*).

Hutt, F. B. and Mueller, C. D. (1942), *J. Heredity, 33:*69–77 (*50*).

Hutt, F. B. and Sturkie, P. D. (1938), *J. Heredity, 29:*370–379 (*54*).

Hwang, J. (1965), *Avian Diseases, 9:*417–422 (*332*).

Hyre, H. M., Patrick, H., and Nestor, K. (1964), *Poultry Sci., 43:*121–124 (*131*).

Ikota, H. and Kondo, K. (1953), *Bull. Nat. Inst. Agr. Sci.* (*Chiba*), *Ser. G., No. 6:*1–7, in Japanese (*33*).

Insko, W. M., MacLaury, D. W., and Baute, E. A. (1943), *Kentucky Agr. Expt. Sta. Bull., 449:*1–16 (*404*).

Insko, W. M., Steele, D. G., and Hinton, C. M. (1941), *Kentucky Agr. Expt. Sta. Bull., 416:*117–138 (*100, 102*).

Insko, W. M., Steele, D. G., and Wightman, E. T. (1947), *Kentucky Agr. Expt. Sta. Bull., 498:*1–23 (*407*).

Jaap, G. R. (1937), *Proc. Oklahoma Acad. Sci., 17:*41–43 (*54*).

Jackson, D. (1965), *Poultry Sci., 44:*887–888 (*384*).

Jeffrey, F. P. (1941), *J. Heredity, 32:*310–312 (*52*).

Jeffrey, F. P., Fox, T. W., and Smyth, J. R. (1953), *J. Heredity, 44*:213–216 (*384, 385, 386*).

Johansen, I. (1961), *Intern. J. Radiation Biol., 3*:411–418 (*117*).

Jones, H. S. (1970), *Am. J. Anat., 127*:89–99 (*320*).

Jull, M. A. (1929), *Poultry Sci., 8*:361–368 (*38*).

Jull, M. A. and Haynes, S. (1925), *J. Agr. Res., 31*:685–694 (*404*).

Jurand, A. (1966), *J. Embryol. Exptl. Morphol., 16*:289–300 (*242*).

Kaestner, S. (1898), *Arch. Anat. Physiol.* (*Anat. Abt.*):81–94 (*390*).

Kaestner, S. (1899), *Arch. Anat. Physiol.* (*Anat. Abt.*):28–32 (*383*).

Källén, B. (1955), *Anat. Record, 123*:169–182 (*12*).

Kallman, R. F., Steele, R. E., Weissbluth, M., and Bagshaw, M. A. (1958), *Radiation Res., 9*:270–277 (*143*).

Kaplan, S. and Grabowski, C. T. (1967), *J. Exptl. Zool., 165*:325–336 (*41, 278*).

Kaplan, S. and Johnson, E. M. (1970), *Teratology, 3*:269–272 (*203, 278*).

Karb, B. (1969), *Tierphysiol. Tiererna. Futtermettel., 25*:43–62 (*186*).

Karnofsky, D. A. (1955), *Cancer Res.* (*Suppl. 3*):83–85 (*255, 260, 262, 265*).

Karnofsky, D. A. and Lacon, C. R. (1964), *Clin. Orthop., 33*:59–70 (*194, 249, 255, 259, 261, 267, 270, 271*).

Karnofsky, D. A. and Lacon, C. R. (1966), *Biochem. Pharmacol., 15*:1435–1442 (*231, 232, 258*).

Karnofsky, D. A., Patterson, P. A., and Ridgway, L. P. (1949), *Proc. Soc. Exptl. Biol. Med., 71*:447–452 (*155, 183*).

Karnofsky, D. A., Patterson, P. A., and Ridgway, L. P. (1950), *Am. J. Roentgenol. Radium Therapy, 64*:280–288 (*112, 114, 118*).

Karnofsky, D. A. and Ridgway, L. P. (1952), *J. Pharmacol. Exptl. Therap., 104*:176–186 (*245*).

Karnofsky, D. A., Ridgway, L. P., and Patterson, P. A. (1950), *Proc. Soc. Exptl. Biol. Med., 73*:255–259 (*42, 116, 194, 195, 196, 249*).

Karnofsky, D. A., Ridgway, L. P., and Patterson, P. A. (1951), *Endocrinology, 48*:596–616 (*284, 285, 286, 288, 289, 314*).

Karnofsky, D. A., Ridgway, L. P., and Patterson, P. A. (1952), *Ann. N.Y. Acad. Sci., 55*(Art. 2):313–329 (*353*).

Karnofsky, D. A., Ross, C., and Leavitt, C. G. (1954), *Fed. Proc. Am. Soc. Exptl. Biol., 13*:1373 (*42*).

Kaufman, L. (1938), *Arch. Geflugelkunde, 12*:65–77 (*398*).

Kaufman, L. (1939), *Proc. World Poultry Congr., 7*:198–200 (*397*).

Kawahara, T. (1956), *Natl. Inst. Genetics* (*Japan*), *Ann. Rept. 6*, pp. 15–16 (*52*).

Kawamura, J. (1967), *Showa Igakukai Zasshi, 27*:418–428 (*268*).

Khera, K. S. (1966), *Toxicol. Appl. Pharmacol., 8*:345 (Abstr.) (*42, 212*).

Khera, K. S. and Bedok, S. (1967), *Food Cosmet. Toxicol., 5*:359–365 (*42, 213, 259, 269*).

Khera, K. S. and Lyon, D. A. (1968), *Toxicol. Appl. Pharmacol., 13:*1–15 (*212*).

Kieler, J. (1954), *Acta Pathol. Microb. Scand., 34:*11–18 (*179*).

Kienholz, E. W., Turk, D. E., Sunde, M. L. and Hoekstra, W. G. (1961), *J. Nutr., 75:*211–221 (*170, 176, 177, 188*).

Kieny, M. (1958), *Arch. Anat. Microscop. Morphol. Exptl., 47:*85–169 (*181*).

Kieny, M. (1960), *J. Embryol. Exptl. Morphol., 4:*457–467 (*11*).

Kirrmann, J. M. (1961), *Arch. Anat. Histol. Embryol., 44:*117–124 (*265*).

Kirrmann, J. M. (1962), *Bull. Biol. France Belg., 96:*365–463 (*126*).

Kirrmann, J. M. and Wolff, Et. (1964), *Intern. Rev. Exptl. Pathol., 3:*365–408 (*123, 125*).

Kitamura, Z. (1957), *Nippon Yakurigaku Zasshi, 53:*449–479 (*272*).

Kiwaki, T. (1956), *Nippon Yakurigaku Zasshi, 52:*186–213 (*268*).

Klein, N. W. (1965), *Technology Principles and Techniques,* Chicago Press, pp. 131–141 (*372*).

Klika, E. and Jelinek, R. (1961a), *Cesk. Morphol., 9:*162–172 (*373*).

Klika, E. and Jelinek, R. (1961b), *Cesk. Morphol., 9:*274–281 (*373*).

Knezevic, Z. and Karanovic, J. (1966), *Bull. Inst. Nucl. Sci., 17:*219–226 (*133*).

Knowlton, F. L. (1929), *Oregon Agr. Exp. Sta. Bull., 253:*1–15 (*54*).

Knox, C. W. and Olsen, M. W. (1936), *Poultry Sci., 15:*345–348 (*366*).

Kocher, W. (1963), *Z. Naturforsch., 18:*1128–1130 (*277*).

Kocher, W. (1965), *Z. Naturforsch., 20:*1136–1137 (*277*).

Kocher, W. (1969), *Arch. Entwicklungsmech. Organ., 162:*161–196 (*277*).

Kocher, W. (1970), *Arch. Entwicklungsmech. Organ., 165:*53–86 (*277*).

Kohn, H. I. and Kallman, R. F. (1956), *Radiation Res., 5:*710–714 (*143*).

Komai, T. and Nakamura, K. (1938), *Mem. Coll. Sci. Univ. Kyoto, Ser. B, 14:*193–242 (*382, 384, 387, 388*).

Kondo, K. and Ikoto, H. (1953), *Bull. Nat. Inst. Agr. Sci. (Chiba), Ser. G, No. 6:*1–7, in Japanese (*387*).

Korotkova, G. P. (1955a), *Bull. Exptl. Biol. Med., 8:*60–61 (*2*).

Korotkova, G. P. (1955b), *Bull. Exptl. Biol. Med., 10:*60–63 (*2*).

Korotkova, G. P. (1957), *Zh. Obshch. Biol., 18*(4):275–287 (*2*).

Kosin, I. L. (1944), *Physiol. Zool., 17:*289–319 (*130*).

Kosin, I. L. and Mun, A. M. (1965), *Poultry Sci., 44:*31–39 (*399*).

Kosin, I. L. and Sato, I. (1960), *J. Morphol., 103:*263–276 (*392*).

Krakoff, I. H. and Meyer, R. L. (1965), *J. Pharmacol. Exptl. Therap., 149:*417–422 (*251*).

Krayevoy, I. (1938), *Akad. Nauk. Ukr. SSR, Mem. Genetics, No. 2:*109–160 (*112*).

Krizenecky, J. (1926), *Zool. Anz., 67:*129–140 (*385, 387*).

Kumanov, S. (1948), *Proc. World Poultry Congr. (Copenhagen), 8*(I):320–326 (*89*).

Kury, G. and Craig, J. M. (1966), *Arch. Pathol., 81:*166–173 (*261*).

Kury, G. and Craig, J. M. (1967), *J. Embryol. Exptl. Morphol., 17:*229–237 (*266*).

Kury, G. and Crosby, R. J. (1967), *Toxicol. Appl. Pharmacol., 11*:72–80 (*261*).

Kushner, H. F. and Panfilova, E. V. (1955), *Zh. Obshch. Biol., 16*:248 (*388*).

Lakshmi, M. S. (1962), *J. Embryol. Exptl. Morphol., 10*:373–382 (*257*).

Lakshmi, M. S. (1967a), *Arch. Entwicklungsmech. Organ., 158*:103–111 (*215, 318, 321*).

Lakshmi, M. S. (1967b), *Arch. Entwicklungsmech. Organ., 158*:112–123 (*265, 289, 292, 315, 317*).

Lamoreux, W. F. (1942), *J. Heredity, 33*:275–283 (*40, 47, 50*).

Lamson, G. H. and Edmond, H. D. (1914), *Conn. (Storrs) Agr. Expt. Sta. Bull., 76*:218–258 (*97*).

Lamson, G. H. and Kirkpatrick, W. F. (1918), *Conn. (Storrs) Agr. Expt. Sta. Bull., 95*:311–351 (*102, 103*).

Landauer, W. (1927), *Arch. Entwicklungsmech. Organ., 110*:195–278 (*179*).

Landauer, W. (1928), *J. Heredity, 19*:453–467 (*41*).

Landauer, W. (1929), *Poultry Sci., 8*:301–312 (*192, 194, 214*).

Landauer, W. (1930), *Proc. World Poultry Congr. (London), 4*:128–133 (*43*).

Landauer, W. (1932a), *Conn. (Storrs) Agr. Expt. Sta. Bull., 179*:195–218 (*145*).

Landauer, W. (1932b), *J. Genet., 25*:367–394 (*43, 50*).

Landauer, W. (1933), *Z. Microscop. Anat. Forsch., 32*:359–412 (*40, 43, 50*).

Landauer, W. (1935), *J. Genet., 31*:237–242 (*50*).

Landauer, W. (1938), *J. Genet., 37*:51–68 (*40*).

Landauer, W. (1939), *Conn. (Storrs) Agr. Expt. Sta. Bull., 232*:1–62 (*43, 44, 46*).

Landauer, W. (1940), *Sigma Xi Quarterly, 28*(4):1–10 (*23*).

Landauer, W. (1941), *Genetics, 26*:426–439 (*50*).

Landauer, W. (1945), *J. Exptl. Zool., 98*:65–77 (*41, 49, 297*).

Landauer, W. (1946), *Am. Naturalist, 80*:490–494 (*41, 295*).

Landauer, W. (1947a), *J. Exptl. Zool., 105*:145–172 (*42, 294, 295, 298, 319*).

Landauer, W. (1947b), *J. Exptl. Zool., 105*:317–328 (*40, 41, 296, 297*).

Landauer, W. (1947c), *Endocrinology, 41*:489–493 (*283*).

Landauer, W. (1948), *Growth Symposium, 12*:171–200 (*48*).

Landauer, W. (1949), *Arch. Anat. Microscop. Morphol. Exptl., 38*:184–189 (*247, 270*).

Landauer, W. (1951), *J. Exptl. Zool., 117*:559–571 (*295, 297, 299, 300, 318*).

Landauer, W. (1952), *J. Exptl. Zool., 120*:469–508 (*191, 192, 243*).

Landauer, W. (1953a), *Growth, 17*:87–109 (*271, 319*).

Landauer, W. (1953b), *J. Exptl. Zool., 122*:469–483 (*239, 241, 271*).

Landauer, W. (1954), *J. Cellular Comp. Physiol., 43*:261–305 (*41, 192, 193, 238, 239, 240, 241, 271*).

Landauer, W. (1956a), *J. Heredity, 47*:57–63 (*40, 46, 51*).

Landauer, W. (1956b), *J. Genet., 54*:199–218 (*54*).

Landauer, W. (1956c), *J. Genet., 54*:219–235 (*43, 53*).

Landauer, W. (1956d), *J. Exptl. Zool., 132*:25–38 (*279*).

Landauer, W. (1956e), *J. Exptl. Zool., 132:*39–50 (*239, 271*).

Landauer, W. (1957), *J. Exptl. Zool., 136:*509–530 (*237, 266, 267, 270*).

Landauer, W. (1959), *J. Heredity, 50:*137–139 (*53*)

Landauer, W. (1960a), *J. Exptl. Zool., 143:*101–105 (*249*).

Landauer, W. (1960b), *J. Exptl. Zool., 143:*107–122 (*268*).

Landauer, W. (1965a), *J. Exptl. Zool., 160:*345–354 (*48, 267*).

Landauer, W. (1965b), *J. Heredity, 56:*131–138 (*46, 47, 53*).

Landauer, W. (1967), *Conn. (Storrs) Agr. Expt. Sta. Monograph, 1:*1–315 (*38, 39*).

Landauer, W. and Baumann, L. (1943), *J. Exptl. Zool., 93:*51–74 (*33, 365, 366*).

Landauer, W. and Bliss, C. I. (1946), *J. Exptl. Zool., 102:*1–22 (*41, 294, 296, 297, 298, 319*).

Landauer, W. and Clark, E. M. (1962a), *J. Exptl. Zool., 151:*245–251 (*295*).

Landauer, W. and Clark, E. M. (1962b), *J. Exptl. Zool., 151:*253–258 (*201, 223, 266, 267*).

Landauer, W. and Clark, E. M. (1964a), *Nature, 203:*527–528 (*397*).

Landauer, W. and Clark, E. M. (1964b), *Nature, 204:*285–286 (*295*).

Landauer, W. and Clark, E. M. (1964c), *J. Exptl. Zool., 156:*313–321 (*266, 267, 274*).

Landauer, W. and Dunn, L. C. (1930a), *J. Genet., 23:*397–412 (*40, 50*).

Landauer, W. and Dunn, L. C. (1930b), *J. Heredity, 21:*290–305 (*54*).

Landauer, W. and Rhodes, M. B. (1952), *J. Exptl. Zool., 119:*221–262 (*318*).

Landauer, W. and Wakasugi, N. (1967), *J. Exptl. Zool., 164:*499–515 (*250, 260*).

Lane, H. K., Darlington, J. M., and Coursin, D. B. (1954), *Proc. Penn. Acad. Sci., 28:*232–242 (*191, 192, 243*).

Lane, H. K. and Lavy, R. C. (1958), *Proc. Penn. Acad. Sci., 32:*261–264 (*243*).

Lane, H. K. and Slepecky, R. A. (1960), *Proc. Penn. Acad. Sci., 34:*179–190 (*229, 230, 256*).

Lasnitzki, I. (1943), *Brit. J. Radiol., 16:*137–141 (*124, 129*).

Lea, D. E. (1941), *Am. J. Roentgenol. Radium Therapy, 45:*605–613 (*114, 115, 142, 143, 144*).

Le Douarin, G. (1960), *J. Embryol. Exptl. Morphol., 8:*130–138 (*126*).

Le Douarin, G. (1963), *Bull. Biol. France Belg., 97:*645–759 (*111, 112, 113, 138*).

Le Douarin, G. (1968), *Ann. Fac. Sci. Univ. Clermont, Biol. Anim., 3*(39):5–31 (*112, 113*).

Le Douarin, N. (1964a), *Bull. Biol. France Belg., 98:*544–576 (*128*).

Le Douarin, N. (1964b), *J. Embryol. Exp. Morphol., 12* (Pt. I): 141–160 (*372*).

Leighton, J., Merkow, L., and Locker, M. (1964), *Nature, 201:*198–199 (*81, 82*).

Lenicque, P. (1959), *Acta Zool., 40:*141–202 (*8, 21, 203, 204*).

Lepkovsky, S., Taylor, L. W., Jukes, T. H., and Almquist, H. J. (1938), *Hilgardia, 11:*559–591 (*162, 184*).

Lepore, P. D., Miller, R. F., and Siegel, H. S. (1962), *Poultry Sci., 41:*1047–1049 (*193, 195*).

Lepori, N. G. (1967), *Monitore Zool. Ital. N. S., 1:*129–148 (*393, 394*).

Lerner, I. M. and Gunns, C. A. (1938), *Growth, 2:*261–266 (*82*).

Levene, C. I. and Gross, J. (1959), *J. Exptl. Med., 110:*771–791 (*224, 225, 253*).

Levi, W. M. (1957), *The Pigeon.* Levi Publishing Co., 667 pp. (*386*).

Lienhart, R. (1937), *Compt. Rend. Soc. Biol., 126:*336–342 (*50*).

Limborgh, J. van and Meenen, F. van (1965), *Ned. Tijdschr. Verlosk Gynaecol., 65:*96–107 (*272*).

Lombardini, L. (1868), *Forme organiche irregulari negli Uccelli e ne'Batrachei.* Pisa, 140 pp. + 2 pl. (*3*).

Loomis, L. N., Cunningham, C. H., Gray, M. L., and Thorp, F. (1950), *Am. J. Vet. Res., 11:*245–251 (*329, 356*).

Lucas, A. H. S. (1890), *Proc. Roy. Soc. Victoria, 2:*111–112 (*390*).

Ludwig, F. and Ries, J. (1944), *Monatsschr. Geburtsh, Gynak., 118:*291–298 (*149*).

Lustig, E. S. de (1954), *Rev. Soc. Arg. Biol., 30:*256–264 (*203, 356*).

Lustig, E. S. de and Fiszer, B. (1956), *Rev. Soc. Arg. Biol., 32:*120–135 (*356*).

Lutz, H. (1948), *Compt. Rend. Acad. Sci., 227:*87–89 (*390, 394*).

Lutz, H. (1949), *Arch. Anat. Microscop. Morphol. Exptl. 38:*79–144 (*8, 393, 394*).

Lutz, H. (1964), *Arch. Anat. Histol. Embryol., 47:*529–538 (*394*).

Lutz, H. (1965), *Bull. Soc. Zool. France, 40:*191–193 (*388*).

Lutz, H., Beaudenon, M. T., and Mesure, M. (1965), *Compt. Rend. Soc. Biol., 159:*646–648 (*8*).

Lutz, H., Bonhomme, C., and Lutz-Ostertag, Y. (1955), *Compt. Rend. Soc. Biol., 149:*1475–1477 (*148*).

Lutz, H., Departout, M., Hubert, J., and Pieau, C. (1963, *Develop. Biol., 6:*23–44 (*7*).

Lutz, H. and Lepy, M. (1958), *Bull. Soc. Zool. (France), 83:*76–81 (*98, 99*).

Lutz, H. and Lutz-Ostertag, Y. (1957), *Arch. Anat. Microscop. Morphol. Exptl. 46:*307–324 (*148*).

Lutz, H. and Lutz-Ostertag, Y. (1959), *Develop. Biol., 1:*364–376 (*386, 389*).

Lutz, H. and Lutz-Ostertag, Y. (1963), *Compt. Rend. Soc. Biol., 157:*1435–1437 (*390*).

Lutz, H. and Lutz-Ostertag, Y. (1964), *Compt. Rend. Soc. Biol., 158:*1504–1506 (*387*).

Lutz-Ostertag, Y., Meiniel, R., and Lutz, H. (1969), *Compt. Rend. Acad. Sci., 268:*2911–2913 (*269*).

Lutz-Ostertag, Y. and Senaud, J. (1966), *Arch. Anat. Microscop. Morphol. Exptl., 55:*363–386 (*33, 359*).

Lyles, S. T., Sharp, M. S., Nabors, R. E., Larsen, S. P., and Potter, P. S. (1965), *J. Infect. Diseases, 15:*400–412 (*349, 359*).

Lyons, M. and Insko, W. M. (1937a), *Kentucky Agr. Expt. Sta. Bull., 371:*63–75 (*42, 187*).

Lyons, M. and Insko, W. M. (1937b), *Science, 86:*328 (*187*).

Machlin, L. J., Pearson, P. B., and Denton, C. A. (1952), *Arch. Ind. Hyg. Occupational Med., 6:*441–444 (*244, 249*).

Mancini, E. (1908), *Intern. Congr. Froid. (Paris), 1*(2):753–754 (*89*).

Mangione, F. (1957), *Arch. Ital. Anat. Embryol., 62:*311–329 (*323*).

Manner, H. W. and Granik, B. H. (1967), *Anat. Record, 158:*371–374 (*398*).

Maraud, R., Coulard, H., and Stoll, R. (1963), *Compt. Rend. Soc. Biol., 157:*1566–1569 (*257*).

Marble, D. R., Harper, J. A., and Hammer, E. V. (1944) *Poultry Sci., 23:*114–117 (*50*).

Marble, D. R. and Morgolf, P. H. (1936), *Poultry Sci., 15:*225–229 (*404*).

Marcacci, A. (1893), *Arch. Ital. Biol., 11:*164–171 (*3*).

Marshall, W. (1951), *Proc. World Poultry Congr.* (*Paris*), *9:*80–87 (*34*).

Martin, J. H. and Insko, W. M. (1935), *Kentucky Agr. Expt. Sta. Bull., 359:*151–177 (*27*).

Martinovitch, P. N., Kanazir, D. T., Knezevitch, Z. A., and Simitch, M. M. (1962), *J. Embryol. Exptl. Morphol., 10:*167–177 (*215*).

Matsumoto, T. (1957), *Nippon Yakurigaku Zasshi, 53:*249–280 (*273, 274*).

Mauro, F. (1923), *Atti Soc. Ital. Sci. Nat., 62:*239–246 (*89*).

May, D. and Glick, B. (1964), *Poultry Sci., 43:*450–453 (*321*).

McCabe, W. R. (1964), *J. Clin. Invest., 43:*2146–2157 (*349*).

McCallion, D. J. and Clarke, R. B. (1959), *Can. J. Zool., 37:*387–390 (*369*).

McCartney, M. G. and Shaffner, C. S. (1949), *Poultry Sci., 28:*223–228 (*311*).

McClymont, G. L. and Hart, L. (1947), *Australian Vet. J., 23:*247–253 (*161, 185*).

McGhee, R. B. (1949a), *J. Infect. Diseases, 84:*98–104 (*352*).

McGhee, R. B. (1949b), *J. Infect. Diseases, 84:*105–110 (*352*).

McKeehan, M. S. (1954), *J. Exptl. Zool., 126:*157–176 (*9*).

McKenzie, J. and Ebert, J. D. (1960), *J. Embryol. Exptl. Morphol., 8:*314–320 (*254*).

McLaughlin, J., Jr., Marliac, J. P., Verrett, M. J., Mutchler, M. K., and Fitzhugh, O. G. (1963), *Toxicol. Appl. Pharmacol., 5:*760–771 (*218, 245, 251*).

McNally, E. H. and Byerly, T. C. (1936), *Poultry Sci., 15:*280–283 (*20*).

Meiniel, R. and Lutz-Ostertag, Y. (1970), *Arch. Anat. Microscop. Morphol. Exptl., 59:*167–183 (*269*).

Meirelles de Miranda, R. (1956), *Publ. 15 Inst. Zootecnia* (*Brazil*)—cited by Landauer, 1967 (*408*).

Menge, H., Calvert, C. C., and Denton, C. A. (1965a), *J. Nutr., 86:*115–119 (*182*).

Menge, H., Calvert, C. C., and Denton, C. A. (1965b), *J. Nutr., 87:*365–370 (*181*).

Menkes, B. and Deleanu, M. (1964), *Rev. Roumaine Embryol. Cytol., Ser. Embryol., 1:*69–77 (*209, 210, 263*).

Menkes, B. and Miclea, C. (1964), *Rev. Roumaine Embryol. Cytol., Ser. Embryol., 1:*79–87 (*263*).

Menkes, B. A., Pavkov, A., and Mircov, O. (1967), *Rev. Roumaine Embryol. Cytol., Ser. Embryol., 4:*35–46 (*263*).

Merkow, L. and Leighton, J. (1966), *J. Cellular Biol., 28:*127–137 (*81*).

Merkow, L. and Leighton, J. (1967), *Am. J. Pathol., 50:*975–991 (*83*).

Meyer, R. K., Rao, M. A., and Aspinall, R. L. (1959), *Endocrinology, 64:*890–897 (*319*).

Miclea, C. (1962), *Studii Cercetari Stiinte Med., 9:*77–83 (*372*).

Miclea, C. and Arcan, A. (1965), *Rev. Roumaine Embryol. Cytol., Ser. Embryol., 2:*37–43 (*277*).

Miclea, C. and Arcan, A. (1966), *Rev. Roumaine Embryol. Cytol., Ser. Embryol., 3:*75–86 (*277*).

Miclea, C. and Arcan, A. (1967), *Rev. Roumaine Embryol. Cytol., Ser. Embryol., 4:*163–171 (*258*).

Miclea, C. and Arcan, A. (1968), *Rev. Roumaine Embryol. Cytol., Ser. Embryol., 5:*151–158 (*264*).

Mitolo, V. (1963), *Acta Embryol. Morphol. Exptl., 6:*109–130 (*370*).

Miyamori, H. (1966), *J. Biol. (Osaka Univ.), 17:*19–39 (*317*).

Monkman, M. (1963), *A Study of the Development of Double-Yolked Eggs.* Inaugural Dissertation, North Illinois, 47 pp. (*386*).

Moog, F. (1962), *Gen. Comp. Endocrinol., 1:*416–432 (*307, 325*).

Moog, F. and Richardson, D. (1955), *J. Exptl. Zool., 130:*29–55 (*285*).

Mookerjee, S. (1953), *Arch. Entwicklungsmech. Organ., 146:*537–543 (*399*).

Morgan, W. C. (1963), *J. Heredity, 54:*116–120 (*54*).

Morgan, W. C. (1966), *Proc. World Poultry Congr. (Kiev), 13:*531–535 (*134*).

Moscona, M. H. and Karnofsky, D. A. (1960), *Endocrinology, 66:*533–549 (*285, 287, 289, 290, 315*).

Moulton, J. E. and Adler, H. E. (1957), *Am. J. Vet. Res., 18:*731–734 (*347, 348, 359*).

Mraz, F. R. (1966), *Poultry Sci., 45:*96–99 (*136*).

Mraz, F. R. (1969), *Poultry Sci., 48:*1460–1462 (*141*).

Mraz, F. R., Wright, P. L., and Ferguson, T. M. (1964), *Proc. Soc. Exptl. Biol. Med., 116:*1114–1118 (*131, 132, 138, 139, 140*).

Muller, H. D. and Moreng, R. E. (1966), *Radiation Res., 29:*85–99 (*116*).

Munro, S. S. (1965), *J. Genet., 56:*285–288 (*388*).

Murillo-Ferrol, N. L. (1963a), *Anales Desar., 11:*63–76 (*272*).

Murillo-Ferrol, N. L. (1963b), *Anales Desar., 11:*391–402 (*272*).

Murphy, M. L., Dagg, C. P., and Karnofsky, D. A. (1957), *Pediatrics, 19:*701–714 (*227, 255, 263, 267, 275, 276*).

Murray, P. D. F. (1926), *Proc. Linnean Soc. N. S. Wales, 51:*187–263 (*9*).

Mussehl, F. E. and Bancroft, P. (1925), *Poultry Sci., 4:*79–81 (*89*).

Naber, E. C. and Blackwood, U. B. (1962), *J. Nutr., 78:*207–214 (*224, 253*).

Naber, E. C., Cravens, W. W., Baumann, C. A., and Bird, H. R. (1954), *J. Nutr., 54:*579–591 (*201, 268, 272*).

Nakamura, T. (1967). *Hiroshima Daigaku Suichikusan Gakubu Kiyo, 7:*25–31 (*321*).

Nalbandov, A. (1942), *J. Heredity, 33:*53–54 (*388*).

Naujoks, H. (1953), *Beitr. Pathol. Anat. Allgem. Pathol., 113:*221–252 (*90, 93*).

Naz, J. F. and Rulon, O. (1946), *Anat. Record, 96:*555 (Abstr.) (*245*).

Needham, J. (1942), *Biochemistry and Morphogenesis.* Cambridge University Press, London, 785 pp. (*27*).

Nelsen, O. E. (1955), *J. Morphol., 96:*359–380 (*106*).

Nelsen, O. E. (1958), *Growth, 22:*109–124 (*90, 99*).

Nelsen, O. E. (1960), *Growth, 24:*361–383 (*90, 92, 372*).

Nemoto, M. (1929), *Tohoku J. Exptl. Med., 14:*1–28 (*88*).

Neuman, R. E., Maxwell, M., and McCoy, T. A. (1956), *Proc. Soc. Exptl. Biol. Med., 92:*578–581 (*273*).

Neuman, R. E. and McCoy, T. A. (1955), *Proc. Soc. Exptl. Biol. Med., 90:*339–342 (*222, 252*).

Newcomer, E. H. (1959), *Science, 130:*390–391 (*41*).

Newcomer, V. D., Wright, E. T., and Tamblyn, E. E. (1952), *J. Lab. Clin. Med.* (*St. Louis*), *39:*303–308 (*214*).

Newman, H. H. (1923), *The Physiology of Twinning.* Chicago University Press, 230 pp. (*382, 390, 391*).

Newman, H. H. (1940), *J. Heredity, 31:*371–378 (*389, 390*).

Noguchi, Y. (1969), *Japan. J. Vet. Sci., 31:*119–134 (*93, 94*).

Norris, L. C. and Bauernfeind, J. C. (1940), *Food Res., 5:*521–532 (*161*).

Nosal, R. A. and Watterson, R. L. (1959), *J. Morphol., 105:*529–567 (*222, 252*).

Noto, T. (1967a), *Sci. Rept. Tohoku Univ., 4th Ser., 33:*51–57 (*193, 243, 246, 254, 279*).

Noto, T. (1967b), *Sci. Rept. Tohoku Univ., 4th Ser., 33:*59–63 (*279*).

Noto, T. (1967c), *Sci. Rept. Tohoku Univ., 4th Ser., 33:*65–69 (*251, 254, 256*).

Nowinski, W. W. and Pandra, J. (1946), *Nature, 157:*414 (*248*).

Nowinski, W. W. and Pandra, J. (1947), *Anat. Record, 98:*84 (Abstr.) (*248*).

Oftedal, P. (1956), in *Progress in Radiobiology,* pp. 131–136, Oliver & Boyd Publishing Co., Edinburgh (*112, 143*).

Ohno, S., Kittrell, W. A., Christian, L. C., Stenius, C., and Witt, G. A. (1963), *Cytogenetics, 2:*42–49 (*41*).

Ojeda, J. L., Bosque, P. G., and Barbosa, E. (1970), *Anales Desar., 15:*241–246 (*276*).

Olcese, O. and Couch, J. R. (1950), *Poultry Sci., 29:*612–614 (*167*).

Olcese, O., Couch, J. R., Quisenberry, J. H., and Pearson, P. B. (1950), *J. Nutr., 41:*423–431 (*167, 185*).

Olsen, M. W. (1962), *J. Heredity, 53:*125–129 (*387, 392*).

Olsen, M. W. (1966), *J. Heredity, 57:*23–25 (*392*).

Olsen, M. W. (1969), *Poultry Sci., 48:*711–713 (*393*).

Olsen, M. W. and Byerly, T. C. (1938), *J. Agr. Res., 56:*221–226 (*365, 366, 367*).

Olsen, M. W. and Haynes, S. K. (1948), *Poultry Sci., 27:*420–426 (*33, 394, 400*).

Olsen, M. W. and Marsden, S. J. (1950), *Poultry Sci., 29:*414–419 (*18*).

O'Rourke, W. F., Bird, H. R., Phillips, P. H., and Cravens, W. W. (1954), *Poultry Sci., 33:*1117–1122 (*174*).

Paff, G. H., Boucek, R. J., Nieman, R. E., and Deichmann, W. B. (1963), *Anat. Record, 147:*379–386 (*86, 149*).

Pantić, V., Kosanovíc, M., and Pavlovíc, M. (1970), *Endocrinology*, 4:95–108 (*315*).

Panum, P. L. (1860), *Untersuchungen über die Entstehung der Misbildungen zunachst in den Eier Vögel*. Reimer, Berlin, 200 pp. (*384*).

Panum, P. L. (1863), *Mitt. Ver. Verbr. Naturw. Kenn.*, *5*:19–37 (*59*).

Papilian, V. and Nana, A. (1932), *Virchows Arch. Pathol. Anat. Physiol.*, *287*:5–9 (*367*).

Papilian, V. and Nana, A. (1933), *Virchows Arch. Pathol. Anat. Physiol.*, *290*:237–241 (*367*).

Parhon, C. I., Laurian, L., Balaceanu, M., and Albu-Aderca, N. (1956), *Rev. Sci. Med., Acad. Rep. Populaire Roumaine*, *1*:5–55 (*292, 293, 319*).

Parker, J. E. (1946), *N. Dakota Agr. Expt. Sta. Bull.*, *9*:6–8 (*404*).

Parodi, A. S., Lajmanovich, S., Pennimpede, F., and Mittelman, N. (1948), *J. Immunol.*, *58*:109–120 (*335, 336*).

Patry, E. and Ferrier, A. (1934), *Compt. Rend. Soc. Biol.*, *116*:928 (*251*).

Paunescu, C. (1964), *Vet. Inst. Pasteur (Romania)*, *3*:115–128 (*333*).

Peers, J., Ransom, S. E., and Huebner, R. J. (1952), *J. Exptl. Med.*, *96*:17–26 (*330*).

Pelagalli, G. V. and D'Angelo, A. (1965), *Atti Soc. Ital. Sci. Vet.*, *19*:367–370 (*320*).

Penionzhkevich, E. E. (1934), *Probl. Inkubatsii*, *1*:60–83 (in Russian) (*400*).

Perdomo, J. T., Harms, R. H., and Arrington, L. R. (1966), *Proc. Soc. Exptl. Biol. Med.*, *122*:758–760 (*174*).

Persaud, T. V. N. and Kaplan, S. (1970), *Life Sci.*, *9*(Part 2):1305–1313 (*262*).

Petersen, C. F., Wiese, A. C., Dahlstrom, R. V., and Lampman, C. E. (1952), *Poultry Sci.*, *31*:129–132 (*166*).

Petersen, C. F., Wiese, A. C., and Lampman, C. F. (1948), *Poultry Sci.*, *27*:471–476 (*177*).

Petersen, C. F., Wiese, A. C., Lampman, C. E., and Dahlstrom, R. V. (1950), *Poultry Sci.*, *29*:618–619 (*166*).

Philips, A. G. and Brooks, F. D. (1923), *Purdue Univ. Agr. Expt. Sta. Bull.*, *275*:3–16 (*69*).

Pictet, R. (1893), *Arch. Sci. Phys. Nat.*, *30*:293–314 (*89*).

Pierro, L. J. (1961), *J. Exptl. Zool.*, *147*:203–210 (*197, 258*).

Pierro, L. J. and Spiggle, J. (1966), *Nat.*, *209*:782–785 (*232, 233, 259*).

Pincus, G. and Erickson, A. E. (1962), *Endocrinology*, *71*:24–30 (*291, 316*).

Piotrowski, J. (1966), *Folia Biol. (Warsaw)*, *14*:205–225 (*285, 317, 320*).

Poley, W. E., Moxon, A. L., and Franke, K. W. (1937), *Poultry Sci.*, *16*:219–225 (*174, 175, 176, 246*).

Prevost, J. L. and Dumas, J. B. (1825), *Ann. Sci. Nat.*, *4*:47–56 (*3*).

Puchkov, V. F. (1959), *Biofizika*, *129*(4):940–943 (*128, 130*).

Puchkov, V. F. (1961), *Radiobiologiya*, *1*(3):437–439 (*125*).

Puchkov, V. F. (1963), *Tr. Inst. Exp. Med., Akad. Med. Nauk, vol. 7–8*(Part I):160–164 (*245*).

Puchkov, V. F. (1965), *Arkh. Anat. Gistol. Embriol.*, *48*(6):11–20 (*125*).

Puchkov, V. F. (1966), *Arkh. Anat. Gistol. Embriol.*, *51*(8):27–35 (*261*).

Puchkov, V. F. (1967a), *Arkh. Anat. Gistol. Embriol.*, *52*(7):37–44 (*121, 122*).

Puchkov, V. F. (1967b), *Byul. Eksperim. Biol. Med.*, *63*(7):99–102 (*254, 262*).

Pun, C. F. (1954), *J. Exptl. Zool.*, *126:*101–134 (*45, 51*).

Purchase, H. G., Cunningham, C. H., and Burmester, B. R. (1966), *Avian Diseases, 10:*162–172 (*49*).

Quan, S. F. (1963), *Zoonoses Res.*, *2:*41–66 (*214*).

Rabaud, E. (1899), *Compt. Rend. Acad. Sci.*, *128:*1183–1185 (*59, 89*).

Randles, C. A. and Romanoff, A. L. (1954), *Poultry Sci.*, *33:*374–377 (*364*).

Rao, M. A., Aspinall, R. L., and Meyer, R. K. (1962), *Endocrinology, 70:*159–166 (*319*).

Rauch, W. and Steinke, L. (1953), *Z. Tierzüchtung Züchtungsbiol.*, *62:*61–66 (*405*).

Rawles, M. E. (1943), *Physiol. Zool.*, *16:*22–43 (*8*).

Réaumur, R. A. F. de (1749), *Art de faire éclore et d'élever en toutes saisons des oiseaux domestiques de toutes espèces, soit par le moyen de la chaleur du fumier, soit par le moyen de celle du feu ordinaire.* Impr. Roy. Paris, Vol. I, 342 pp.; Vol. II, 339 pp. + 15 pl. (*102, 364*).

Rector, G. H. M. and Eastwood, D. W. (1964), *Anesthesiology, 25:*109 (*101*).

Reczko, E. (1955), *Z. Immunitaetsforsch.*, *112:*459–469 (*336*).

Rendel, J. M. (1943), *Biometrica, 33:*48–56 (*405*).

Reyss-Brion, M. (1956), *Arch. Anat. Microscop. Morphol. Exptl.*, *45:*342–353 (*115, 120, 121*).

Riddle, O. (1923), *Am. J. Anat.*, *32:*199–252 (*33, 387*).

Riddle, O. (1930), *Am. J. Physiol.*, *94:*535–547 (*27*).

Ridgway, L. P. and Karnofsky, D. A. (1952), *Ann. N.Y. Acad. Sci.*, *55*(Art. 2):203–215 (*243, 244, 246, 247, 249*).

Rigdon, R. H. (1968), *Avian Diseases, 12:*202–207 (*38*).

Rizzoli, C., Dessi, P., and Cestari, A. (1959), *Arch. Ital. Sci. Pharmacol.*, *9:*76–81 (*260*).

Robertson, G. G., Williamson, A. P., and Blattner, R. J. (1955), *J. Exptl. Zool.*, *129:*5–44 (*340, 341, 358*).

Robertson, G. G., Williamson, A. P., and Blattner, R. J. (1960), *Yale J. Biol. Med.*, *32:*449–463 (*338*).

Robertson, G. G., Williamson, A. P., and Blattner, R. J. (1964), *Am. J. Anat.*, *115:*473–485 (*339, 358*).

Robertson, I. S. (1961a), *J. Agr. Res.*, *57:*49–56 (*364*).

Robertson, I. S. (1961b), *Brit. Poultry Sci.*, *2:*49–58 (*29*).

Rogers, K. T. (1964), *Am. J. Anat.*, *115:*487–508 (*373, 374*).

Rogler, J. C., Parker, H. E., Andrews, F. N., and Carrick, C. W. (1959), *Poultry Sci.*, *38:*398–405 (*171, 172, 173, 187, 324*).

Rogler, J. C., Parker, H. E., Andrews, F. N., and Carrick, C. W. (1961), *Poultry Sci.*, *40:*1554–1562 (*172, 187*).

Rogulska, T. (1965), *Zool. Polon., 15*:65–76 (*378*).

Rogulska, T. (1968), *J. Embryol. Exptl. Morphol., 20*:237–245 (*272, 273*).

Rolle, G. K. and Bevelander, G. (1966), *J. Morphol., 118*:317–329 (*198, 199, 275*).

Romanoff, A. L. (1928), *Effect of Humidity on the Growth and Calcium Metabolism of the Chick Embryo.* Inaugural Dissertation, Cornell University, Ithaca, N.Y., 213 pp. (*102*).

Romanoff, A. L. (1929a), *Biol. Bull., 56*:351–356 (*404, 406*).

Romanoff, A. L. (1929b), *J. Exptl. Zool., 54*:343–348 (*102*).

Romanoff, A. L. (1929c), *Science, 70*:484 (*115*).

Romanoff, A. L. (1930a), *Cornell Univ. Agr. Expt. Sta. Mem., 132*:1–27 (*97, 102, 105*).

Romanoff, A. L. (1930b), *J. Morphol. Physiol., 50*:517–525 (*96, 97*).

Romanoff, A. L. (1931), *Anat. Record, 48*:185–189 (*367, 368*).

Romanoff, A. L. (1933a), *Poultry Sci., 12*:305–309 (*24, 26*).

Romanoff, A. L. (1933b), *Poultry Sci., 12*:388–389 (*367, 368*).

Romanoff, A. L. (1934), *Cornell Univ. Agr. Expt. Sta. Bull., 616*:1–39 (*25, 27, 67, 69, 102, 103*).

Romanoff, A. L. (1935a), *J. Agr. Sci., 25*:318–325 (*67, 69, 71, 78*).

Romanoff, A. L. (1935b), *Science, 81*:536–537 (a review of *Problems of Ionification*, 1933, by A. L. Tchijevsky, Vol. 1, 487 pp., Voronezh, USSR) (*149*).

Romanoff, A. L. (1936a), *Poultry Sci., 15*:311–315 (*59, 61, 66, 68, 69, 71, 75, 77, 78*).

Romanoff, A. L. (1936b), *Feathered World, 95*:375 (*28*).

Romanoff, A. L. (1938), *Cornell Univ. Agr. Expt. Sta. Bull., 687*:1–30 (*25*).

Romanoff, A. L. (1939), *Proc. World Poultry Congr.* (*Cleveland*), *1*:184–186 (*67, 68*).

Romanoff, A. L. (1940a), *Food Res., 5*:291–306 (*397*).

Romanoff, A. L. (1940b), *Poultry Sci., 19*:360 (Abstr.) (*21, 24*).

Romanoff, A. L. (1941), *J. Cellular Comp. Physiol., 18*:199–214 (*21, 24, 96, 104, 302*).

Romanoff, A. L. (1942), *Anat. Record, 84*:16 (Abstr.) (*145*).

Romanoff, A. L. (1943a), *Poultry Sci., 22*:148–154 (*25, 27, 67, 69, 102, 103*).

Romanoff, A. L. (1943b), *J. Cellular Comp. Physiol., 21*:123–127 (*145, 149*).

Romanoff, A. L. (1944), *Biodynamica, 4*:329–358 (*110*).

Romanoff, A. L. (1946), *Anat. Record, 96*:568–569 (*149*).

Romanoff, A. L. (1949), *Auk, 66*:264–270 (*27*).

Romanoff, A. L. (1951), *Anat. Record, 111*:549–550 (Abstr.) (*48, 73, 74*).

Romanoff, A. L. (1960), *The Avian Embryo: Structural and Functional Development.* Macmillan, New York, 1305 pp. (*viii, 8, 17, 24, 38, 44, 60, 61, 62, 63, 88, 89, 205, 312, 345, 367, 388, 390, 392, 393, 400*).

Romanoff, A. L. (1967), *Biochemistry of the Avian Embryo: A Quantitative Analysis of Prenatal Development.* Wiley-Interscience, New York, 398 pp. (*viii, 23, 44, 96, 104, 115, 170, 173, 227, 231, 283, 302, 365*).

Romanoff, A. L. (unpublished) (*9, 18, 22, 23, 25, 27, 28, 59, 61, 62, 75, 76, 81, 83, 102, 103, 171, 286, 368, 402, 403, 406, 407*).

Romanoff, A. L. and Bauernfeind, J. C. (1942), *Anat. Record, 82:*11–23 (*42, 160, 161, 162, 163, 184*).

Romanoff, A. L. and Bauernfeind, J. C. (unpublished) (*162, 164*).

Romanoff, A. L. and Bless, A. A. (1942), *Proc. Natl. Acad. Sci. U.S., 28:*306–311 (*110, 112, 133, 149*).

Romanoff, A. L. and Cottrell, C. L. (1939), *Proc. Soc. Exptl. Biol. Med., 42:*298–301 (*110, 149*).

Romanoff, A. L. and Faber, H. A. (1933), *J. Cellular Comp. Physiol., 2:*457–466 (*71*).

Romanoff, A. L. and Frank, K. (1941), *Proc. Soc. Exptl. Biol. Med., 47:*527–530 (*149*).

Romanoff, A. L. and Klevay, L. M. (unpublished) (*384, 385, 386*).

Romanoff, A. L. and Laufer, H. (1956), *Endocrinology, 59:*611–619 (*305, 306, 307, 308, 309, 325*).

Romanoff, A. L. and Ludlam, H. W. (1943), *Anat. Record, 87:*41 (Abstr.) (*145*).

Romanoff, A. L. and Romanoff, A. J. (1933), *Cornell Univ. Agr. Exptl. Sta. Mem., 150:*1–36 (*96, 97*).

Romanoff, A. L. and Romanoff, A. J. (1949), *The Avian Egg.* Wiley, New York, 918 pp. (*viii, 2, 17, 367, 384, 387, 397, 403, 405*).

Romanoff, A. L., Smith, L. L., and Sullivan, R. A. (1938), *Cornell Univ. Agr. Expt. Sta. Mem., 216:*1–42 (*25, 58, 61, 64, 65, 66, 67, 69, 70, 72, 73, 75, 76, 77, 83, 84, 85*).

Romanoff, A. L. and Sochen, M. (1936), *Anat. Record, 65:*59–68 (*86*).

Romijn, C. (1967), *Tijdschr. Diergeneesk., 91:*1474–1484 (*104, 105*).

Romijn, C., Fung, K. F., and Lokhorst, W. (1952), *Poultry Sci., 31:*684–691 (*301, 302, 321, 324*).

Rosenberg, E. E. (1957), *Nature, 180:*706–707 (*252*).

Rossi, A., Filoni, S., Bartoli, G. M., and Stefanelli, C. A. (1966), *Atti. Accad. Nazl. Lincei, Rend., 40:*313–320 (*248*).

Ruch, J. V. (1962), *Arch, Anat. Histol. Embryol., Norm. Exptl., 45:*64–129 (*388*).

Ruch, J. V. (1964), *Compt. Rend. Soc. Biol., 158:*2460–2464 (*316*).

Ruch, J. V. and Kremer, O. G. R. (1962), *Compt. Rend. Soc. Biol., 156:*1508–1509 (*207*).

Ruch, J. V. and Kremer, O. G. R. (1963a), *Compt. Rend. Soc. Biol., 157:*2291–2293 (*207*).

Ruch, J. V. and Kremer, O. G. R. (1963b), *Bull. Anatomists Assoc.* (*Brussels, Belgium, 11th Congr.*), *2:*368–374 (*206*).

Ruch, J. V., Kremer, O. G. R., Rumpler, Y., and Gerlinger, P. (1965), *Therapie, 20:*1585–1592 (*205, 206, 207, 257, 275*).

Ruch, J. V., Kremer, O. G. R., Schneegans, E., and Rohmer, A. (1961), *Compt. Rend. Soc. Biol., 155:*1109–1112 (*204, 206*).

Ruch, J. V., Kremer, O. G. R., Schneegans, E., and Rohmer, A. (1962a), *Compt. Rend. Soc. Biol., 156:*379–382 (*205, 206*).

Ruch, J. V., Kremer, O. G. R., Schneegans, E., and Rohmer, A. (1962b), *Compt. Rend. Soc. Biol., 156:*930–932 (*206*).

Ruch, J. V. and Rumpler, Y. (1964), *Arch. Anat. Microscop. Morphol. Exptl.*, *53*:301–323 (*290, 291, 316*).

Ruch, J. V., Zahnd, M., Kremer, O. G. R., Rumpler, Y., and Gerlinger, P. (1965), *Arch. Biol. (Liege)*, *76*:25–48 (*207, 257, 275, 316*).

Rumpler, Y. (1966), *Compt. Rend. Soc. Biol.*, *160*:2426–2428 (*323*).

Rüst, W. (1932), *Arch. Gefluegelk.*, *6*:110–116 (*50*).

Ruth, E. S. (1916), *Philippine J. Sci., B. Trop. Med.*, *11*:109–120 (*387*).

Ruth, E. S. (1918), *Philippine J. Sci. (Sec. B)*, *13*:201–205 (*387, 390*).

Rychter, Z. (1963), from *Advances in Morphogenesis, Vol. 2*:333–371 (*373*).

Sadler, W. W. (1955), *Poultry Sci.*, *34*:754–760 (*21*).

Salazar, M. R., de (1950), *Rev. Clin. Espan.*, *37*:400–405 (*185*).

Salzgeber, B. (1962), *Compt. Rend. Acad. Sci.*, *254*:3591–3593 (*236, 264*).

Salzgeber, B. (1966), *Rev. Franc. Etudes Clin. Biol.*, *11*:60–68 (*236, 264*).

Salzgeber, B. (1968), *Ann. Embryol. Morphogen.*, *1*:313–331 (*264*).

Salzgeber, B. and Salaün, J. (1965), *J. Embryol. Exptl. Morphol.*, *13*:159–170 (*241, 275*).

Sames, G. L. and Leathem, J. H. (1951), *Proc. Soc. Exptl. Biol. Med.*, *78*:231–232 (*315*).

Sanctuary, W. C. (1925), *Poultry Sci.*, *4*:141–143 (*28, 29*).

Sandor, S. T. and Elias, S. T. (1968), *Rev. Roumaine Embryol. Cytol., Ser. Embryol.*, *5*:51–76 (*220, 221, 251*).

Sandvik, O. (1958), *Hereditas*, *44*:407–414 (*133, 134*).

Saunders, J. W. and Gasseling, M. T. (1959), *J. Exptl. Zool.*, *142*:553–559 (*372*).

Saunders, J. W. and Gasseling, M. T. (1963), *Develop. Biol.*, *7*:64–78 (*373*).

Schneller, Sister Mary Beatrice (1951), *J. Morphol.*, *89*:367–395 (*111, 114, 117, 120*).

Schowing, J. (1961), *J. Embryol. Exp. Morphol.*, *9*:326–334 (*374*).

Schowing, J. (1965), *J. Embryol. Exp. Morphol.*, *14*:255–263 (*374*).

Schowing, J. (1968a), *J. Embryol. Exp. Morphol.*, *19*:9–22 (*372, 374*).

Schowing, J. (1968b), *J. Embryol. Exp. Morphol.*, *19*:23–32 (*372, 374*).

Schowing, J. (1968c), *53rd Congr., Tours., 1968, Bull. Assoc. Anat., No. 143*:1519–1526 (*372, 375*).

Schowing, J. and Celers, J. (1965), *Compt. Rend. Acad. Sci.*, *260*:5104–5107 (*373*).

Schryver, H. F. (1966), *J. Exptl. Zool.*, *162*:81–88 (*398*).

Scott, H. M. (1933), *Poultry Sci.*, *12*:49–54 (*400*).

Sell, J. L., Hajj, R., Cox, A., and Guenter, W. (1967), *Brit. Poultry Sci.*, *8*:55–64 (*173, 175*).

Serebrovsky, A. S. and Petroff, S. G. (1929), *J. Heredity*, *20*:331–332 (*54*).

Shear, H. H., Heath, H. D., Imagawa, D. T., Jones, M. H., and Adams, J. M. (1955), *Proc. Soc. Exptl. Biol. Med.*, *89*:523–528 (*357*).

Sheard, C. and Higgins, G. M. (1930a), *J. Exptl. Zool.*, *57*:205–222 (*145*).

Sheard, C. and Higgins, G. M. (1930b), *Proc. Soc. Exptl. Biol. Med.*, *27*:467–474 (*145*).

Sherbet, G. V. and Lakshmi, M. S. (1968), *Nature*, *217*:1257–1258 (*318*).

Shibko, S. I., Arnold, D. L., Morningstar, J., and Friedman, L. (1968), *Proc. Soc. Exptl. Biol. Med., 127:*835–839 (*219, 251*).

Shipkowitz, N. L. and Clarke, M. K. (1955), *Science, 122:*969–970 (*194, 249*).

Siegel, B. V., Smith, M. J., and Gerstl, B. (1957), *A.M.A. Arch. Pathol., 63:*562–570 (*294*).

Silver, P. H. S. (1962), *J. Embryol. Exptl. Morphol., 10*(II):423–450 (*372, 374*).

Simon-Reuss, I. and Spear, F. G. (1947), *Brit. J. Radiol., 20:*63–70 (*139*).

Sinha, M. D., Ringer, R. K., Coleman, T. H., and Zindel, H. C. (1959), *Poultry Sci., 38:*1405–1409 (*310, 324*).

Sittmann, K. (1967), *Genet. Res., Camb., 10:*229–233 (*53*).

Sittmann, K. and Abplanalp, H. (1965), *J. Heredity, 56:*220–223 (*53*).

Sittmann, K., Abplanalp, H., and Fraser, R. A. (1966), *Genetics, 54:*371–379 (*39*).

Sittmann, K. and Craig, R. M. (1967), *J. Heredity, 58:*44–46 (*51*).

Sittmann, K., Richards, W. C. P., and Abplanalp, H. (1965), *Can. J. Genet. Cytol., 7:*636–640 (*54*).

Sittmann, K., Wilson, W. O., and McFarland, L. Z. (1966), *J. Heredity, 57:*119–124 (*54*).

Slonim, D. and Roslerova, V. (1965), *Acta Virol., 9:*473–480 (*332*).

Smith, B. E., Gaub, M. L., and Moya, F. (1965), *Anesthesia Analgesia, Current Res., 44:*726–732 (*4, 101*).

Smith, R. T. and Thomas, L. (1956), *J. Exptl. Med., 104:*217–231 (*346, 347, 349*).

Smith, S. I., Weber, C. W., and Reid, B. L. (1970), *Toxicol. Appl. Pharmacol., 16:*179–185 (*212, 213*).

Snell, E. E. and Cravens, W. W. (1950), *Proc. Soc. Exptl. Biol. Med., 74:*87–91 (*199, 253*).

Spira, L. (1956), *Arch. Intern. Pharmacodyn., 108:*55–61 (*247*).

Spratt, N. T. and Haas, H. (1960), *J. Exptl. Zool., 145:*97–137 (*7, 8, 393*).

Stanzhevskaya, T. I. (1966a), *Byul. Eksperim. Biol. Med., 61*(4):85–88 (*202, 272*).

Stanzhevskaya, T. I. (1966b), *Arkh. Anat. Gistol. Embriol., 51*(7):7–12 (*241, 275*).

Starr, D. and Weston, J. (1962), *Anat. Record, 142:*328 (Abstr.) (*316*).

Stearner, S. P. and Christian, E. J. B. (1968), *Radiation Res., 34:*138–152 (*112, 127, 128*).

Stearner, S. P., Sanderson, M. H., Christian, E. J., and Tyler, S. A. (1960), *Radiation Res., 12:*286–300 (*133, 135, 136*).

Stephan, F. and Sutter, B. (1961), *J. Embryol. Exptl. Morphol., 9:*410–421 (*278*).

Stiles, K. A. and Watterson, R. L. (1937), *Anat. Record, 70:*7–12 (*366*).

Stockard, C. R. (1914), *Anat. Record, 8:*33–41 (*220*).

Stocum, D. L. (1966), *J. Morphol., 118:*183–196 (*325*).

Stoll, R., Bodit, F., and Maraud, R. (1968), *Compt. Rend. Soc. Biol., 161:*1680–1684 (*223, 252*).

Stoll, R., Coulaud, H., Faucounau, N., and Maraud, R. (1966), *Arch. Anat. Microscop. Morphol. Exptl., 55:*59–76 (*302, 304, 321*).

Strudel, G. (1953), *Ann. Sci. Nat. Zool., 15*:251–329 (*11*).

Strudel, G. (1971), *Compt. Rend. Acad. Sci., Ser. D, 272*:673–676 (*237*).

Sturkie, P. D. (1941), *J. Morphol., 69*:517–535 (*54*).

Sturkie, P. D. (1942), *J. Heredity, 33*:202–208 (*54*).

Sturkie, P. D. (1946), *J. Exptl. Zool., 101*:51–64 (*394*).

Suchetet, A. (1897), *Des Hybrides à l'état sauvage. Règne Animal. I. Classe des Oiseaux.* Paris, 1001 pp. (see Romanoff, 1960, pp. 109–111) (*38*).

Sumulong, M. D. (1925) *Philippine J. Sci., 28*:549–557 (*385*).

Sunde, M. L., Cravens, W. W., Elvehjem, C. A., Halpin, J. G. (1950), *Poultry Sci., 29*:696–702 (*156, 183*).

Szarski, H. (1948), *Bull. Intern. Acad. Polon. Sci. Lettres, Ser. B., Sci. Nat., Nos. 7–10*:133–144 (*74, 87, 88*).

Taglioretti, D. and Eridani, S. (1958), *Atti. Soc. Lombarda Sci. Med. Biol., 13*:445–447 (*267*).

Tannreuther, G. W. (1919), *Anat. Record, 16*:355–367 (*383, 390, 391*).

Taylor, A. and Carmichael, N. (1949), *Cancer Res., 9*:498–503 (*353*).

Taylor, A., Thacker, J., and Pennington, D. (1941), *Science, 94*:542–543 (*158*).

Taylor, L. W. (1934), *Poultry Sci., 13*:378 (*33*).

Taylor, L. W. and Gunns, C. A. (1947), *J. Heredity, 38*:66–76 (*40, 46, 51*).

Taylor, L. W. and Kreutziger, G. O. (1965), *Poultry Sci., 44*:98–106 (*96, 98*).

Taylor, L. W. and Kreutziger, G. O. (1966), *Poultry Sci., 45*:867–884 (*96, 98*).

Taylor, L. W. and Kreutziger, G. O. (1969), *Poultry Sci., 48*:871–877 (*96, 98*).

Taylor, L. W., Kreutziger, G. O., and Abercrombie, G. L. (1971), *Poultry Sci., 50*:66–78 (*98*).

Taylor, L. W., Sjodin, R. A., and Gunns, C. A. (1956), *Poultry Sci., 35*:1206–1215 (*90, 96, 98*).

Tazelaar, M. A. (1928), *Quart. J. Microscop. Sci. (London), 72*:419–446 (*61, 63, 85*).

Tchijevsky, A. L. (1933), *Problems of Ionification*, Vol. I, 487 pp. (in U.S.S.R., Voronezh), translated by A. L. Romanoff, 1935, *Science, 81*:536–537 (*149*).

Tengerdy, R. P., Fitch, J. W., and Moreng, R. E. (1969), *Poultry Sci., 48*:751–753 (*105*).

Teriukhanov, A. B., Pogoniailo, G. F., and Tivelev, P. G. (1965), *Sb. Tr. Nauch.-Issled. Inst. Bolez. Ptits., Issue 1*(12):209–212 (*2*).

Thompson, J. N., Howell, J. McC., Pitt, G. A. T., and Houghton, C. I. (1965), *Nature, 205*:1006–1007 (*164, 185*).

Titus, H. W. (1953), *Poultry Sci., 32*:593–601 (*26*).

Tobin, M. J. and O'Brien, J. P. (1952), *Wasmann J. Biol., 10*:1–6 (*118*).

Tur, J. (1904), *Compt. Rend. Soc. Biol., 57*:236–238 (*119*).

Tur, J. (1908), *Arch. Entwicklungsmech. Organ., 25*:615–631 (*59*).

Twiesselmann, F. (1935), *Compt. Rend. Soc. Biol., 119*:1169–1172 (*393*).

Twiesselmann, F. (1938), *Arch. Biol. (Liege), 49*:285–367 (*393*).

Tyler, S. A. and Stearner, S. P. (1960), *Radiation Res., 12*:303–316 (*113, 135*).

Ummersen, C. van (1961), in *Biological Effects of Microwave Radiation.* Plenum Press, New York, pp. 201–220 (*146, 147, 149*).

Vadehra, D. V., Baker, R. C., and Naylor, H. B. (1970), *J. Food Sci., 35:*5–6 (*2*).

Vannini, E. (1947), *Boll. Soc. Ital. Biol. Sper., 23:*240–242 (*315*).

Venzke, W. G. (1941), *Am. J. Vet. Res., 2:*281–284 (*320*).

Vidal, A. (1952), *Ann. Endocrinol.* (*Paris*), *13:*982–990 (*306, 325*).

Vidal, A. (1953a), *Ann. Endocrinol.* (*Paris*), *14:*437–443 (*306, 307, 322, 325*).

Vidal, A. (1953b), *Ann. Endocrinol.* (*Paris*), *14:*444–449 (*301, 303, 322*).

Vogel, H. H. and Jordan, D. L. (1959), *Radiation Res., 11:*667–683 (*143, 144*).

Vogel, N. W. (1965), *Proc. Penn. Acad. Sci., 39:*57–60 (*375, 376*).

Volskii, M. I. (1959), *Dokl. Akad. Nauk SSSR, 128:*857–859 (*99*).

Waddington, C. H. (1941), *J. Heredity, 32:*268–270 (*389*).

Wagley, P. F. and Morgan, H. R. (1948), *Arch. Pathol., 46:*441–450 (*253*).

Waite, R. H. (1919), *Maryland Agr. Expt. Sta. Bull., 233:*1–101 (*399*).

Warner, N. L. and Burnet, F. M. (1961a), *Australian J. Biol. Sci., 14:*580–587 (*320*).

Warner, N. L. and Burnet, F. M. (1961b), *Australian J. Exp. Biol. Med. Sci., 39:*235–248 (*320*).

Warren, A. E. (1934), *Am. J. Anat., 54:*449–485 (*372*).

Warren, D. C. (1934), *Kansas Agr. Expt. Sta. Bull., 37:*1–42 (*400*).

Warren, D. C. (1941), *J. Heredity, 32:*3–5 (*54*).

Warren, D. C. (1942), *Kansas Agr. Expt. Sta. Tech. Bull., 52:*1–44 (*37*).

Warren, D. C., Mueller, C. D., and Hutt, F. B. (1944), *J. Heredity, 35:*354–358 (*54*).

Warren, S. and Dixon, F. J. (1949a), *Radiology, 52:*714–729 (*137*).

Warren, S. and Dixon, F. J. (1949b), *Radiology, 52:*869–880 (*137, 138*).

Warynski, S. (1886), *Rec. Zool. Suisse, 3:*261–312 (*391*).

Waters, N. F. and Bywaters, J. H. (1943), *J. Heredity, 34:*213–217 (*53*).

Watterson, R. L., Fowler, I., and Fowler, B. J. (1954), *Am. J. Anat., 95:*337–400 (*11*).

Weber, A. (1906), *Arch. Zool.* (*Paris*), *Ser. 4, 5:*lxxxi–xc (*59*).

Wedlock, D. E. and McCallion, D. J. (1968), *Experientia, 24:*620–621 (*11*).

Weiss, H. S., Wright, R. A., and Hiatt, E. P. (1965), *Aerospace Med., 36:*201–206 (*99*).

Weiss, P. and Matoltsy, A. G. (1959), *Develop. Biol., 1:*302–326 (*13*).

Wendt, E. (1966), *Westdeutscher Verlag*, pp. 679–699 (*118*).

Wheeler, R. S. and Hoffmann, E. (1949), *Endocrinology, 45:*208–209 (*171, 172, 246*).

Wilburg, J. (1937), *Bull. Intern. Acad. Polon. Sci. Lettres, Ser. B* (*II*)*:*131–137 (*13*).

Wilder, O. H., Bethke, R. M., and Record, P. R. (1933), *J. Nutr., 6:*407–412 (*170*).

William, C., Gopalakrishnan, C. A., and Raman, K. S. (1969), *Indian Vet. J., 46:*696–699 (*33*).

Williamson, A. P., Blattner, R. J., and Lutz, H. R. (1963), *Proc. Soc. Exptl. Biol. Med., 112:*1022–1025 (*242, 275, 370*).

Williamson, A. P., Blattner, R. J., and Robertson, G. G. (1953), *J. Immunol., 71:*207–213 (*340*).

Williamson, A. P., Blattner, R. J., and Robertson, G. G. (1967), *Proc. Soc. Exptl. Biol. Med., 124:*524–532 (*370*).

Williamson, A. P., Blattner, R. J., and Simonsen, L. (1957), *Proc. Soc. Exptl. Biol. Med., 96:*224–228 (*338, 339, 358*).

Willier, B. H. (1924), *Am. J. Anat., 33:*67–103 (*372*).

Willier, B. H. (1954), *J. Cellular Comp. Physiol., 43:*307–317 (*10*).

Wilson, C. W. (1935), *Acta Radiol., 16:*719–734 (*114, 139*).

Windle, B. C. A. (1895), *J. Anat. Physiol., 29:*346–351 (*3*).

Witebsky, E., Peck, S., and Neter, E. (1935), *Proc. Soc. Exptl. Biol. Med., 32:*722–725 (*205, 207, 208*).

Wolff, Em. and Lutz-Ostertag, Y. (1961), *Arch. Anat. Microscop. Morphol. Exptl., 50:*439–468 (*388*).

Wolff, Et. (1933), *Compt. Rend. Soc. Biol., 114:*1030–1032 (*126*).

Wolff, Et. (1936a), *Les bases de la tératogenèse expérimentale des vertébrés amniotes, d'après les résultats de méthodes directes.* Inaugural Dissertation, Univ. Strasburg, 375 pp. (*33, 123, 125, 409*).

Wolff, Et. (1936b), *Arch. Anat. Histol. Embryol, 22:*1–382 (*3, 33, 123, 125, 127*).

Wolff, Et. (1937), *Titres et Travaux Scientifiques.* E. Dietz, Haguenau, 36 pp. (*123*).

Wolff, Et. (1938), *Compt. Rend. Soc. Biol., 127:*440–442 (*372, 373*).

Wolff, Et. (1948), *Compt. Rend. Soc. Biol., 142:*1282–1306 (*8, 393, 394*).

Wolff, Et. (1966), *Cell Differentiation and Morphogenesis.* North-Holland Publishing Co., Amsterdam, pp. 1–23 (*11*).

Wolff, Et. and Wolff, Em. (1936), *Compt. Rend. Soc. Biol., 123:*1191–1193 (*314*).

Wolff, Et. and Wolff, Em. (1937), *Compt. Rend. Soc. Biol., 124:*367–369(*314*).

Wolff, Et. and Wolff, Em. (1951), *J. Exptl. Zool., 116:*59–98 (*372*).

Wright, P. L. and Mraz, F. R. (1965), *Proc. Soc. Exptl. Biol. Med., 118:*534–539 (*140, 141, 248*).

Wright, R. A., Lessler, M. A., Weiss, H. S., and Hiatt, E. P. (1965), *Aerospace Med., 36:*311–314 (*99*).

Wucher, O. (1900), *Anat. Hefte, 15:*91–121 (*391*).

Yamamoto, R. and Bigland, C. H. (1966), *Am. J. Vet. Res., 27:*326–330 (*352*).

Yamamoto, R. and Ortmayer, H. B. (1966), *Avian Diseases, 10:*268–272 (*352*).

Yamamoto, S. and Noguchi, Y. (1964), *Nippon Juigaku Zasshi, 26:*181–190 (*90, 93, 94*).

Youngner, J. S., Ward, E. N., and Salk, J. E. (1950), *Proc. Soc. Exptl. Biol. Med., 75:*157–161 (*233, 235, 262*).

Yushok, W. D. (1950), *The Relationship of Thyroid Activity to the Growth and the Cytochrome* c *Content of Chick Embryos and their Organs.* Inaugural Dissertation, Cornell University, Ithaca, N.Y., 86 pp. (*307, 310, 324*).

Zagaevsky, J. S. and Lutikova, P. O. (1944), *U.S. Egg Poultry Mag., 50:*75–77, 88–90 (*2*).

Zahnd, J. P. and Rumpler, Y. (1961a), *Compt. Rend. Soc. Biol., 155:*650–653 (*312, 323*).

Zahnd, J. P. and Rumpler, Y. (1961b), *Compt. Rend. Soc. Biol., 155:*1566–1569 (*312, 323*).

Zanini, A. C. (1965), *Rev. Fac. Farm. Bioquim.* (*Sao Paulo*), *3:*199–240 (*244*).

Zartman, D. L. and Jaap, R. G. (1968), *J. Reprod. Fertility, 17:*215–217 (*40*).

Zinn, K. M. (1970), *Invest. Ophthamol., 9:*165–182 (*375*).

Zusman, I. N. (1968), *Zool. Zh., 47:*407–417 (*18*).

Zwilling, E. (1942), *J. Exptl. Zool., 91:*453–463 (*12*).

Zwilling, E. (1945), *J. Exptl. Zool., 98:*237–247 (*372*).

Zwilling, E. (1949), *J. Exptl. Zool., 111:*175–187 (*53*).

Zwilling, E. (1955), *J. Exptl. Zool., 128:*423–441 (*372*).

Zwilling, E. (1956a), *J. Exptl. Zool., 132:*157–171 (*52, 372*).

Zwilling, E. (1956b), *J. Exptl. Zool., 132:*173–187 (*372*).

Zwilling, E. and DeBell, J. T. (1950), *J. Exptl. Zool., 115:*59–81 (*216, 217, 230, 274*).

Subject Index

Boldface entries represent the titled sections of the text. Boldface numbers are page references to numbered tables. Boldface italic numbers are page references to graphic illustrations.